D1297307

Evolution under the microscope

QH
366.2
.S95
2002

Evolution under the microscope

*a scientific critique of the
theory of evolution*

David W. Swift

Leighton

Nyack College
Eastman Library

Published by
Leighton Academic Press
Stirling University Innovation Park
FK9 4NF
United Kingdom

Copyright © David W. Swift, 2002
All rights reserved.
The moral right of the author has been asserted.

Apart from any fair dealing for the purposes of research or private study, or criticism or review, as permitted under the UK Copyright Designs and Patents Act 1988, this publication may not be reproduced, stored or transmitted, in any form or by any means, without prior permission in writing of the publishers, or in the case of reprographic reproduction only in accordance with the terms of the licences issued by the Copyright Licensing Agency in the UK, or in accordance with the terms of licences issued by the appropriate Reproduction Rights Organization outside the UK.

This book is sold subject to the condition that it shall not, by way of trade or otherwise, be lent, resold, hired out, or otherwise circulated without the publisher's prior consent in any form of binding or cover other than that in which it is published.

ISBN 0 9543589 0 2

A catalogue record for this book is available from the British Library.

Cover photograph: a crystal of DNA,
viewed through a microscope with crossed polarized illumination.
Copyright © Michael W. Davidson and The Florida State University Research Foundation.

Printed by The Inglewood Press Ltd, FK10 2HU

Sit down before fact as a little child,
be prepared to give up every preconceived notion,
follow humbly wherever and to whatever abysses nature leads,
or you shall learn nothing.

THOMAS HUXLEY, 1860

Contents

Subsidiary text boxes

Figures

ACKNOWLEDGEMENT

Figure 4 is reproduced from Figure 3 of Baba *et al.* (1981) by kind permission of Springer-Verlag GmbH & Co. KG.

Preface

The theory of evolution is the leading explanatory principle of biology. Its chief strength is that it provides a unifying rationale for many diverse biological phenomena, and it is also of value to biologists as a working hypothesis which stimulates research. Since being proposed by Darwin nearly 150 years ago, the theory has been eminently successful and, whilst recognising there are differences of opinion over details, it is now widely accepted as fact – as factual as any other established scientific theory. Indeed it is affirmed to such an extent that those who do not share this conviction are likely to be considered ill-informed and/or anti-science.

However, there are scientists who in the course of their work have come across what they see as substantial if not insurmountable obstacles. This is my own position. Because evolution so permeates our education, it is not surprising that from an early age I totally accepted the whole story of evolution from simple forms of life, and probably of the origin of life itself. But in studying for a degree in natural sciences – in which the theory of evolution formed a backdrop to almost every aspect of the curriculum – I was prompted to question whether the biochemical structures and mechanisms which were being discovered at the time could really have arisen in an opportunistic evolutionary manner: and it seemed to me that they could not. Over the years since then, mostly spent as a research scientist, this view has been strengthened as we have learned more of how biological systems work; and I have become well aware that similar issues have been raised by others, including experts in their field. Eventually, the opportunity arose to research the subject more fully, and this book is the outcome.

It is primarily and predominantly an examination of evolution as a scientific theory: how it arose, the evidence on which it is based, the extent to which the theory of evolution is a satisfactory explanation of that evidence – and, importantly, the facts that are inconsistent with an evolutionary explanation and consequently undermine the theory as a whole. A central issue is that whilst I accept the principle of natural selection it cannot account for the formation of biological macromolecules, though this is the usual evolutionary explanation for them. Evidence for the operation of natural selection at the level of the whole organism, and evidence from the fossil record do not answer this objection, in fact I show that they add weight to it.

The book is written firstly for biologists because it is primarily they whom I want to challenge to take a fresh look at the facts. However, because of the widespread interest in and acceptance of evolution, I have sought to make the book accessible to a much wider readership. To this end the early chapters describe how the theory of evolution arose (Darwinism), became well established once we understood hereditary processes (Neo-Darwinism), and finally assimilated our modern understanding of the biochemical nature of genetic mechanisms (the modern synthesis). The aim of these chapters is to provide an adequate understanding of the science of evolution to enable subsequent discussion to be followed readily by those with a basic knowledge of biology, and hopefully by many who may not even have that. For example, evolution has become so generally accepted as a factual biological theory that the principle of evolution has been assimilated by many other disciplines – such as sociology, philosophy and even theology. I trust the book will be of interest, indeed of value, to non-biologists from disciplines such as these, who need to hear from a scientist of the substantial difficulties with the theory – that it is nothing like so secure as they have probably been led to believe.

So far as any 'science versus religion' debate is concerned, I write entirely as a scientist and my purpose here is not to offer any sort of reconciliation or accommodation between the two camps. To present the scientific advances of the 16th and 19th centuries in a proper light it is necessary to describe some of the religious background; but there is no discussion here of Genesis 1 or any other religious text. My only comment in this area relates to the changing attitude to the concept of a supernatural God who intervenes in the affairs of the world, as natural science developed and the laws of science were discovered.

First and foremost I am indebted to my family and friends who have supported and encouraged me throughout this project – which, like so many others, turned out much longer than anticipated – with special mention of my brother for reading the manuscript. Thanks, too, to Professor David Bebbington, Dr Adrian Newton and an anonymous reviewer for their valuable comments and constructive criticisms.

June 2002
D. W. S.

1

Trial and Error

A common theme of many of the events, exhibitions and the like organized to celebrate the millennium was that of man's achievements thus far, with confident anticipation of further success in the future. It was an appropriate theme: science and technology are advancing rapidly and much of our everyday lives is now dependent on them. Those alive just a century ago, even the Victorians who were so impressed with progress made in their time, could scarcely have imagined how we would be living now. It is not only the technological advances that have mechanized many domestic chores or enabled rapid transport, which they might reasonably have foreseen; but whereas powered flight was unknown at the start of the century, it is already a generation since man walked on the moon. Similarly, the use of electricity was in its infancy and electronics was totally unknown; but now the computing power in a handheld calculator far exceeds that of a roomful of hardware of the 1950s; and with a laptop PC or even just a mobile phone we can link up to a world-wide network of computers which allows us to access information from, or send messages to, almost anywhere on earth. So it is not surprising that our world-view has become dominated by science and technology, to the extent that perhaps we take much of it for granted.

Undoubtedly an aspect of this world-view is the perception of evolution: it is felt that our technological progress is a natural extension of the biological process(es) that led to our existence in the first place. In fact, with the capability of genetic engineering, we are even beginning to consider the possibility of directing our future biological evolution, and the recent success of sequencing the human genome makes that day look all the closer. I am not suggesting that embracing the theory of evolution has accelerated progress, but it has provided a philosophical backdrop which helps us to make sense of the exponential rate of change of our everyday lives.

Further, it is widely recognised that formulation of the theory of evolution in the middle of the 19th century was not merely an explanation of biological development, but it led to a radical impact on man's view of himself. Because it relates to our origins, it has philosophical, theological and sociological implications; and this, of course, is why it is such a contentious subject – why the theory of evolution is still debated well over a century after Darwin. Indeed, whilst this book focuses on scientific challenges to evolution, I also recognise that it is not just a dispassionate scientific theory but, because of its wider implications, it carries non-scientific

1

baggage which hinders a completely objective assessment.

If the general principle of evolution teaches us anything, it is that what we are today is the product of what happened yesterday. So it should come as no surprise that any proper study of current views regarding evolution needs to examine how those views evolved! Hence, to appreciate modern attitudes to evolution it is important to see how the theory emerged out of wider scientific progress of the preceding centuries, and has its roots even further back. It is partly for this reason that I start with a historical approach. My second reason is that it provides a convenient and, I trust, interesting way of introducing the technical background that is needed to understand the science of evolution.

There are many excellent histories of science available, ranging from broad overviews to detailed accounts of a particular branch of science or an especially important period such as the 'scientific revolution' of the 16th and 17th centuries. Not being a historian, it is certainly not my intention here to duplicate such efforts, and I do not presume to offer anything new. Indeed, to many readers the outline I give here may be very familiar and no doubt some will feel I am oversimplifying. However, with today's specialisations – particularly between the arts and sciences, and even different disciplines within the latter – all too easily we can focus on our own area and know little outside of it. So for the benefit of readers who, like me, went through secondary and higher education and perhaps even scientific training without learning much of the history of science, I think it is instructive to outline the rise of modern science, to set the scene for the emergence of evolutionary biology. It also serves to introduce some of the other themes relevant to my later discussion.

The Dawn of Science

There were various civilisations in the 'fertile crescent' of the Middle East from at least as early as 3000 BC. Notable among them were the Sumerians in Mesopotamia who were remarkably cultured and technically advanced; for example, they invented writing and the wheel. Yet despite their progress they were not really scientific: theirs was primarily a practical knowledge rather than theoretical – one might call them artisans rather than scientists. 'Science' started with the Greeks.

Classical Greece

What is known as ancient Greece comprised various city-states extending over an area that included southern Italy and the west coast of Turkey as well as the Balkan Peninsula and islands of modern-day Greece. As with Mesopotamia, there had been civilisations in the region for a long time before, but from about 600 BC there was a dramatic rise of Greek culture and intellectual achievement.

They recognised that much of their knowledge came from earlier cultures, such as astronomy from Babylonia and geometry from Egypt, but it was the Greeks who first sought to piece together a coherent philosophy of nature. The world seems changeable and even erratic; but the Greeks believed that behind this there must be consistent patterns and principles; and they sought to elucidate that order – to discover the fundamental laws of the universe – which is the essence of science. At least in the West, a scientific perception so dominates our world-view that almost without thinking we ask 'How does this work?'. So it is hard for us to appreciate just how radical a concept this was for their time, how much of a break from previous ways of thinking. But for a culture steeped in mythology, it was a remarkable intellectual advance; and it started mankind along the road from perceiving nature as being at the mercy and whim of the gods, to that of the universe being governed by regular non-personal laws. It is true that, because of their limited knowledge, their early ideas were very speculative – and all too often they did not realise how speculative – but they did begin to think of the world as a material entity in which there is a consistent operation of cause and effect.

Before turning to their science, it is worth mentioning Socrates (c.470–399 BC). He was sentenced to death, accused of promoting atheism and corrupting the minds of youth, because his teaching was perceived as undermining the traditional beliefs of his contemporaries. Socrates was more concerned with ethics than science, so it would be misleading to portray his conflict as that between science and religion. But it was between the status quo and change, and is an early example of the way in which the establishment can react to new ideas that it fears may challenge its authority. Struggle against the establishment in one form or another has been a common theme in the development of new ideas, whether scientific or otherwise. And, of course, this was an important consideration in the development of evolutionary theories during the 19th century (AD).

The Pythagoreans

One of the first advances made by the Greeks was in mathematics. The Babylonians and Egyptians had of course used arithmetic for their astronomy and geometry, some being quite sophisticated such as enabling the solution of quadratic equations. But it was the Pythagoreans who began to develop mathematics as a science in the sense of introducing rigorous methods and proofs. The essence of a mathematical proof is that, starting with premises (called axioms) which are considered to be self-evidently true, by a series of logical deductions from these it is possible to demonstrate the truth of a more complex conclusion – a conclusion which might or might not have seemed right intuitively. For example, based on the axioms of two-dimensional geometry, the Pythagoreans developed the well-known theorem that for a right-angled triangle the square of the hypotenuse equals the sum of the squares of the other two sides. This rigorous deductive approach was a major contribution to the development of mathematics and is still followed in mathematical proofs today.

However, their success in logic, mathematics and geometry, which are non-empirical disciplines in that they do not rely on observation, led them to think that a similar approach could be equally successful with, and should be applied to, the natural sciences. One of the attractions of the deductive method is that it appears to be a route to certain knowledge because, provided the axioms are true and all the logical steps of the proof are valid, one can be certain the conclusion is definitely true. So the Pythagoreans and other Greek natural philosophers after them adopted a similar approach to the study of nature. That is, they attempted to decide from reason alone (*a priori*) what must be the fundamental truths about nature (corresponding to the axioms of mathematics), and then proceeded to deduce, or 'prove', how nature should be in the light of those underlying principles. We need to see that this was their approach in order to make sense of their conclusions.

For instance, the Pythagoreans were so taken with the elegance of numbers that they believed numbers must be at the heart of the universe, even in some cryptic sense comprising its very substance. This view was supported when they discovered that musical notes are related in simple mathematical ratios. Hence, they believed that nature would follow numerical or geometrical patterns, and especially the circle which they considered to be the perfect geometrical figure. So in their model of the universe they assumed that the planets follow circular orbits, and they placed them at distances from the earth corresponding to musical ratios. They reasoned it had to be that way for the cosmos to be in harmony.

So, although the classical Greeks were the first to try to understand nature, their approach was flawed. They thought that starting from basic presuppositions or principles it is possible to deduce wider truths about nature. It meant they were more concerned with thinking about how things ought to be rather than looking to see how things actually are; and, of course, the conclusions one comes up with depend on one's premises – how you think the universe should be, or how you think a god would make the universe. This philosophical approach, in which reasoning was more important than observation, dominated science for a long time.

Whilst many individuals made valuable contributions to the birth of natural philosophy (or natural science), two figures stand out in view of their substantial influence on later western thought – Plato and Aristotle.

Plato

Plato (*c*.428–*c*.347 BC) was born into an aristocratic family and at first had political ambitions but later abandoned these, partly due to the execution of his friend Socrates, and turned to learning. In 387 BC he founded the Academy in Athens, which provided a comprehensive curriculum including such subjects as mathematics, astronomy, biology and political theory as well as philosophy, and might be regarded as the first European university.

In his natural philosophy Plato mainly followed the Pythagorean deductive approach of deciding what are the basic principles and then reasoning how things must be in the light of those. This is shown in his important Theory of Forms (or Theory of Ideas). He reasoned that if something is perfect it must be unchanging (as it cannot become more or less perfect), and conversely anything that changes must be imperfect. Because any physical object is subject to change, it must be imperfect; in fact the whole material world is imperfect, and to find perfect reality we must perceive beyond material things: any physical object is but a pale reflection of a perfect archetype. So, for example, the *essence* or fundamental Idea or Form of a rose or a horse is more important, even more real, than any physical rose or horse. Clearly this sort of attitude to the natural world reinforced the non-empirical approach typical of his predecessors.

Plato was primarily a philosopher, his comments on science being limited compared with his works on other subjects such as ethics, but he wrote a treatise on the natural world, called *Timaeus*. In keeping with his general philosophy of nature, in this he describes how the material world was formed (by a 'Demiurge' or Maker) as an imperfect temporal copy of perfect eternal realities. Plato's philosophical way of perceiving the natural world – very unscientific in modern terms – retarded the advance of science for many centuries because it was his works that were most readily available in the West during the early Middle Ages.

Aristotle

Aristotle (384–322 BC) spent about 20 years at Plato's Academy, first as a student and then a teacher, and subsequently established his own institution, the Lyceum. Much of what we have of Aristotle's works is based on his lecture notes, and they include a large amount of information on astronomy, meteorology, plants and animals. For a while he was tutor to the young Alexander the Great.

Whilst following much of Plato's philosophy, Aristotle increasingly adopted an independent position regarding natural science. In particular, he rejected Plato's Theory of Forms with its emphasis on non-material essences, instead maintaining that the natural world is real and meaningful and a proper subject for investigation, and insisting that careful observation must have an important part in the study of nature. So, whilst he did not reject the Pythagorean deductive approach, he maintained it was insufficient to rely on this exclusively, and that deductive conclusions must be compared with what we actually see in nature. He also advocated the principle of induction – of seeking general patterns or regularities from a set of observations – which is the first step in identifying scientific laws. Aristotle was therefore particularly important in the development of science: he was the first philosopher of science as well as pioneering the practice of empirical science.

Another important aspect of his philosophy was his understanding of 'causes' or explanatory factors. He maintained that everything has four causes: the *material cause* which is the substance out of which a thing is made; the *efficient cause* which is the agent of production or change; the *formal cause* which specifies what kind or type of a thing it is; and the *final cause* which is its purpose. For example: the material cause of a statue is the stone or other substance of which it consists; the efficient cause is the sculptor; the formal cause is its recognisable shape, e.g. a personage; and the final cause is to be a work of art.

Aristotle's causes led to some important consequences for an understanding of the natural world. First, he was convinced that nothing arises by itself (everything must have an efficient cause), so this reinforced a cause-and-effect appreciation of nature, and an expectation that natural laws will operate in the world. However, he recognised that ultimately there must be an uncaused First Cause or 'prime mover'. Second, because he believed that everything has a 'final cause' in the sense of a purpose, it follows that the First Cause activates the world with a purpose. In other words, it was evident to Aristotle that the world has been designed, and his philosophy of causes codified this teleological view of nature – that everything is formed with its end purpose in view. Aristotle and Plato were agreed on this, and their opinions had much influence through the Middle Ages and after. But this view was not shared by all the classical Greek philosophers. The debate about design in the universe is not new, and still rages today; and obviously this is a particularly important issue when it comes to biology, as we shall see in due course.

With his philosophy of science and the beginnings of an empirical approach Aristotle made significant advances over the work of earlier philosophers. Unfortunately, he was just about the last of the great classical Greek philosophers, and although his work was influential for a time in Alexandria, it was then largely ignored if not forgotten for about a thousand years until the revival of learning in Europe.

Alexandria

When Alexander the Great died in 323 BC the Greek empire was divided among his generals. This marked the demise of classical Greece, and our focus of attention moves to the city of Alexandria which had been founded in 332 BC on one of his early campaigns. On his death, one of the generals became governor of Egypt, and made Alexandria his capital. He became Ptolemy I, founder of the Ptolemaic Dynasty which lasted until the death of Cleopatra and her son when Egypt was taken over by Rome under Octavian, later Emperor Augustus.

Ptolemy I (and/or II) established a museum and library in Alexandria along the lines of the Lyceum in Athens. At its height the library boasted about half a million manuscripts and was responsible for preserving most of the ancient texts that have survived. It was the literary and scientific centre of the ancient world until the Roman Empire declined.

Historians generally consider this Alexandrian period to be less intellectually accomplished than that of the earlier Athenian Greeks. Nevertheless, the Alexandrians performed the invaluable role of collating the work of the earlier natural philosophers, and occasionally supplementing it with their own work. Euclid is a prime example of this: he founded a school of mathematics at Alexandria and compiled all the previous work on geometry, and probably added some proofs of his own.

Also, perhaps because the museum was established by one of his pupils, its focus tended to be along Aristotelian rather than Pythagorean lines. That is, they were more empirically minded, with due regard for observation and experiment, practising applied science compared with the more theoretical or philosophical emphasis of Athens. Archimedes typifies this: he was an exceptional mathematician, but also investigated physics, and applied his mathematics and physics to engineering. He is famous for discovering the density of a substance and how to measure it; and he invented various military machines for use in defence of his native city, Syracuse, against the Romans.

Alexandria's prominence as a centre of learning lasted for about 300 years, and for a time it was also an important crossroads between East and West. However, the city was subjected to various attacks by the Romans and Arabs, and its importance declined with the rise of Byzantium (later called Constantinople, now Istanbul) and Cairo.

The Middle Ages

By the time Augustus became the first Roman emperor, Greek science had declined and there was then no major advance in science for more than a thousand years. A number of reasons have been suggested for this dismal state of affairs.

First of all, it is recognised that the classical Greeks were exceptional in their perception and creative thinking. Even the Alexandrians who followed in the Greek tradition are seen as lacking originality – primarily having developed earlier Athenian ideas rather than formulating new ones.

As for the Romans, it is well known that they were excellent engineers, but theirs was strictly applied science and they made hardly any significant contribution to scientific thought or progress. According to the historian George Sarton 'the stolid indifference of the Romans to intellectual or scientific matters stifled science more than the barbarian invasions' (Introductory Chapter), and to some extent they reverted to religious or mythological explanations of the natural world, which set back science even more. However, there were a few Roman writers (often referred to as the Latin Encyclopaedists) who produced compilations of the available knowledge on various subjects, and these came to be standard textbooks through the Middle Ages. Undoubtedly the most important of these was *Natural History* by Pliny the Elder who had a keen interest in the natural world, an interest which unfortunately resulted in him being killed when observing the eruption of Vesuvius that destroyed Pompeii. It was the largest collection of facts about the natural world available in the Middle Ages, and the principal source for many later writers, at least until the 12th century. But Pliny was credulous and uncritical, including many fanciful ideas and myths; so, whilst this work helped to preserve the earlier knowledge, it also perpetuated some erroneous ideas about nature which took a long time to be thrown off. Eventually the Roman Empire collapsed in the face of barbarian invasions and internal strife; and the ensuing political instability further inhibited intellectual activity.

A major setback at the time was that most of the works of the ancient Greeks were lost so far as western Europe was concerned; more survived in the Eastern Empire, centred on Byzantium, but this was practically inaccessible to the West. In particular, it seems likely that none of Aristotle's works on natural science, with their empirical outlook, was available, and possibly the only work on natural philosophy was part of Plato's *Timaeus*. The inaccessibility of the classical works was accentuated by the transition under the Romans from Greek to Latin as the language of learning. Further, the intellectual mood of the times was to venerate the ancients, and scholars were primarily concerned with preserving the old knowledge rather than inquiring for themselves. Plato's deductive and non-empirical philosophy did nothing to change this attitude, but rather reinforced it; and his influence persisted right through the Middle Ages.

Christianity

And then there was Christianity. Some writers have blamed the intellectual decline on this new religion because of its focus on another world. But it is evident that the decline had already taken place before Christianity had any substantial impact – it was still very much in the minority before the 4th century.

However, the early church fathers sought to enhance the credibility of Christianity by comparing it with classical Greek philosophies. By far the most important of these was Augustine (354–430), an intellectual who had explored other religions and philosophies before becoming a Christian, and who saw in Plato much that is consistent with Christian teaching. In particular, he was happy to align Plato's assertion of a Maker with the Christian Creator, Plato's Forms became eternal ideas in the mind of God, and the associated derogation of material things was consistent with the Christian emphasis on spiritual values. This approval of Plato might have suited the early church, but it was unfortunate for science, as it reinforced adherence to Plato's non-empirical methods. In addition, it was instrumental in giving theology a dominating role over all branches of knowledge, including science:

Plato's deductive philosophy emphasized the importance of having the right fundamental principles (axioms), from which wider conclusions about nature are derived, and the view that it was more important that any theory should be consistent with those principles than with actual observation. It was evident to the Christian church that theology must be the source of those principles, and the foundation stone for all branches of knowledge; and so theology became supreme – the chief if not sole arbiter of what was and what was not acceptable. Nowadays, with our much clearer distinction between religion and science, it is hard for us to appreciate that religious views could so dominate science; but this is where we should recognise that our current viewpoint is an outcome of the scientific revolution: before that it was very different. During the medieval period, theology became the 'Queen of Sciences' – valued above all other disciplines, and the filter through which everything else was interpreted. This role of theology lasted for more than a millennium and was the chief source of conflict in later centuries when advances in science challenged old ideas about the world. Its position lasted for so long not only because of the role of Christianity, but also because of the medieval frame of mind which looked back to the old authorities rather than forward; and the domination by theology would not be overthrown until this scholastic attitude (see below) was abandoned in favour of a more empirical approach – one based on the observation of nature.

On the other hand, throughout the early Middle Ages the church fulfilled a valuable role of preserving and transmitting knowledge, especially through its monasteries and abbeys which were the chief centres of Christian life and learning. There is also no doubt that its teachings and practices had a beneficial stabilising influence through the turbulent years of the 'Dark Ages' before new European nations emerged. But it should be noted that almost exclusively education was based on studying works from the past – there was little breaking of new ground.

Probably the most regrettable impact of Christianity, at least so far as scientific investigation is concerned, arose from the amalgamation of church and state. Inevitably, when Constantine established Christianity as the state religion in the 4th century, the church became politicized, with bishops having civil authority as well as their ecclesiastical roles; and with the weakness of nations during the Middle Ages, the church acquired a dominant role in society.

One of the chief areas where the church exerted its authority was in the area of knowledge. This had adverse consequences for science because the association of church and state blurred the distinction between theological and secular knowledge. The church, which was so involved with learning, teaching both theology and the natural sciences (indeed, there was not this clear distinction), came to attach the same importance to secular knowledge as it did to tenets of the Christian creed. So any departure from accepted secular teaching might be seen as heresy in much the same way as unorthodox religious views; and this inhibited the acquisition and dissemination of new knowledge. Christianity did not cause the de-

cline in science, but it resisted recovery from it – and continued to retard
scientific advance right through the Renaissance and scientific revolution,
even to the 19th century. Possibly one of the chief successes of the theory
of evolution was finally to throw off the church's hold over science, which
I shall discuss further in due course.

Islamic science

The restoration of learning in the West, in particular recovery of the semi-
nal works of classical Greece, came from the Arab world. Islam was
founded at the beginning of the 7th century. At first its impact on science
was negative: the violent expansionist policies of the early Muslims, and
their respect for the Koran to the exclusion of all other books, led to the
loss of much of the classical learning in the Middle East, notably final de-
struction of the library in Alexandria, and isolated the West from such as
remained. But gradually the mood changed, there was a rise in intellectual
activity, and the ancient texts became valued once again. Baghdad was
founded in 762 and rapidly became a centre of learning, including translation of many classical works, notably those of Aristotle. By the 10th century, nearly all the texts of Greek science that were to become known to
the western world were available in Arabic, and from the 11th century this
learning was transmitted to Europe. It was this influx of the classical
works that fostered the Renaissance.

As well as translating the classical works, some Arabic scholars added
commentaries to their translations. A particularly important figure to
emerge was the philosopher Averroës (Ibn Rushd, 1126–1198) who lived in
Cordoba, the capital of Spain and a major centre for culture and learning.
To a large extent he adopted Aristotle's philosophy and expounded his
own opinions in his commentaries on Aristotle's works. The significant
views he held were that the universe has always been in existence, which
removed the need for a creator, and although he accepted a prime mover,
this was somewhat remote compared with the traditional view of God. He
also maintained that reason took precedence over religious dogma. Not
surprisingly these ideas were seen as challenging traditional Islamic beliefs, not least in undermining the concept and role of God, and provoked
opposition from some Muslim religious leaders. Averroës was exiled for a
time and his works burned, and a general ban was imposed on studies that
challenged orthodox Islam.

Revival of learning

By the end of the first millennium AD western Europe had regained some
political stability, and along with this came increased population and developing urban life. Cathedral and monastic schools prospered, and there
was a general reawakening of cultural interest. The first modern universities, notably Oxford and Paris, were established in the 12th century, and
included specialist schools to focus on specific subjects. Gradually there

was a shift away from monasteries to universities as the chief centres of learning – but this slackened the church's hand on learning only a little.

There was also an expansion of trade which led to increased contact with neighbouring countries. This was extended to the Middle East by the Crusades, which led to direct contact with Arabs, as did the progressive reclaiming of Spain from the Moors which took place at about the same time. Most of the key documents had been received in the West by the end of the 12th century, and in the next many of the translations were revised, either with a better knowledge of Arabic or directly from Greek, and by its close virtually all of Aristotle's surviving works had been acquired.

Scholasticism

We now need to consider how this influx of classical learning was received in the West. As outlined above, intellectual life, indeed virtually all of life, was dominated by the church. For most people, what the church said really mattered, and only a tiny proportion of the population would have been equipped to challenge it even if they had wanted to. So, to a large extent, the church's response to the newly available material was crucial to its wider acceptance. To understand this we need to look at scholasticism and the prevailing frame of mind of the scholars at the time.

I have already mentioned that centuries earlier Augustine had grappled with the interrelation of Christianity and philosophy. Part of the revival of learning that took place from about the 11th century was that a number of intellectuals arose, Anselm (1033–1109) being the first significant figure, who began to reconsider this issue, especially in the light of the classical literature which was beginning to reach the West (and to which Augustine had not had access). Their purpose was to reconcile Christian doctrine with secular philosophy: so far as possible to prove what they could of the truth of Christianity, and failing that at least to demonstrate that it was reasonable. Their approach was essentially that of studying in great detail and comparing 'authorities' – on the one hand the Bible and teachings of the early church fathers, especially of Augustine, and on the other the teachings of the Greek philosophers, which increasingly meant those of Aristotle. Scholasticism refers to this sort of reasoning and debate that took place through to about the time of the scientific revolution. In practice, much of their approach was deductive, and in time their deliberations came to focus increasingly on sterile arguments over logic.

In the short term, whilst Aristotle's information about the natural world was welcomed, the initial reception of his philosophy was hostile. This was because, from the early Middle Ages, the church had had an established philosophy of the natural world, essentially Augustine's synthesis of Plato and the Bible, where God had a prominent role as creator and sustainer of the universe. But Aristotle, with his emphasis on the operation of natural processes (recall his four 'causes'), maintained that the material universe must always have been in existence, i.e. could not have had a beginning,

so there was no need of a creator. Whilst his mechanical understanding of
the universe required a first cause to get things moving, thereafter natural
laws applied in terms of a continuous series of causes and effects – a determinism which left little room for God to be actively involved in the
world's affairs, contrary to Christian claims. Also, it should be remembered that the classical texts were received first in Arabic and accompanied by commentaries of the Arabic philosophers, especially of Averroës
who highlighted differences between Aristotle's teaching and traditional
religious views, and placed reason above religious revelation as a source
of knowledge. Averroës had, of course, challenged Islamic doctrine, not
Christian, but the church recognised that his arguments also attacked the
Christian teaching of God's role in nature. Indeed, a faction of Averroists
arose who vociferously promoted Averroës' views. The church's response
to this attack was, by the Council of Paris in 1210, to ban Aristotle's natural philosophy and threaten to excommunicate anyone who studied it.

However, the church found a champion in Thomas Aquinas (1225–74).
Born in Italy and first educated at the University of Naples, he then studied at Cologne under Albertus Magnus (see below), and subsequently
taught in Paris. Like Augustine nearly a millennium before, Aquinas was
an astute intellectual, with a thorough grasp of philosophy and theology,
and a lucid teacher and writer. Over a period of about 25 years he wrote
voluminously, providing a renewed synthesis of natural philosophy and
religion, seeing secular knowledge and religious revelation as two sides of
the same coin. Like Augustine, he insisted that secular and religious truth
must ultimately be consistent, which was in sharp contrast to Averroës'
view of 'double truth' where philosophy and religion could hold opposing
views (with philosophy having precedence). Where possible, Thomas
adopted the Aristotelian view, notably interpreting his First Cause as God,
Aristotle's earth-centred universe could be seen as illustrating the central-
ity of mankind in Christian theology, and his hierarchy of creatures (*scala
naturae*, see Ch. 4) supported the Christian teaching of man as ruler of
nature.

Superficially, the church's philosophy of nature at the end of the 13th
century was not very different from that at its beginning; but the important
change over this period was that it had taken a step towards accommodating a scientific view of nature, and of seeing God as acting through natural
laws rather than supernaturally.

With this resolution, the way was open for Aristotle's works to be embraced by western scholars. By the end of the 13th century he was *the*
'Philosopher' (with Averroës the 'Commentator') – the most important of
the classical thinkers, and his works were central to any university teaching. Dissemination of Aristotle's works certainly stimulated renewed interest in science and philosophy, and it might have been expected that his
emphasis on an empirical scientific method, on the need to check theory
with observation, would have led to a more empirical outlook and practice.

But this did not happen – at least not in the short term.

It seems the reason for this was the medieval frame of mind. They had an excessive sense of respect for the early works (both secular and religious), seeing them as unassailable 'authorities'. The scholastics used these works as premises from which they deduced conclusions by logical argument. Once they had Aristotle, they simply adopted his works as another of those authorities, and proceeded to apply their deductive methods to them as well. It seems their outlook was so much that the ancients knew it all, that they were simply rediscovering knowledge, that it did not occur to them that they should or even could investigate for themselves.

Further, the attitude of the church towards secular knowledge remained unchanged. It absorbed Aristotle's views into its teaching and in time attached to them the same importance as to its theological doctrines. It is hard to know whether they had this attitude because they felt particular theories of the natural world were important to their theology, or simply because they could not see the distinction between them.

Empirical science emerges

Albertus Magnus

A leading figure responsible for disseminating and promoting classical and Arabic learning in the West, especially the natural science of Aristotle, was Albertus Magnus. He was born around the year 1200 in Bavaria, and studied at the University of Padua in Italy where he became acquainted with the classical Greek and Arabic works. He spent much of his life at Cologne (where he taught Thomas Aquinas) and some time as Professor of Theology at Paris, but also travelled widely through western Europe, teaching and researching. Whereas Thomas' strength was in philosophy and theology, Albert excelled in the natural sciences; and in order to explain Aristotle's teaching to his students, he paraphrased all of Aristotle's works that were available at the time. The name by which he came to be known (Albert the Great) reflects the fact that, even in his lifetime, he became highly respected for the breadth and depth of his knowledge and teaching.

Albertus Magnus was somewhat anomalous because, although a leading scholar, he did not conform to the usual scholastic mould. As we have seen, the typical scholars' approach was exclusively to respect and interpret the ancient authorities, but never to challenge them. Although Albert greatly valued Aristotle's work, he did not accept it blindly or uncritically; rather, he weighed what Aristotle said in the light of other information available to him, often from his own observations. He wrote extensively on many topics such as mineralogy, botany and zoology, and whilst he usually followed Aristotle, he occasionally criticised him. Indeed, in his *Summa theologica* he argued strongly against the Averroists' view that Aristotle had been infallible.

Development of the scientific method

Contemporary with Albert were two English scholars, Robert Grosseteste who became first Chancellor of the University at Oxford, and his somewhat better known pupil Roger Bacon. They also helped to publicize Aristotle's works, but they focused on the physical sciences rather than natural science, especially optics and astronomy. Probably more important than their writings on specific scientific subjects were their views on and development of the scientific method, as they criticised the scholastic practice of relying exclusively on books, and clearly advocated an experimental approach. When discussing Aristotle, I mentioned his introduction of the inductive method to science – of examining a body of data to discern patterns, leading to the identification of scientific laws. Grosseteste and Bacon progressed this inductive approach by formulating rules to assist the detection of patterns, especially for identifying important factors and discarding irrelevant ones. Perhaps more important still, Roger Bacon insisted that the conclusions determined by induction should be tested, not only by reference to the initial observations, but by carrying out experiments specifically devised to test the validity and scope of the proposed scientific principle. This was a valuable methodological insight; it is getting close to the general scientific procedure of proposing a hypothesis and then seeking additional data to substantiate (or refute) it.

However, Bacon was ahead of his time and his scientific method was not adopted. He was also censured by the ecclesiastical authorities, partly it seems because of his criticism of their scholastic attitude, and also due to his dabbling in astrology and alchemy, for which he was better known at the time.

End of the Middle Ages

So the introduction of Aristotle's works brought something of a revival to science during the 13th century, and by its end his physics and natural science had been assimilated by the West. Also, a few individuals had begun to appreciate and develop his empirical approach. Eventually this stimulated scientific inquiry, but it was a slow process. For a long time, perhaps a couple of centuries, the main attitude remained that of rediscovery: it was assumed that the ancients had had comprehensive knowledge and all that needed to be done was to rediscover it – there was no need to discover anything new, indeed there was nothing new to discover. No doubt the Black Death of the 14th century and the 100 years war between England and France, which spanned the 14th and 15th centuries, also retarded progress.

Nevertheless, despite the scholastics and their medieval mentality, empirical science gradually began to emerge from its stagnation. At this time the remaining important classical Greek scientific works became available, such as Archimedes' mechanics and Ptolemy's *Almagest* (see Ch. 2). Also, during the 15th century printing was invented in the West, greatly facilitating dissemination of not only the classics, but also contemporary ideas. All of these factors continued to promote scientific inquiry, and with that to challenge the medieval and scholastic mindset.

As scientific thought began to develop, a paradoxical outcome was that, although Aristotle had advocated an empirical and questioning approach to natural science, Aristotelianism came to be seen as anti-science because of its association with scholasticism. As we have seen, the scholars had assimilated Aristotle's science, his physics and cosmology, and it had become part of the orthodox teaching. But they had incorporated this with the old scholastic mindset, accepting the science on the basis of Aristotle's authority, and they ignored his teaching that theories must be reviewed in the light of observation. Because they believed that the classical philosophers knew everything, they assumed that Aristotle's physics must have been right. However, evidence began to emerge that showed Aristotle's physics was wrong; and because the church dogmatically insisted it was right, it ended up inhibiting science.

An influential figure, at least in England, who helped to overcome the shackles of Aristotelian scholasticism, was Francis Bacon (1561–1626, not to be confused with the earlier Roger Bacon). Whilst primarily a politician, in later life he turned to the philosophy of science and wrote, among many other things, *The Advancement of Learning* (1605) and *Novum Organum* (1620) in which he set out his methodology for science. He emphasized the importance of observation as the basis for scientific knowledge, and of induction for the elucidation of scientific laws. He advocated that scientists should consider a set of data without preconceptions, letting the facts speak for themselves, and enumerated procedures for drawing out inductive inferences. Although modern historians do not regard Bacon's ideas as particularly innovative, there is little doubt that his writings catalysed progress through the scientific revolution, and still carried considerable weight in Darwin's day.

Eventually the rise of empirical science was unstoppable, and the 16th century saw the beginnings of the scientific revolution. Rumblings between the establishment and new ideas persisted and eventually confrontation was inevitable. There was discontent on several fronts, but the first major battleground was astronomy. Science's success here led to a break from the old ways and to the rise of modern scientific methods, rapidly leading to achievements in other areas. So astronomy merits special attention and I shall turn to this and related issues now. I shall follow this with geology which, with its interpretation of fossils and extension of the perceived age of the earth, had a direct bearing on the formulation and acceptability of the theory of evolution.

2

Revolutions and Revolutionaries

It may seem strange to have a chapter on astronomy in a book primarily about biological evolution; and the reason is not that I will be supporting some theory of evolution from space such as has been proposed by various authors from time to time. Rather, it is because the development of astronomical ideas, especially the radical change from a geocentric (earth-centred) system to heliocentric (sun-centred) marked a turning point in the history of science. It was much more than merely a change in the prevailing theory of the cosmos. There was a major advance in scientific method, and a rise in the standing and influence of science itself as it was seen to triumph over the dogma that had dominated intellectual pursuits for so long previously. It marked the demise of the old world order and was the vanguard of the scientific revolution.

Undoubtedly the change in cosmology arose from the steady accumulation of astronomical observations, and a growing commitment (at least by some) to the principle that theories about the world must fit observed facts. So in the first instance it was a triumph for empirical science.

But it did not stop there: the developing ideas of astronomy and mechanics culminated in the late 17th century with Newton's work on physics, and gravitation in particular, which had a substantial influence on subsequent scientific thinking and practice. His work epitomized what science is about and how it should be done, and served as a stimulus for further scientific research. It led to progress in many fields, and certainly was part of the motivation for Darwin's work on evolution in the 19th century. Also, the ensuing scientific progress led to the industrial revolution and subsequent technological advances. Newton's work is still cited for its pioneering value, even though his astronomy, mechanics and theories about gravity have all now been superseded.

Then, of course, the advances in astronomy were much more than merely a stimulus to dispassionate science. Because the old ideas were held dogmatically by the establishment, adopting new ideas did not just mean replacing one scientific theory by another, but led to conflict with the authorities themselves, principally with the church which had such a prominent role in society. And because the establishment lost – ultimately it had to lose because its ideas about astronomy were wrong – it lost credibility; and at the same time the credibility of science increased and the opinions of scientists gained influence.

Also, along with the new astronomy came a better appreciation of the scale of the universe and that the earth is not at its centre, indeed that the universe may not have a 'centre' at all. This not only changed man's view of himself and his role in the universe, but was seen to challenge the anthropocentricism of established religion. And in similar vein, as the more fundamental theories of physics began to emerge, this further emphasized the operation of natural laws, and at the same time undermined the supernatural.

All of these factors – the emphasis on nature's laws, stimulus to scientific inquiry, and challenge to the old dogmas – helped to pave the way for the theory of evolution. Further, along with a better understanding of the heavens as they are, came ideas of evolutionary cosmology – of how the universe had developed – and we will see that these had a direct bearing on the formulation of evolutionary ideas for biology.

There is one other reason why I have included an account of the astronomical revolution. It is remarkable how often, even today, if some sort of criticism of evolution is expressed, a retort is something like 'What about Galileo?'. Galileo is seen by many as the champion of science against religious dogma; and a popular perception is that any rejection of the theory of evolution must be on grounds of religious dogma, or at least be for nonscientific reasons. One of my chief aims in writing this book is to challenge that perception.

Early Astronomy

Babylonians

The Babylonians were accomplished astronomers. From at least as early as the second millennium BC they made observations of the sun, moon, Venus and some stars, and began compiling records of their movements. Such observations were useful to them for regulating their agricultural activities, as part of their religion and the growing practice of astrology.

It is obvious even to a casual observer that the night-time sky moves westwards in a similar way to the sun's daytime movement. It is also readily apparent that from night to night the stars advance a little, so that progressively new constellations come into view (just before dawn) while others are no longer visible, having passed below the horizon by the time the sun has set. Over the course of a year the sequence is completed, with the same constellations being visible, at the same time of the night, as had been the case a year before. The net effect of this is that the sun appears to move eastwards across the starry background, its track through the constellations being known as the ecliptic. The moon follows a similar pattern, but much more rapidly, completing a cycle in a month; and its altitude varies above and below the ecliptic. Very early on the Babylonians were able to predict lunar and possibly solar eclipses.

By about the 6th century BC their observations included all five of the planets that are visible to the naked eye – Mercury, Venus, Mars, Jupiter and Saturn. These 'wanderers' meander through the starry background –

usually eastwards like the sun and moon, but occasionally moving backwards (westwards) for a while before resuming their easterly course. The
retrograde movements last for from a couple of weeks for Mercury up to
several months for Saturn, and are associated with changes in altitude
such that the planets trace out a slow and rather flat 'loop the loop' in the
night sky. The Babylonians devised mathematical relationships for describing the observed positions and for predicting future movements, with
remarkable precision by about 500 BC onwards. Yet, despite the immense
effort in observation and arithmetic, it seems all of this was without any
attempt to conceive how the sun, moon and planets are spatially arranged
and move in relation to each other and the earth. That was taken up by the
Greeks.

Classical Greeks

It is evident from Homer's writings that, like the Babylonians, the early
Greeks held mythological views about the origin and structure of the earth
and heavens. But with the rise of classical Greece, natural philosophers
began to theorize as to the true nature of the cosmos. It is all too easy for
us to criticise their theories, but for their time they represented a genuine
attempt to explain what they saw; their theories were speculative hypotheses based on inadequate information, but not irrational, and were a 'giant
leap for mankind' over the previous fanciful ideas.

For example, the earth seemed to them to be massive – large and
heavy, with no sensation of movement; and it appears to be surrounded by
a giant rotating spherical dome which bears the celestial bodies and carries them around the earth. Also, they still considered man to be of prime
importance in the universe and in the plans of the gods. For these reasons
then, it was only natural to assume that the earth was stationary in the
centre of a spherical universe which revolved around it.

This sort of idea started with the Pythagoreans, who were probably the
first to realise that the earth is a sphere rather than being flat; and, as
mentioned previously, consistent with their ideas of mathematical perfection they presumed that the sun, moon and planets followed circular orbits
around it, and at distances corresponding to musical notes. Plato's ideas
are based on this. He suggested that the movements of the sun, moon and
planets can be accounted for by combining two circular movements
around the earth – one centred on the earth's axis to account for the diur-
nal rotation (which included the 'fixed stars'), and one centred on the
poles of the ecliptic which gave their gradual movement against the background stars.

Although I have emphasized their philosophical approach to science, it
should be noted that the Greeks were not just theorists. Through careful
observation they learned that the sun and moon move at variable rather
than constant speeds, and their apparent sizes vary, suggesting that their
distance from the earth changes. These features are inconsistent with a

uniform circular motion about the earth; but, although they realised this, the Greeks were committed to circular motion for philosophical reasons. This commitment to the circle and emphasis on ideal geometrical forms may seem odd to us; but really it is no different from the desire today for so-called 'elegance' in scientific theories – to the Greeks, a circle was elegant.

Despite its failings, because it was described in his *Timaeus*, it was Plato's model that dominated throughout the subsequent Middle Ages in Europe. There is little doubt that Plato was well aware of its inadequacies and, although he did not offer any further explanation himself, he challenged others to devise a better system. In other words: given the earth, sun, moon and five planets; what motions in relation to each other can account for the observed movements of the other bodies as seen from the earth? This exercise came to be known as 'saving the appearances'.

The challenge to propose a better model was ably taken up by Eudoxus, who accompanied Plato for a few years. For each planet he proposed a combination of four circular motions. Two of these corresponded to those of Plato's model. Then there were two inner circular movements having the same rate of rotation but set at an angle to each other: together these superimposed a to-and-fro and up-and-down oscillation on the basic motion provided by the outer two circles, resulting in the planet's retrograde movement and changes in altitude. This was a major improvement on Plato's model, but, despite its complexity, because it was based on circular movements centred on the earth, it still did not account for the variable speed and apparent size of the sun and moon.

We can hardly ignore Aristotle in view of his importance and the influence of his works at the time of the Renaissance, but in fact he made little contribution to astronomy. In effect he adopted Eudoxus' model, but gave it a more physical basis by proposing that the circular movements were achieved by transparent crystalline spheres rotating within each other at appropriate speeds and axial tilts with respect to each other, each sphere transmitting its movement to all those within it. He also distinguished between the sublunary region (within the orbit of the moon, and of course including the earth itself) and the celestial region beyond. The sublunary region was imperfect and changeable, composed of the four elements – earth, water, air and fire – and subject to terrestrial laws of physics; but the celestial region was perfect and unchangeable, composed of ether, and a supernatural physics could operate there.

Alexandria

It was during the Alexandrian period that more complex geometrical solutions were introduced to the problem of 'saving the appearances'. Appolonius of Perga lived in the 2nd century BC and is best known for his work on geometry, especially on conic sections (e.g. ellipse, parabola), but he also made valuable contributions to astronomy. For the sun and moon he proposed eccentric circular motion, i.e. following circular orbits but not centred exactly on the earth, as this could explain how their distance from the earth, and their apparent velocity when viewed from the earth, could vary. For the planets he proposed an epicycle which is illustrated in Figure 1. In this system the planet follows a small circular orbit (the epicycle), but the centre of that follows a large circular orbit around the earth, called a deferent. He realised that by choosing a suitable epicycle in terms of its radius and rate of rotation compared with the deferent, this system could produce retrograde motion. Following Appolonius' work, epicycles became the principal means for modelling planetary movements. Notably, they were adopted by Hipparchus, who also lived in the 2nd century BC and, although not well known, was possibly the greatest ancient astronomer in terms of the quality of his observations and theoretical analysis.

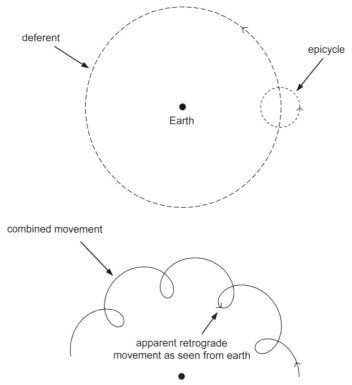

FIGURE 1. Epicycle.

The astronomer Ptolemy (not one of the earlier ruling Ptolemies) lived during the 2nd century AD by which time Alexandria was in decline. As mentioned previously, in general the Alexandrians were not pioneers, but they collated earlier work, and to some extent developed it. In typical Alexandrian fashion, Ptolemy compiled an encyclopaedia of astronomy, which came to be known as the *Almagest*. In this he brought together all the earlier work, especially that of Hipparchus, and refined the epicycles to obtain better agreement with observations. In total, he used 54 epicycles to describe the movements of the planets, and eccentric circles for the sun and moon. He shows his commitment to the old ideas by commenting that circular motion is 'in agreement with the nature of Divine Beings', although his application of this principle was a bit dubious sometimes.

What is striking about the *Almagest* is its strictly theoretical or mathematical nature. Ptolemy was interested only in devising geometrical schemes for correctly explaining the observations and reliably predicting future movements. He even used different schemes to explain different phenomena of the same body, such as one to account for changes in apparent velocity and another for changes in apparent size of the moon. It is clear that he did not worry about what physical reality might be behind his schemes. In a later work he did consider the physical nature of the cosmos, but essentially followed Aristotle's crystalline spheres and did not try to relate them to the epicycles.

Early heliocentric theories

Not all the ancient astronomers followed the usual arrangement of their time of having an immovable earth at the centre of the universe. Once it was recognised that the earth was not flat but a globe, it was apparent to some that an alternative to the heavens revolving daily around a fixed earth, was that the earth could rotate on its axis within a fixed heavenly sphere. Aristotle considered this option but rejected it because he was so convinced the earth was immovable – that if it rotated it would fly apart.

The most progressive idea came from Aristarchus (*c*.310–250 BC) from the Island of Samos. Although we do not have his own account, we know from Archimedes that he proposed a heliocentric universe i.e. with all of the planets, including the earth, orbiting the sun. But his views were perceived as demoting the earth and thereby offending the gods, so were rejected as heretical, and Aristarchus was persecuted. This reaction inhibited further ideas along these lines, and was a clear forerunner of the conflict that ensued when heliocentric theories were revived.

Before condemning this treatment of Aristarchus we should recognise that at the time there were 'scientific' reasons for rejecting a heliocentric universe. The most important of these was that if the earth orbits the sun, this implies substantial movement of the earth within the universe (from one side to the other of its orbit), so our perspective of the stars will change during the course of a year[1], but any such parallax was not evident. This implied that the fixed stars were *very* much more distant than was then supposed (so any parallax was imperceptible) and the universe vastly greater than seemed possible – or that Aristarchus was wrong.

---

[1] An analogy to this would be the changing view of the seating in a sports stadium as one runs around the track. Or simply binocular vision – that the view from each eye is slightly different because they are not at exactly the same place.

Arabic astronomy

Apart from their invaluable role of preserving the early Greek works, including the *Almagest*, the Arabs made a number of contributions to astronomy. They continued with detailed observations and, by comparing their results with early data, especially those of Hipparchus, they had access to about a thousand year's of observations, much more than had been available to Ptolemy. This enabled them to refine some astronomical features, e.g. the length of the solar year and rate of precession, and identify more slowly varying ones such as those due to periodic changes in the earth's orbit and rotation.

All their theoretical astronomers followed Ptolemy's representation of planetary movement based on epicycles. They learned improved methods of trigonometry from India and using these greatly simplified the associated geometrical calculations. In so far as they gave any thought to the physical structure of the universe, they mainly adopted Aristotle's crystalline spheres, even though these are difficult to reconcile with epicyclic motions.

The revival of learning

As mentioned in Chapter 1, the Renaissance was stimulated by the introduction of ancient Greek learning into western Europe. These of course included Greek writings on astronomy, and at the same time the West gained access to astronomical tables constructed by the Arabs. Also Gerbert, later Pope Sylvester II (died 1003), brought knowledge of the abacus and astrolabe from Spain, the latter especially aiding the revival of astronomy. As a result Western scholars started to use the Arab astronomical tables, and to make their own observations for comparison.

Aristotle's cosmology was of course very similar to Plato's in the *Timaeus* which had prevailed in Europe throughout the Middle Ages. Once his philosophy and views became influential, this served to reinforce adherence to the traditional view of a central earth surrounded by spheres which carried the heavenly bodies. This arrangement was underpinned by Aristotle's mechanics which emphasized the heaviness and immovability of the earth, and the 'naturalness' of circular movement around it.

However, Latin translations of Ptolemy's *Almagest* became available during the 13th century. It was evident that epicycles gave a much more accurate account of observations than simple circles, and they were readily adopted by mathematicians and practising astronomers, especially for constructing astronomical tables. But others, principally philosophers who were not satisfied with mere mathematical devices for 'saving the appearances', challenged the epicycles on the grounds that they lacked a physical basis, especially because it could be seen that different mathematical models could give the same results. For example, Thomas Aquinas was inclined to accept the crystalline spheres but doubted the reality of the epicycles and wondered if there could be some other explanation for the phenomena they accounted for. In time, as more observations accumulated, further refinements were needed to the epicycles in order to give good agreement. To the mathematicians this was their strength – the epicycles were adaptable; but to others this flexibility simply accentuated doubts about their reality. The basic problem was that a simple circular model was woefully inaccurate, but epicycles seemed too theoretical and lacking real substance.

The controversy continued through the 14th century, during which many more of the early Greek works were recovered, including those from other schools of thought than Plato and Aristotle. Also, the general revival of enquiry and discussion of specific questions regarding the physical basis of the astronomical schemes reopened some of the old questions. There was renewed debate as to whether the earth rotated rather than the stars, and whether the universe is bounded or infinite – it was recognised that if the universe is infinite then it is meaningless to talk of the earth being at its centre. And all of this argument took place against a philosophical background which was steeped in theology – where for many the theological implications of any theory were paramount, but also where the omnipotence of God was used by some to argue that almost anything could happen.

In the 15th century there was a concerted effort to address the astronomical issues, partly prompted by an increasingly urgent need to reform the Julian calendar which had become significantly out of phase with the natural year. The stage was set for someone to propose radical ideas: that someone turned out to be Copernicus.

THE COPERNICAN REVOLUTION

Copernicus

Nicholaus Copernicus was born in 1473, into a wealthy Polish family, and was educated at the University of Cracow where he learned astronomy. Subsequently he studied law at other universities, but throughout this time maintained an interest in astronomy, including making observations firsthand, and lectured on the subject in Rome in 1500. In due course, he held various ecclesiastical and diplomatic positions, but in later life turned increasingly to scientific pursuits.

Around 1514 he wrote an essay, generally referred to as *Commentariolus*, in which he set out the basic ideas of his heliocentric theory. However, it did not start to circulate until about 1530, and then only in manuscript copies, but it did gain the attention of scholars and ecclesiastics. In response to this he was asked to write a fuller account, which emerged as *De revolutionibus orbium coelestium* (On the revolutions of the celestial spheres), but it was not published until his death in 1543. In this Copernicus refers to classical Greek precedents for a moving earth, to some extent portraying his planetary scheme as a rediscovery of earlier theories. This approach was of course typical of the times, and no doubt Copernicus was seeking to demonstrate that he had Greek authority for his proposal; but he carefully avoided any reference to Aristarchus.

The principal features of Copernicus' system are that the earth rotates on its axis, and that the earth and all the other planets orbit the sun which takes over the central place in the universe. The moon, of course, continues to orbit the earth. The sequence of planets from the sun was in the order of their times to complete an orbit (i.e. the length of their 'year'), which meant that Mercury and Venus were nearer to the sun than the earth, and explained why they are always in roughly the same direction as the sun when viewed from the earth. Copernicus' theory also stated that the fixed stars were stationary, their apparent daily movement resulting from the earth's rotation, but were much more distant than previously supposed, to account for the lack of observed parallax.

The chief attraction of the heliocentric system was that it readily, even naturally, explains the retrograde motion of the other planets as seen from the earth: it arises automatically when the earth laps one of the outer planets, or is lapped by an inner one. It was this elegant feature that persuaded Copernicus and his early followers of the truth of this system; by comparison, the Ptolemaic system, which required epicycles to generate retrograde motion, seemed contrived. But it should be noted that a basic heliocentric system only explains the retrograde movements qualitatively – predictions of planetary positions based solely on circular orbits (even around the sun) were much less accurate than using Ptolemy's system. So Copernicus still had to introduce many epicycles and the like to give good agreement with observations; in fact, with 34 epicycles it was not much less complex than the preceding Ptolemaic systems. Copernicus also retained the crystalline spheres.

The initial response to Copernicus' ideas was limited. He had good contacts within the church authorities, and it seems he had made every effort to get the ecclesiastics on his side, discussing his ideas before publication, and dedicating his *De revolutionibus* to the Pope, Clement VII. Also, his early ideas had received little publicity, and he was no longer around once his main work was published. An important reason why there was so little impact to begin with is that the Preface said the work merely presented an alternative approach for calculating the planetary positions, useful for compiling astronomical tables, but was not advocating an actual sun-centred physical system, and presumably the retention of some epicycles aided this impression. However, the Preface was written by someone else and there is little doubt that Copernicus himself saw it as a real planetary system. Nevertheless it gave the Catholic Church reason for not taking *De revolutionibus* too seriously, and it did not object to Copernicus for over 70 years. In the sixteenth century, his views were opposed more by Protestants, including Luther (1483–1546). For the first hundred years after its publication, astronomers who openly adopted a heliocentric system were very much in the minority, but gradually evidence emerged to support it.

Tycho Brahe

Undoubtedly the principal practising astronomer who opposed Copernicus, and proposed an important alternative theory, was the Dane, Tycho Brahe (1456–1601). In 1576 he set up an observatory on an island near Copenhagen and spent the next 20 years carrying out many high quality observations. Telescopes had not yet been invented, but he used large pointers and sighting tubes which enabled positions to be determined to just a few minutes of arc (much more accurate than Hipparchus). Also, unlike most of his predecessors, he plotted planetary movements throughout their orbits rather than focusing only on particular parts; these data later proved invaluable to Kepler.

Brahe was convinced by the arguments for an immovable earth. It was not only the lack of parallax, but if Copernicus were right then not only was the earth rotating, but it was also moving at very high speed in its orbit around the sun. It was felt that there should be some sense of a wind due to these movements. Also, it was thought that, if the earth was moving, falling objects on earth would not fall straight down, but should land somewhat to the side because of the distance moved by the earth's surface in the time the object was falling. (Now, of course, we understand why these do not occur.) But, on the other hand, Brahe recognised the conceptual benefits of the planets orbiting the sun, especially in terms of explaining retrograde motion.

Consequently, Brahe proposed a compromise solution in which the sun (and moon) orbited a central and stationary earth, but all the planets then orbited the sun. The starry sphere also revolved around the earth. Hence, it satisfied the traditional view of a central earth, and in terms of explaining observations from earth it is mathematically identical to Copernicus' system, so it was adopted by many astronomers at the time. It is interesting to note that in his system the orbit of Mars around the sun crossed the orbit of the sun around the earth; this would seem impossible if the celestial bodies were carried on crystalline spheres, and clearly suggests that Brahe had abandoned that concept.

Brahe recognised that even his system did not give accurate agreement between theory and observation, but died before he could develop the scheme further. He charged his assistant Kepler with this task, and we will come to his contribution in due course.

Bruno

First, however, must be mentioned the severe conflict between advocates of the Copernican system and the establishment which developed, especially in Italy, towards the end of the 16th century and reached its height early in the next. A key figure in this was Giordano Bruno (1548–1600). Whereas Copernicus had endeavoured to present his ideas in the least antagonistic manner, it seems Bruno went out of his way to be as provocative as possible.

Bruno was a Dominican monk and could be regarded as a scholar, but he was not a scientist, and he embraced the Copernican theory for philosophical rather than scientific reasons. He focused on its non-anthropocentricism, i.e. that mankind is no longer the centre of interest in the universe. He reasoned that if the earth is like the other planets (in that it orbits the sun), then presumably the other planets are like the earth. In particular, he assumed they were populated, and provoked awkward questions about the standing of their inhabitants in relation to Christian teaching about mankind. Similarly, he reasoned that, just as the earth is not a special planet, the solar system is not special either, but there must be many other inhabited systems in the universe; in fact he used Copernicus' system to advance the idea that the universe is infinite, so there were innumerable populated solar systems. And he argued that if the universe is infinite then it must be equated with God, and this led him to adopt pantheistic beliefs. He presented his views in *On the Infinite Universe and Worlds* (1584), written while in London, and his ideas were no doubt reinforced if not formed by his contact with English scientists. He returned to Italy, and in 1592 was handed over to the Inquisition, charged with heresy, blasphemy and immorality – and ultimately burned at the stake.

By way of contrast to Bruno should be mentioned the English mathematician Thomas Digges (?1546–95). He strongly supported the Copernican system and explicitly proposed an infinite universe with the stars at various distances from the earth rather than carried on a celestial sphere. Digges published his views in 1576, after the supernova of 1572 (see below), almost a decade before Bruno's *Infinite Universe*. Digges' work, *A Perfect Description of the Celestial Orbes...*, was very popular and went to many editions; but how many today have heard of Thomas Digges? This is a clear indication that Bruno was condemned more for his outspoken and overtly anti-Christian views, than for his support of heliocentric cosmology. It probably also reflected a greater toleration of unorthodox secular views by the Church of England.

However, there is no doubt that the extreme and disruptive views of Bruno became associated, at least in the minds of the Catholic Church, with Copernican cosmology and provoked stronger resistance to heliocentric views as a result, especially in its persecution of Galileo.

Galileo

Galileo Galilei (1564–1642) was born near Pisa and educated at the university there. He first studied medicine but soon turned to science. For a time he was Professor of Mathematics at Pisa, but lost favour with the university authorities because he questioned Aristotle's physics, and moved in 1592 to a similar position at the University of Padua where he began his astronomical work. From 1610 he held the position of mathematician-in-residence to the Grand Duke of Tuscany. It is worth noting that Galileo was foremost a mathematician and physicist, rather than an astronomer, and that he antagonized his academic peers over Aristotelian physics long before he turned to astronomy.

In the latter years of the 16th century a number of events occurred which began to shake confidence in Aristotelian cosmology. Supernovas are 'exploding stars', which increase dramatically in brightness in just a few days and then fade over the following months. Occasional supernovas had been observed since antiquity, but had been assumed to be within Aristotle's sublunary region where change was permissible. However, in 1572 one occurred which was so bright that initially it outshone Venus and then lasted for about 18 months, during which careful observation by various astronomers showed that it was as far away as the fixed stars, which clearly challenged Aristotle's idea of the perfect and unchanging celestial region. Shortly after, in 1576, a comet appeared, and again observation showed it was beyond the moon, giving further evidence of change in the heavens; and the observations suggested that its trajectory took it through several crystalline spheres, and moved in a parabolic course rather than circular which heavenly bodies were supposed to follow. There was another supernova in 1604, and it seems to have been this that sparked off Galileo's interest in astronomy.

Early in the 17th century the telescope was invented, and in 1608 Galileo made his own and began to examine the sun, moon and planets, publishing his early results in 1610 as the *Starry Messenger*. He revealed various phenomena that challenged the Aristotelian view of the universe: there were sun spots which showed that the sun was not a perfect unblemished heavenly body; evidence for a heliocentric solar system came from realising that Venus passes through phases like the moon; and he also found that there were very many more stars than previously thought.

Undoubtedly his most famous astronomical discovery was that Jupiter has moons. The significance of this was partly that it contradicted the long-held philosophical notion that there could be only seven heavenly bodies (apart from the stars); but more importantly it showed unequivocally that bodies could orbit something other than the earth. This latter point challenged the central role of the earth, and it refuted Aristotelian physics which stated that the only natural movement for a body was either towards the centre of the earth or in a circular orbit around it. Because Aristotle's physics underpinned geocentric theories of the universe, proving his physics to be wrong, and wrong in a celestial context, seriously weakened them.

Of course, what made Galileo's discoveries especially famous was the establishment's stubborn refusal to accept them, to the point that some individuals would not even look through Galileo's telescope to see the evidence for themselves, maintaining that it was only an artefact if it could not be seen with the naked eye. This reaction would be inexcusable today because, although we know that our senses can sometimes be deceived, on the whole we give great weight to information from our senses, especially to what we can see, even through an instrument. But it was very different in Galileo's time: then, with Plato's philosophy still influential, and the deductive approach to science still dominant, they gave more weight to getting their premises right and deducing the implications, and they stressed that the senses are fallible. If observations did not agree with what was expected, it was they that must be questioned first rather than their deductions.

In the following few years Galileo continued to publish on astronomical and physical subjects, such as mechanics and hydrostatics, contesting Aristotle's teaching and supporting Copernican cosmology. There was strong resistance by the academic and ecclesiastical authorities, who refused to question the basis for their own ideas and persistently argued against Galileo on the basis of Aristotle's natural philosophy. Eventually in 1616 the Catholic Church, incited by various academics and in the person of Cardinal Bellarmine, pronounced that a moving earth was heretical. It forbade any further teaching or writing in support of Copernican cosmology, and *De revolutionibus* was placed on the *Index Librorum Prohibitorum* (the Papal Index of Forbidden Books, which carried the sanction of excommunication for offenders).

Bellarmine died in 1621; and in 1623 Cardinal Barberini, who in earlier years had been a friend of Galileo, became Pope Urban VIII. Galileo obtained his permission to produce a supposedly impartial assessment of the Copernican versus Ptolemaic cosmologies. When this emerged in 1632 as *Dialogues Concerning the Two Chief World Systems* it was but a thinly veiled support of the heliocentric system. Galileo's opponents used this to bring him before the Inquisition, where Galileo recanted and was punished by house imprisonment. But there he wrote his *Discourse on Two New Sciences* which further undermined Aristotelian physics.

It is often portrayed that the establishment resisted change solely on the basis of protecting its authority and dogmas. However, this view suggests that the church knew Galileo was right but just did not want to accept it, whereas it is more likely that most of the participants genuinely believed in the truth of Aristotelian physics and in their deductions from theology regarding the natural world. They were conscious that their beliefs had a long, well-established pedigree, and they were not going to rush to abandon them just because one or two recently observed phenomena did not seem to fit.

Galileo is justifiably acclaimed for the many contributions he made to the advancement of science. First, of course, was his open support for the Copernican heliocentric system, which helped to win over supporters, although it was not generally accepted until later in the 17th century, especially after Newton. He also made persistent attacks on various aspects of Aristotle's physics; and, by doing this, not only did he undermine geocentric astronomy, he also helped pave the way for further scientific advances in physics and mechanics – in effect he laid the groundwork for Newton. He criticised Aristotelian physics simply because it did not agree with observed facts, and Galileo was a strong advocate of empirical science – that science must be based on experiment, and not on deduction from *a priori* premisses.

It was this latter point, of course, which brought him into direct confrontation with the establishment. In effect, it was the synthesis of Christian theology and Aristotelian science that had been formulated by the 13th-century scholastics which became its downfall. Had it not been for this synthesis then it seems the church could have taken an entirely neutral stance towards the new astronomy; but because of it, the church was reactionary and tried to suppress Galileo. What began to emerge from about this time was a separation of secular from sacred. Galileo encapsulated this in the well-known phrase – that the Bible tells us how to go to heaven, but not how the heavens go – and an increasing number of both scientists and theologians after him agreed with this sentiment.

THE NEWTONIAN REVOLUTION

After Galileo the two principal rival cosmologies were those of Copernicus and Brahe which are mathematically equivalent in terms of astronomical observations, so there was nothing to choose between. Brahe's had the advantage of following the orthodox geocentric approach, but it seemed somewhat contrived to have all the planets orbiting the sun which then orbited the earth; but the problem with the heliocentric system was the failure to identify any hint of the parallax that should result from the earth moving around its orbit. Astronomers had to wait for Newton before the debate could be settled. But first we need to mention Kepler.

Kepler

If I had adopted a strictly chronological approach then I would have considered Kepler, Tycho Brahe's successor, before Galileo. However, although Kepler made valuable contributions to cosmology, they were largely overlooked until Newton made use of them much later in the 17th century.

Johannes Kepler (1571-1630) was born in Germany and trained at the University of Tübingen, primarily in theology and the classics, but he was also introduced to astronomy and adopted heliocentric cosmology at an early stage. Yet in many ways he was one of the old school, having pronounced Pythagorean ideas, especially sharing their fascination with mathematics and believing in the importance of numbers and geometrical shapes for understanding the make-up of the universe.

In 1600 he became assistant to Brahe, gaining access to his extensive high quality data – data which Kepler could not have obtained for himself because he did not share Brahe's observational skill. However, with his interest and ability in mathematics, Kepler set about using Brahe's data to examine planetary orbits in detail. To begin with, faithful to his Pythagorean and Christian principles, he remained committed to the concept of a circle and sought the 'right' combination of epicycles and eccentric deferents that could accurately model the observations. But after five years of assiduous effort he failed. Previously, data had been sufficiently imprecise for satisfactory agreement to be possible with systems based on epicycles, but Brahe's data was so good that it highlighted the deficiencies in that sort of a construction.

But Kepler did not give up. Instead he abandoned the circle and went on to investigate a wide range of alternative geometrical figures in search of one that could fit the data. Eventually he determined that the planetary orbits are elliptical, with the sun at one focus, and that such an arrangement provides very accurate agreement with observations without needing embellishment with e.g. epicycles. However, the planets do not traverse their orbit at a uniform speed, but are faster when near the sun and slower when further away. Kepler worked out his system first for Mars and published these results in 1609 in *A new astronomy...*, and then demonstrated the applicability of his theory to all the planets, including the earth, in 1619.

As mentioned earlier, despite the relatively early date of Kepler's work, it was largely ignored for half a century, even by Galileo who corresponded with Kepler. It seems that despite his innovative thinking in other areas, Galileo was committed to planetary movement being based on circles, and was more interested in advocating the heliocentric concept than finer details of the orbits. Also, even though Kepler's scheme was framed in terms of a heliocentric system, it could be applied to Brahe's combination of heliocentric and geocentric motion, so it still did not resolve between the two systems.

Newton

Isaac Newton (1643-1727) graduated from Trinity College, Cambridge, in 1665, and after a two-year interval at home because of the plague, returned to Cambridge for most of his working life. He became recognised as a pre-eminent scientist in his own time, including being appointed President of the Royal Society.

Many of Newton's predecessors, working in the physical sciences, had been handicapped by a lack of mathematical techniques for developing and applying the physical phenomena and concepts they were grappling with. One of Newton's early achievements was in mathematics where he devised a form of calculus which later enabled him to progress his ideas on mechanics.

When it came to mechanics, Newton completely overturned the Aristotelian concept of motion. Whereas Aristotle had thought it was necessary to keep pushing something to keep it moving, Newton realised it was the other way around – a force is needed to start an object, but once moving it will keep on doing so, and in a straight line, unless a force acts on it to alter that motion, whether to speed it up, slow it down, or alter its direction of movement. This was Newton's first law of motion, and the second was that the acceleration of an object was proportional to the force acting on it, and inversely proportional to its mass.

Newton applied his laws of motion to the subject of gravity. Galileo had shown that falling objects move with uniform acceleration, i.e. their speed increases by the same amount each second. Obviously the motion of falling was attributed to gravity, and Newton proposed that the constant acceleration was due to the constant pull (force) of the earth's gravity on the falling object. He also reasoned that the force of gravity is proportional to the mass of the object, because all objects fall at the same rate (neglecting air resistance) whatever their weight. [2]

He then applied his laws of motion and his understanding of gravity to the question of planetary orbits, which had been brought to his attention by Edmund Halley (1656–1742, after whom the well-known comet is named). From his first law of motion, Newton knew that circular motion is not a stable 'natural' motion as Aristotle had thought, but requires continual application of a force to effect the continuous change in direction from a straight line.[3] He saw that the earth's gravity could provide the force necessary to keep the moon in orbit around the earth, and extrapolated from this that the sun's gravity could keep the planets in their orbits around it. He went on to show that gravity could explain all of these – the rate at which something falls at the earth's surface, the orbit of the moon around the earth, and the various orbits of the planets around the sun – if the force of gravity exists between all objects (universal gravitation) and decreases in proportion to the square of the distance separating the relevant objects.

What was particularly striking, and carried much weight in demonstrating the truth of his laws of mechanics and theory of gravitation, was the result when he looked at orbits in more detail. What he found was that, when one body is orbiting another under the influence of an attractive force such as gravity which varies inversely with the square of distance, it will *necessarily* adopt an elliptical orbit with the principal body at one focus. Even more convincing was that the speed at which it moves around

---

[2] That is, the gravitational force on an object is proportional to its mass, but the acceleration is inversely proportional to its mass, so the net acceleration is independent of the mass.

[3] To appreciate this, consider moving a weight in a circle around oneself, attached to a string. The pull on the string (so-called centrifugal force) is what makes the weight move in a circle; let go of the string (i.e. remove the force), and the weight will appear to fly off, but actually it is simply carrying on in a straight line, along a tangent to the original circular motion.

the orbit varies – it slows down as it moves away from sun, so is slowest when it is furthest away, and accelerates towards the sun, so it is fastest when nearest; and this theoretical variation of speed fitted in exactly with what Kepler had found.

Newton published his theories regarding mechanics and gravitation in *Philosophiae Naturalis Principia Mathematica* (usually known as Newton's *Principia*) in 1687, and its importance was recognised immediately. Newton's work finally confirmed the heliocentric view of the solar system because with his mechanics the lighter body orbits the heavier, not the other way around, so it was no longer tenable to maintain that the sun orbited the earth. (It might be said that *final* confirmation came in 1838 when the astronomer and mathematician Friedrich Bessel (1784–1846) identified the long-sought-after parallax of a relatively nearby star, 61 Cygni; but by then very few if any astronomers doubted the heliocentric system, and it was considered just a matter of time before this observational proof was found.)

The Scientific Revolution

Superficially, all that has been said in this section so far has been but the final stage of the Copernican Revolution – the widespread adoption of a heliocentric view of the solar system (and an appreciation that the universe is much larger than the solar system). But Newton's work on mechanics and astronomy did much more than merely provide the final piece in that jigsaw. It is important to gain some appreciation of how his work was significant in other ways too. He had a major impact on the nature of science as a whole, and how it is perceived; so much so, that even modern-day scientists and philosophers of science cite Newton for his pioneering work. In fact the influence of his work extended well beyond science.

Scientific method

First to be mentioned is his contribution to the scientific method. We have already come across deduction and induction; Newton showed how these two cognitive processes should be used in scientific inquiry. To his predecessors, induction was essentially the identification of patterns in a set of data, but Newton extended this concept to the hypothesizing of underlying laws to account for those patterns. He then tested his hypotheses by deducing from them some implication or consequence that could be verified by experiment: if the observations were found to agree with the predictions then the theory was strengthened, if not then it should be rejected or at least modified. This is the basis of what is called the hypothetico-deductive method of doing science, which is generally considered to be the modern approach to science, though I have more to say on that later. In effect, Newton established a model for subsequent scientific research.

Scientific explanation

Perhaps the most important contribution Newton made to the scientific enterprise related to what might be called the level of scientific enquiry. Before Newton, astronomers from the Greeks onwards had focused exclusively on devising a spatial arrangement of the earth, moon, sun and planets which could account for their observed movements: they had been 'saving the appearances'; this was also true of Kepler who determined the correct form of planetary orbits. But Newton was different, radically different: he did not merely confirm that the planets move in elliptical orbits, he provided an underlying explanation for why they do so. This explanation was based on a more fundamental property of nature – that of gravity, and the laws of motion – i.e. his explanation was at a more basic level. In other words, Newton's work represented a radical change in the nature of science: from merely identifying and *describing* phenomena to *explaining* them. Indeed, he changed the *purpose* of science from that of merely describing the universe to that of explaining it in terms of its fundamental laws. Following Newton's lead, scientists in every field set out to determine the fundamental principles – the scientific laws – pertaining to their particular discipline. This was not only in the physical sciences, but in chemistry and biology as well; and there is no doubt that Darwin saw natural selection as a law of biological science, analogous to gravity being a law of physical science. Further, that he saw his theory of evolution by natural selection as 'explaining' the diversity of biology in this Newtonian sense.

The mechanical universe

Another major impact of Newton's work, arising especially from its emphasis on mechanics, and that the laws of mechanics and gravitation appeared to pervade the universe, was that it reinforced perception of the universe as a large and complex machine – a machine in which all its parts interact consistently in obedience to natural laws. Actually, Newton had realised that a planet's normal elliptical orbit about the sun would be perturbed by the gravitational attraction of other planets and comets, and had thought that God would need to nudge them back into their proper course every now and then to avoid the solar system becoming unstable. It was about a century later that the mathematician Pierre-Simon, Marquis de Laplace (1749–1827), showed that, whilst the perturbations are real enough, they eventually cancel each other out and need not be a cause for concern.

Hence, it became increasingly recognised that the world, indeed the universe in so far as we know it, consistently follows the operation of natural laws. This contrasted with the traditional view that God was intimately involved with the universe. At least for most 17th-century scientists, notably for Newton, God was still unquestionably the architect of the universe, but he had instituted natural laws and delegated its day-to-day

running to the operation of these laws. In other words, whilst God remained the First Cause, the universe operated in terms of regular secondary causes. This was an important consideration so far as science was
concerned, because if there is the possibility that God will meddle with
nature then trying to determine regular cause and effect is fraught with
uncertainty; but if nature is free of divine interference then it means that
its laws will operate consistently and hence be amenable to scientific investigation. So this further spurred on efforts to elucidate nature's laws.

The consistent operation of cause and effect has other important implications. On one hand it leads to questions of determinism – that the future
is entirely determined by the present state of affairs – which is a problem
for philosophers *vis-à-vis* free will. But the scientists turned the concept
around and realised that, if the universe runs and has always run like a
great machine, following unchanging natural laws, then it may be possible
to retropolate (extrapolate backwards) from the state of affairs today to
determine what the situation had been in the past. Hence arose the concept of evolution, at least in the sense of the physical universe, and the rise
of cosmogonies – theories about the origin of the cosmos.

Obviously these ideas challenged the traditional Christian view of the
earth having been made in the beginning essentially as it is now. And from
the idea of the earth developing from some primitive form, it is not too big
a step to think of how life on earth might have developed too – leading to
the idea of organic evolution. But what evolutionary processes need is
time, and plenty of it – which is the subject of my next chapter.

Authority and Reason

However, we can hardly leave the scientific revolution without a few comments on its wider implications, because the revolution was not only about
how science is carried out, but also about the standing of science itself. In
particular, there was from this time a marked decline in the authority of
the church and a corresponding rise in the weight given to the voice of
science and reason.

We have seen how, from the amalgamation of church and state early in
the Middle Ages, the church became the principal authority in Europe, and
exercised authority over what it considered to be orthodox secular knowledge as well as orthodox theology. Arising from its attempts to present
Christianity favourably compared with contemporary knowledge and philosophy, it devised a synthesis first with Platonism (chiefly by Augustine)
and then with Aristotelianism (chiefly by Thomas Aquinas). The close association of secular and theological ideas, coupled with the medieval
mindset and scholastic attitude which venerated the ancient authorities,
made the church suspicious of new ideas and reactionary towards the
growth of knowledge. This showed up most clearly in its dealings with
Galileo: it was evident to many that the church had suppressed Galileo
simply by weight of authority and not by reason or cogency of argument.

The subsequent growth of science through the 17th century, especially the work of Newton, left no doubt that Aristotelian physics was incorrect and that the church had been trying to defend an erroneous position. It was partly because the church's view had been wrong, but perhaps mainly because of the authoritarian way in which it had sought to impose its view, that its credibility was seriously undermined. Superficially, in the short term, it appeared to win; but the fact that it won only by brute force and not by force of argument, meant that in the long run it lost; first it lost credibility, and this meant that in due course it also lost its authority. The standing of the church had, of course, also been eroded by the Reformation of the 16th century. So one side of the coin was the waning authority of the church: before the scientific revolution its voice went almost unchallenged; but that was severely eroded by its stance towards the advance of science.

The other side of the coin was the growing confidence in science. In the Middle Ages, the abundance of mythical ideas, pseudo-science, and general mistrust of the senses, meant that science was scarcely discernible as a distinct voice at all. Then, beginning with the Renaissance, and especially through the 16th and 17th centuries, science began to emerge as a coherent discipline – with weight given to observed facts and to reasoned conclusions based on those facts. This was clearly seen in the evident coherence of Newton's mechanics – the way it extended from commonplace falling objects on earth to describing how the heavens moved, and the impressive way in which theory and observation agreed. Science was perceived to be objective, free of religious or philosophical bias. Increasingly, truth was perceived to lie in the laws of science: people were prepared to listen to science even when it seemed at odds with established views, and science began to emerge as the voice of reason and objectivity.

In fact the scientific revolution of the 17th century ushered in what is called the Age of Enlightenment which flourished during the 18th, especially on the continent. This was a time of reaction against any form of traditional authority, including the classical philosophers as well as traditional Christian teaching, and of increasing reliance on reason alone as the basis for knowledge.

The church has never recovered from its defeat at the hands of Galileo and Newton – even though neither had intended it – which is shown by the events of the 17th century still being cited so often today. The voice of established religion was not silenced, but it was seriously weakened: no longer could it make pronouncements regarding the natural world and expect people to take notice. This can be seen when the issue of the age of the earth arose, and the church's stance was much more to try to argue its case than pontificate.

3

Time and Tide

This chapter is primarily about man's perception of time and the age of the earth. It has an emphasis on geology because it was largely the rise of geological science that led to an overturning of traditional beliefs in a young or recently formed earth and the birth of modern ideas regarding its antiquity. It is well known that this much greater age of the world provided Darwin with the time he needed for evolution to take place. But there was also a significant change in the perception of time itself – from a static view of the universe to one in which everything is developing – which contributed to the emergence of evolutionary ideas.

Ancient beliefs about the age of the earth

An aspect common to many ancient civilisations, in the East and West, was a cyclical perception of time. Being aware of the periodicity of nature – months, seasons, years – and the cycle of repeated generations, it was only natural to believe that time itself recurred, even to the point that history repeated itself. So this became part of many early religions and philosophies, often tied up with recurring celestial phenomena. Many of the classical Greeks, including Plato and Aristotle, followed this view that life on earth consisted of a series of repeated cycles. However, whereas Plato held that the universe had been formed at some time in the past, Aristotle strongly opposed any idea of creation and maintained the view that time is infinite and the world is eternal.

A cyclical view of time confers a sense of not getting anywhere, that ultimately there is no progress – which is quite alien to our modern concept, and probably would not be tenable in an age when we see so rapid an expansion in our knowledge of the world around us and such technological advance. But by the time the Roman Empire stretched across Europe there was already a sense that the golden age of Greece had been lost, and a cyclical view supported the hope that one day it may be restored. Indeed, a feature of many ancient religions and philosophies was that the current era is one of degeneracy from a preceding ideal age that had existed at the beginning of things – which is quite the opposite of our modern forward-looking perspective.

However, these views had only limited influence on the subsequent development of Western civilisation; this was dominated by the Christian rectilinear (straight-line) view of time which took over Western thought with the Christianisation of Europe, and which has shaped our present outlook.

THE CHRISTIAN ERA

By about the middle of the second century AD Christianity had spread to much of mainland Europe, and had come into contact with its cultures – predominantly those of Greece and Rome with their many established deities. Not surprisingly, this new religion was seen by them at first as a minor sect – no more than a newfangled and transient cult or superstition, arising from an inconspicuous part of the Empire.

To counter this, and to establish credentials for their fledgling faith, one approach adopted by the early church fathers was to emphasize that Christianity, although recently founded as such, had its roots in the much older and well-established Judaism. In fact they pointed out that most of the Jewish histories and prophets even predated the classical Greeks, and were at least comparable with the earliest available histories of neighbouring ancient nations such as Egypt and Babylon. To substantiate this claim they set out to determine the chronology of the Old Testament, based on which creation was dated at about 5000 BC. These chronologies were used by the historian Eusebius for his *Chronicles* and *Church History* which were widely disseminated and provided the dating system that was accepted by the western world for at least the next thousand years.

Augustine made three significant points concerning time: First, he stated that time itself started with the creation – the universe was created with time, not in time; so there was no 'before' the beginning. Then, in keeping with the Christian chronologies just mentioned, he maintained that the creation had been relatively recent, not more than 6000 years before his time, in contrast to the very ancient earth and recurring ages of Plato. And, importantly, Augustine clearly portrayed a strong historical or directional perspective of time. Due to his recognised intellectual and philosophical ability, taken with his prominent position as bishop of Hippo, his teaching became adopted by the church at large, and so became part of the generally accepted view in western Christendom.

So the Christianisation of Europe became a major turning point in the development of a sense of time and history. Cyclical views were largely abandoned in the West; and, probably of greater significance, there was widespread acceptance that the earth was no more than a few thousand years old. The purposefulness of time and history which Augustine communicated in due course helped to stimulate enquiry and promote a scientific approach, because 'cause and effect' needs the concept of time to have a sense of direction (called the 'arrow of time'); but first came the barren years of the Middle Ages.

It was during the Middle Ages that the current system of dating from the birth of Christ was adopted. The idea was introduced in the 6th century, became widely known through Bede's writings, such as *On the Reckoning of Time*, and came into general use in Europe by the 11th century.

As mentioned previously, probably the most significant impact of the revival of learning so far as science was concerned was the dissemination of Aristotle's work in the West. After the initial reaction, Aristotle's ideas were received with much enthusiasm, and this extended to his cosmogony which included an eternal earth. As a result, despite the fact that Albertus Magnus and Thomas Aquinas reasserted Augustine's teachings about time, there was some revival of the old ideas of time being cyclical or recurring.

However, as we have seen, as scientific inquiry progressed over the ensuing two to three centuries, there was a gradual erosion of confidence in Aristotle's physics and cosmology, culminating in acceptance of a heliocentric view of the universe and Newton's mechanics. With this rise of science, cyclical views of time again fell from favour and a rectilinear view was restored.

Also, the 16th century saw the Reformation with its renewed emphasis on the authority of Scripture, which re-emphasized traditional biblical chronology and time-scale. This was reinforced by the French scholar Joseph Scaliger (1540–1609) who learned several ancient languages, and studied Eusebius' works. He showed that, contrary to what had become the popular view, ancient history was not confined to the Greeks and Romans, but included Egypt, Babylon and Persia, and he compiled a chronology of these nations' histories. His chronology and the events of these nations paralleled those described in the Bible. He even proposed that the era based on the Julian calendar should be considered as having started on 1 January 4713 BC as the various cycles used in antiquity coincided at that time. All of which gave further support to the biblical scale of time.

It was in the middle of the 17th century that Bishop James Ussher produced his *Annals of the World* in which he constructed a chronology, based primarily on the Bible, which gave the date of creation as 4004 BC. This date is well known for it was printed in the margins of Bibles until quite recently, and served to popularize the view of a very young earth.

However, almost as soon as this concept of the young age of the earth was re-established it was challenged by the general stimulus to enquiry that was part of the scientific revolution. One of the subjects demanding investigation was fossils: they had been known for a long time, but still lacked a satisfactory explanation. As Cuvier (see below) said in his *Essay on the Theory of the Earth*, it is only due to the existence of fossils that we even suspect the earth has a past.

THE FOSSIL ENIGMA

Fossils are of considerable interest nowadays, especially with the popularity of dinosaurs and the like, and in recognising their importance for deciphering geology and the course of evolution. However, until about the 16th century or even later there was nothing like such attention given to them; in so far as people thought about them at all they were regarded merely as curiosities along with various other strange objects that were occasionally unearthed. There was speculation about them and ideas as to their origin, from as early as the Greeks, but there was no general consensus as to their nature. One of the fundamental questions which persisted unanswered for perhaps 2000 years was whether fossils, given their resemblance to organisms, were in some way of organic origin or merely inorganic artefacts. This was a key issue to resolve, because it was only when they were clearly recognised as being derived from living things that it became necessary to consider what geological forces or processes, and acting over what sort of time period, could account for their existence.

Early ideas about fossils

The early Greek natural philosophers knew of fossils, and their historian Herodotus (?484–425 BC) was one of those who, recognising fossilized marine shells hundreds of kilometres from the sea, concluded that this showed such land had once been submerged. But this view was by no means shared by all. An alternative, also consistent with an organic origin, was that they were carried by a deluge such as the Deucalion flood (an account in Greek mythology with similarities to Noah's flood in Genesis). Some thought of fossils as unsuccessful attempts at life, such as Xenophanes who envisaged that life emerged from mud; whilst others reckoned they had nothing to do with living creatures at all, they just happened to look like them.

Aristotle's views on fossils are somewhat anomalous. In the first place, given his extensive and detailed interest in natural history, it is surprising that he says so little about them. Secondly, although, along with his belief in timeless ages, he proposed that land is being elevated or new land emerging from the sea to compensate for erosion, he did not see fossils as a consequence of this or cite them in support of such a scenario. Rather, he seems to have thought of fossils primarily as originating *in situ* in the same sort of way as other objects in the ground, perhaps resulting from the action of celestial forces. Perhaps he saw them as part of the *scala naturae* which included a gradation from inanimate objects to living organisms, mentioned in the next chapter.

Augustine rejected any idea of fossils resulting from the action of celestial forces, as not being consistent with the Christian creator. Instead, he proposed that at creation God had planted seeds in the ground which subsequently developed into fossils (which resembles ideas put forward by Theophrastus who succeeded Aristotle at the Lyceum). This is an early example of attempts made by various theologians to propose solutions to harmonize observations of nature, especially geology, with their understanding of what the Bible teaches on such matters.

Perhaps if Aristotle had clearly seen fossils as arising from living creatures, then with the recovery of and prominence given to his work this view would have become generally accepted from that time. But he did not do so, and debate as to their nature persisted for centuries, which is why the existence of fossils did not challenge the traditional time-scale for so long. As late as the 16th century, the term 'fossil' could still mean any sort of object dug from the ground, rather than its modern meaning as a relic of past life. There just was not the same distinction between living and non-living as is so fundamental to the modern understanding, nor between objects formed from living things and those resulting from solely physico-chemical processes. For example, Konrad von Gesner (1516–65), a biologist whose work I will refer to in the next chapter, wrote *On Fossil Objects* (1565) which is well-known for its extensive illustrations of fossils, and in which he compared the appearance of various fossils with living organisms, but nevertheless did not assume that fossils were once living.

Organic origin

As already indicated, a few of the Greek natural philosophers had believed that fossils were derived from living things. So did some of the Arabs who preserved and transmitted their work, notably Avicenna (Ibn Sina, 980–1037), who thought that fossils had originated as plants and animals, but had been buried in mud and transformed into the element earth by some sort of mineralising or petrifying force. His views were substantially adopted by Albertus Magnus, who thus disseminated them in the West through his own writings.

At about the same time there was renewed interest in the possibility that fossils resulted from the burial of plants and animals by the biblical flood. This had been suggested first by Tertullian, one of the early church apologists, and was popularized in Renaissance Italy by Ristoro d'Arezzo in his *Composition of the World* (1282) which describes a wide range of natural phenomena. With the Reformation, the Genesis flood became established in the West as the leading explanation for fossils. This view remained dominant through to the 19th century, and is still maintained by some, notably many 'creation scientists', today.

So the general uncertainty about the origin of fossils – especially whether organic or not, and even if organic then it seemed they could be explained by the Genesis flood – meant that for a long time they could readily be accommodated within traditional beliefs, and there was no need to question the generally accepted biblically-based age of the earth.

Beginnings of the modern view

Modern ideas regarding fossils can be traced to Leonardo da Vinci (1452–1519), although it is doubtful how influential his views really were, as his relevant notes on the subject were not published until the end of the 18th century. Nevertheless, with an artist's keen eye he recognised the close similarity of some fossils to living creatures, down to fine details of structure, and he totally rejected any suggestion of a non-organic origin, especially ridiculing the notion of formation by plasticizing or celestial forces.

He concluded that the creatures must have been buried rapidly by silt-laden water, and in time the mud turned to stone. Because the fossils occur in orderly layers he concluded that they were not deposited by a flood, but progressively. This also meant that the fossils could not have been carried to the mountains by a deluge, and he proposed that these occurred where fossil-bearing ground was elevated.

The two who really introduced modern thinking about fossils were the Englishman, Robert Hooke (1635–1703), and Dane, Niels Stensen (1638–86), who spent most of his life in Italy and is often known by his Latin name of Steno.

Hooke was a scientist and for over forty years held the position of curator of experiments at the Royal Society through which he published much of his work. Perhaps he is best known for his work on the elasticity of materials, but he also made a major contribution to microscopy. He manufactured his own microscopes and studied a wide range of living organisms, publishing his observations in *Micrographia* (1665), which was the first book to be devoted to microscopy and included many detailed illustrations – well worth a look if you get the opportunity. In this he discussed fossils, although much of his more detailed work came later. From close observation it was evident to Hooke (as it had been to Leonardo) that fossils must in some way be copies of or derived from living creatures and not be some inorganic artefact. He concluded that the material surrounding the fossilized object must have been fluid when the object was buried in order to take on its image – comparing it with sealing wax taking the impression of a seal.

Steno trained as a physician and acquired an interest in fossils when asked to examine a shark's head by his patron Duke Ferdinand II and he recognised that the fossils long-known as tongue stones must be derived from shark's teeth. This prompted him to carry out various studies of the rock formations in Tuscany, and as a result of his researches he wrote a *Dissertation concerning a Solid Body enclosed by a Process of Nature within a Solid*. A significant contribution made by Steno were his comments about how fossil-bearing rocks are laid down in horizontal layers, and particularly in enunciating the principle of superposition – that lower strata must have been laid down before higher ones and must, therefore, be older. This was especially important in introducing the concept of time, and that different fossiliferous strata could be interpreted as snapshots taken at different periods.

So by the end of the 17th century it was becoming generally accepted that fossils were in some way derived from living organisms. But this conclusion immediately raised difficult questions in the minds of most scientists. First was the time required for their formation, including the indication that they had been deposited in various strata at different times, perhaps over a protracted period, rather than in one event which could be attributed to a flood. Second, whilst some fossils bore a close resemblance to living organisms, it was evident that many did not, which implied that some organisms had become extinct, and even raised the spectre of some form of evolution. At the time there remained the possibility that creatures known only from fossils may be found alive in some as yet unexplored recess of the earth; but as time went by and exploration widened, it became increasingly clear that many fossils do represent the remains of creatures which are now extinct.

The dilemma of the scientific community in general was felt keenly by a contemporary of Hooke and Steno, the Rev. John Ray (1627–1705). Ray was a keen naturalist and made outstanding progress on the classification of organisms which will be mentioned in the next chapter. He totally accepted traditional biblical chronology and also believed that species cannot change – after the initial creation no more were made *and none lost* – he believed that for species to become extinct would be inconsistent with the providence of God, which is reminiscent of Plato's ideas. But he was confronted with fossils. As a naturalist he was an expert observer, and he could not deny that fossils were almost certainly of organic origin, describing them as the petrified remains of organisms. Yet he was faced with the two problems: first that here was evidence of creatures which were no longer existing, and second the time-scale. The latter was typified by his sight of fossilized trees at Bruges which clearly appeared to result from processes taking far longer than the traditional biblical time-scale of 6000 years, or perhaps produced by processes which acted far more dramatically in the past than they do at present.

We should note that on this issue the church was no longer in a position to impose any particular interpretation of nature as it had over Galileo regarding astronomy. Rather, individual scientists had their own religious convictions which seemed at odds with what was being discovered about the world. Many had a genuine belief that what the Bible says is important and should be respected; it could not simply be ignored, and they searched for ways in which they could explain the occurrence of fossils by natural processes which could also be consistent with the biblical narrative. For many the Genesis flood continued to be the accepted explanation which could account for fossils within a biblical framework; but for others, as more discoveries were made, it looked ever less plausible. The difficulty remained unresolved for at least another century, but eventually progress was made through the rise of theories concerning the formation of the universe and earth – ideas which had been prompted by the advances made in astronomy outlined in the preceding chapter.

COSMIC EVOLUTION

One of the major consequences of the Copernican revolution, when it became apparent that the stars were immensely more distant than had been supposed previously, was the way it expanded man's view of the universe;

and we have seen how Bruno and others took this to its extreme and asserted that the universe is infinite. However, whereas Copernicus maintained the old view that the universe was created essentially as it is today, in the 17th and 18th centuries ideas emerged of the universe not being static, but that it, and especially the earth, developed from some primeval form to its present structure, and may still be developing; and with these theories came the concept of an expanded time-scale.

Descartes

This radical step was taken first by René Descartes (1596–1650), who is better known for his work as a philosopher and mathematician. Although it is commonly said that Descartes merely paid lip-service to the church's views in his writings, Bertrand Russell's opinion was that Descartes was a sincere Catholic who wanted to persuade the church to be more tolerant of modern science, rather than that his orthodoxy was merely politic. Descartes studied but never practised law, and followed a military career before turning to philosophy, mathematics and science. He spent much of his adult life in the Netherlands, and carried out most of his writing there. In the 17th century, this was the only European country where the state permitted reasonable freedom of thought, and other philosophers such as Locke and Hobbes took refuge there too.

Descartes accepted the Copernican heliocentric model of the solar system and propounded a theory for its formation. He first described this in 1633 in *Le Monde, ou Traité de la Lumière* (The World...), but withheld it from publication because of the condemnation of Galileo's work. However, the essence of his cosmogony is described in his *Principes de la Philosophie* which was published in 1644, and *Le Monde* was eventually published posthumously.

Descartes adopted an atomist-type view in which the universe is full of corpuscles (there was no empty space), and change is due to physical interaction between them. At the beginning, the corpuscles were set into motion by God and their subsequent movement formed vortices which eventually condensed to produce the stars and planets. It is evident that his cosmogony is little better than the classical Greeks', being highly speculative and based on very little science; but he preceded Newton's fundamental work on mechanics and astronomy. His theory is important because of the ideas behind it: first, its mechanistic character, that once God set the universe into motion it followed natural 'scientific' laws without further divine intervention; and second that along with this came the idea of the universe developing, evolving, progressing.

He also believed that in its early state the earth was hot, and then as it cooled the surface cracked to produce the surface features we see today. In this we see an early attempt to explain the formation of geological structures. Interestingly, Descartes retained a time-scale for his cosmogony of just five to six thousand years.

Burnet

Thomas Burnet (*c*.1635–1715) was a fellow at Cambridge and a clergyman who took the basic idea of Descartes' cosmogony and endeavoured to build a comprehensive and explicitly Christian history of the world around it, and published his ideas in *A Sacred Theory of the Earth* (1684).

One of his prime objectives (following Augustine and Aquinas) was to demonstrate harmony between the Bible and natural science, and he said that where such harmony was not apparent it was because of a misunderstanding of one or the other. For example, he challenged the traditional interpretation of the Bible that the earth has always been as it is, arguing from biblical texts that the earth has undergone radical changes in the past, and will undergo further substantial changes in the future. In particular, he asserted that the world had been a very different place before the Genesis flood, and that the processes which led to the flood also caused many of the earth's present physical features.

Burnet followed Descartes' view that God was the First Cause who had started the universe, but that since then the universe has run according to natural laws or secondary causes, without God's direct intervention. Some felt that distancing God in this way demoted him, but Burnet considered that the reliable running of the universe showed how capable the Creator was:

> We think him a better Artist that makes a Clock that strikes regularly every Hour from the Springs and Wheels which he puts in the Work, than he that hath so made his Clock that he must put his Finger to it every Hour to make it strike. [*Sacred Theory*, Book 1]

In this way Burnet provided a mechanistic rationalisation of Genesis which helped people to see the formation of the world as resulting from the outworking of natural processes, rather than necessarily by a supernatural fiat. Also, the apparently abrupt epochs of the Genesis 'days' of creation gave way to the idea of gradual progression and even geological evolution; and this also meant that the processes which moulded the earth could be susceptible to scientific investigation. So, overall, Burnet seemed to provide an account that was eminently reasonable science (for his day), and consistent with Genesis; and his *Sacred Theory* became very popular.

In *Sacred Theory* Burnet conforms to a time-scale of five to six thousand years, arguing for a young earth on the basis that the mountains have not been eroded away, earth is not overrun by man, and extant histories go back only a few thousand years. However, he subsequently accepted criticism from some quarters that the processes he described would take much longer than the six days mentioned in Genesis and he produced a less known work *Archaeologiae philosophicae* (1692) in which he allowed a greater span of time and treated Genesis as an allegory rather than so literally. But the wide-spread criticisms arising from this publication led to his removal from office! It is evident that even at the end of the 17th century the biblical time-scale was still held strongly by public and scientists alike; it would take another century before an extended time-scale was widely accepted. In the meantime, the *Sacred Theory* continued to be popular and served to perpetuate the diluvial (Genesis flood) explanation for fossils.

Newton

At about this time, there was a significant spin-off from Newton's work on gravitation. Close observation of Jupiter showed its image is not a true circle but is wider at the equator than at its poles i.e. Jupiter is an oblate sphere. A number of terrestrial phenomena suggested that this may also be true of earth, and it was confirmed by comparing the distances along the earth's surface of a given angle of latitude near the poles and at the equator. Newton showed that an oblate sphere is the shape that would be adopted by a rotating ball of liquid. On the one hand this provided further support for his theory of gravity; but, of significance here, it was also consistent with the earth having cooled from a liquid state. Consequently, after Newton, most serious cosmogonic theories were based on the earth forming as a rotating molten ball and then cooling.

De Maillet

A significant turning point in ideas about the formation of the earth came in the early 18th century with Benoît de Maillet (1656–1738). For a while he was French consul in Egypt and there came across old records showing how the level of the Nile had varied over long periods of time, indicating an overall fall. De Maillet extrapolated from this and other observations that the whole world had once been covered by water which had gradually receded, leaving sediments behind on the sides of the early mountains, and these had in turn been eroded to produce secondary deposits. De Maillet thus saw the fossil-bearing strata as resulting from natural processes operating over a long period of time, rather than a brief catastrophe: that is, unlike most of his predecessors he completely ignored the biblical time-scale and flood. For fear of censure at his radical ideas, de Maillet published his ideas anonymously in a book called *Telliamed* (his name spelt backwards) which was first circulated clandestinely and not published until after his death. As expected, this work came in for a great deal of criticism, but it introduced the idea of an initial world-wide ocean which, as we will see, was an important feature of one early school of geological thought; and it opened up the way for others to break from the shackles of Genesis.

We should note that de Maillet lived in the first half of the 18th century, during the Enlightenment when, in response to the advances of science in the preceding century, thinkers began to throw off the constraints imposed by the old authorities and exercise a new intellectual freedom. De Maillet sought to provide a rational basis for the formation of the earth

which completely disregarded biblical 'revelation' and the associated concepts of teleology (final causes) and the supernatural. The fact that de
Maillet felt he had to publish anonymously shows how real the constraints
were, and this is reinforced by the resistance to Buffon who expressed his
own ideas more openly.

Buffon

Georges Louis Leclerc (1707–88) was born into a French aristocratic family and became Comte de Buffon, by which name he is better known, in
1773. He trained as a lawyer, but had much more interest in the natural
world and was keeper of the royal zoological gardens in Paris for most of
his life.

Buffon was particularly impressed with Newton's achievements in
identifying fundamental laws behind the physical world, and he endeavoured to do the same for the world of nature. He set out to produce a comprehensive account of the natural world, presented from this 'scientific'
viewpoint, in his *Histoire Naturelle* which issued in 36 volumes between
1749 and 1789, and embraced mineralogy, botany and zoology. In the introductory volume he outlined his theory for the formation of the world,
but because it ignored biblical chronology he was censured by the Sorbonne, in the face of which he retracted and incorporated a statement in
his fourth volume to the effect that he totally accepted what Scripture had
to say on the matter. But much later, in 1779, he presented a fuller version
of his ideas in a supplementary volume entitled *Epochs of Nature*.

His cosmogony drew on ideas from his predecessors, but he combined
them into a coherent scheme and, importantly, extended the time-scale.
He proposed that the material which formed the planets was knocked off
the sun by a passing comet, and coalesced into discrete balls which then
gradually cooled. As we have seen, others had already suggested that the
earth started as a molten ball – but what was new with Buffon was his attempt to employ a scientific basis to estimate how long it would take to
cool. Newton had made his own estimate, but because it indicated a period
of many thousands of years, which was much greater than the biblical
time-scale to which he adhered, he had not pursued this. Buffon took
Newton's work and carried out experiments of his own on the rate of cooling of small iron balls, and came up with a time of about 75,000 years for
the earth to cool to its present temperature. With our modern knowledge it
is evident that his extrapolation was naive and seriously flawed – but that
is not the issue. The important point is that he made a genuine objective,
scientific estimate of the age of the earth based on the time required for a
physical process; and that he gave weight to this rather than to the time
specified by some ancient 'authority', whether that authority be biblical or
otherwise. In other words, Buffon's cosmogony represents a significant
advance, not so much in that he got it right, but that he consciously tried
to apply a scientific method to the question of the formation and develop-

ment of the early earth.

The other important point of course is that, as a result of his method, he completely shattered the traditional time-scale. His figure of 75,000 years may seem as nothing to us now, but it was an order of magnitude greater than that believed previously. To reconcile his theory with Genesis (no doubt for the benefit of the Sorbonne rather than for himself), he argued that the days of Genesis should be interpreted symbolically, and he divided the formation of the world into six 'epochs' corresponding to the 'days' of the creation week, with the present age being the seventh 'day'. Most of the time involved in the formation of the earth occurred before man arrived on the scene, and Buffon still placed that at only six to eight thousand years ago.

Buffon also proposed that the hot earth was surrounded by water vapour which condensed as the earth cooled sufficiently, resulting in a world-wide ocean. The impact of the initial rainfall and subsequent recession of the sea led to various episodes of erosion and deposition which resulted in the sedimentary strata. In Buffon's scheme, most of these episodes occurred in the early epochs, before man was on the scene; the biblical flood was but a minor subsequent event.

GEOLOGY COMES OF AGE

Geology was born as an independent science in the latter half of the 18th century and the first half of the 19th, the latter period commonly known as the heroic age of geology because so much was achieved, especially the assembling of a reasonably complete stratigraphical column which is still the basis of that used today (see Appendix 1). Its parents were the two schools of geological thought known as Neptunism and Plutonism which developed throughout this period.[1] Neptunism, so called because of its emphasis on the role of water, was associated by some with traditional views, including adherence to some sort of biblical interpretation, and employing catastrophes such as the flood in its explanations. In contrast, Plutonism emphasized the role of subterranean forces such as earthquakes and volcanic activity in the shaping of the earth's crust; and out of this school arose the doctrine of uniformitarianism which insisted on the steady action of terrestrial processes, rather than extreme or catastrophic events. To a large extent the two views developed side by side, but I think it is more straightforward in a short account such as this to outline them separately. I shall start with the Neptunists who follow on naturally from the earlier cosmogonies.

---

[1] Neptune was the Roman god of the sea, and Plouton the Greek god of the underworld.

Neptunism and Catastrophism

The German mineralogists

Geology took an important step forward with the German mineralogical school which grew up in the mining areas of Saxony. The industrial revolution had prompted exploration for coal and other minerals, which encouraged geological investigation, and there was a shift in emphasis – people became much more interested in classifying rocks in terms of their mineralogical nature and economic value rather than with cosmogony. Nevertheless, to explain the formation of the various strata, this group adopted the view proposed by de Maillet that the early earth had been covered by a world-wide ocean and, recognising that granite is crystalline, believed it had formed by chemical precipitation from this primeval ocean which they suggested had had a high concentration of dissolved minerals. The ocean covered the earth completely and resulted in a layer of granite over the whole earth surface; mountains and valleys occurred due to irregularities in the initial crust. As the ocean receded the mountains were uncovered and exposed to weathering and erosion, and material was washed into the sea where it deposited to form sedimentary layers; as the ocean receded further, some of these early sediments were themselves eroded and redeposited to produce more recent strata. The mineralogists introduced a stratigraphical classification: the crystalline basal rocks (principally granite) were denoted primary, with secondary and tertiary being subsequent sedimentary layers. These groups form the basis of the modern classification, although now with many subdivisions and a quaternary group for recent superficial deposits.

Undoubtedly the best-known advocate of this view for the formation of the various types of rock and their classification was Abraham Werner (1750–1817) who spent most of his life teaching at the Freiburg Mining Academy. Although he wrote little, he was a popular teacher, with many visiting students, and became influential through his lectures. Being much more interested in the mineralogical value of rocks rather than how they had originated, whilst he adopted the view I have just described, he did not explain where the water had come from or went to, he did not necessarily associate the ocean with any 'Deluge', and did not seem to care too much about how long the processes had taken.

On the other hand, others, especially some geologists in Britain such as Kirwan and Deluc, interpreted the world-wide ocean as the Genesis flood, assigned a short time-scale, and saw Werner's scheme as a ready way of harmonizing science and Scripture. Given the popularity and high regard for Werner's scheme, it provided them with an up-to-date, 'scientific', natural explanation for the world consistent with the biblical account – it could be seen as a contemporary version of Burnet's scheme.

Index fossils

In terms of their mineralogical composition, many sedimentary rocks are rather similar, even if from very different periods; but what began to emerge in the 18th century was the use of fossils to characterize different strata. Steno had laid the foundation for this with his principle of superposition, and others such as Werner used fossils to some extent to distinguish between various strata. But the person usually credited with pioneering the use of index fossils because of his extensive work on stratigraphy is the engineer and surveyor, William Smith (1769–1839). In the course of building canals in Britain he became well aware of the strata in a given region and that each tended to be characterized by a particular assemblage of fossils, and he used such correlations to trace strata from one region to another, which was especially useful for locating important coal measures.

Smith's prime interest was in making sense of the various strata; but in so far as he considered the time aspect he followed a biblical chronology. His work was used by the Rev. Joseph Townsend to produce a book entitled *The Character of Moses established for veracity as an historian, Recording events from Creation to the Deluge* (1813–15), in which he attacked the view's of James Hutton (see below).

Cuvier

While Smith was working in England, major advances in the interpretation of fossils were made in France by Baron Georges Cuvier (1769–1832). He was a prominent figure who established his reputation in science as a zoologist and is recognised as the founder of comparative anatomy; he held various academic and public positions, including Professor of Zoology at the Museum of Natural History in Paris, and was very influential.

He acquired an interest in fossils when asked to look at bones from Paraguay, which, with his skill in comparative anatomy, he was able to recognise as the almost complete skeleton of a giant sloth, which he named *Megatherium*, and which was evidently now extinct. This prompted him to undertake investigations in the Paris Basin, first in superficial alluvia and then in gypsum quarries excavated from Tertiary rocks. He was able to reconstruct and identify many types of animals (mostly mammals), and demonstrated that fossils in upper layers resembled modern creatures, but at increasing depth they were progressively more different, and many of the lowest were totally unlike modern ones. He concluded that many ancient animals had become extinct, in fact whole fauna had occurred in the past that had now been lost. Cuvier's work was a significant advance over that of Smith who had used fossils only to distinguish between different strata, without inferring a progressive change in the fossils.

Cuvier recognised that the fossil-bearing rocks occurred in discrete strata, with distinct changes in the nature of the fossils between them, including some which he identified as alternating between fresh and marine forms of life. He was satisfied that each stratum could be built up by normal sedimentary processes, but believed that normal processes could not account for the transitions in between and that extreme events must have caused them. He accepted that such 'catastrophes' could have been of short duration, with the biblical flood having been the last of these events, perhaps only a few thousand years ago; but suggested that hundreds of thousands of years would have been required to build up the sedimentary layers.

Along with a colleague, Alexandre Brongniart, Cuvier wrote *Researches on the fossil bones of quadrupeds* (1812), in which they discussed the progression of life through successive strata, and for which they are considered major founders of palaeontology. This work was published in England as *Essay on the Theory of the Earth*, which was mentioned earlier, where its references to catastrophes continued to support a biblical perspective on geology through the first part of the 19th century.

Buckland

The last person to mention here is the Rev. William Buckland (1784–1856) who, as well as being a clergyman, was an accomplished naturalist and made important contributions to both geology and palaeontology. He graduated from Oxford, became ordained in the Church of England, and then returned to Oxford, being its first reader in geology from 1818 to 1845. It was in this latter post that he was most influential.

From early in his career Buckland accepted the growing evidence that the earth had (or at least appears to have had) a remote past – a geological history far greater than the biblical few thousand years. However, on the other hand, he was adamant that the geological strata evidenced a catastrophic and widespread flood, and in the not too distant past. He presented his argument that geology demonstrates the biblical flood in *Reliquiae diluvianae*, in 1823.

> The grand fact of an universal deluge at no very remote period is proved on grounds so decisive and incontrovertible, that had we never heard of such an event from Scripture or any other Authority, geology itself must have called in the assistance of some catastrophe to explain the phenomena of diluvial action which are universally presented to us, and which are unintelligible without recourse to a deluge exerting its ravages at a period not more ancient that that announced in the Book of Genesis.

So the harmony, or perhaps compromise, between Scripture and geology that Buckland formed (and, because of his standing, many followed his lead) was to accept biblical chronology at least since the flood, but to be open to the possibility of a much more ancient earth before that. The flood also marked a significant watershed in the operation of processes which shaped the earth: that exclusively natural processes had occurred since the flood, but perhaps supernatural processes might have occurred beforehand. The more dramatic operation of forces in the distant past could, of course, be used to promote the idea of the earth appearing older than it actually is. This last point of view was by no means followed by all, and indeed was the very issue that the uniformitarians challenged.

Plutonism and Uniformitarianism

One of the problems with Werner's scheme, even evident to some of his pupils, was that although granite is crystalline it seemed increasingly unlikely that it could have arisen by precipitation from water, and much more likely that it had crystallized by slow solidification from a molten state. In the middle of the 18th century two French geologists had recognised the volcanic nature of the Massif Central, and this was a clear pointer to the origin of granite as solidification of magma, and an early indication of the action of subterranean forces in moulding earth's surficial features.

Hutton

At least in terms of the development of modern geology, the person usually credited with advocating the role of such forces, and of them acting steadily over a protracted period of time, is James Hutton (1726–97). He was born in Edinburgh and trained as a physician but never practised as such, working instead as an agricultural chemist and then as a mineralogist and geologist. It was towards the end of his life that he formed his ideas about earth processes, and these were published first as his *Theory of the Earth* in the Proceedings of the Royal Society of Edinburgh in 1788.

Unfortunately, in the enthusiasm to credit Hutton with formulating the uniformitarian principle, what is often overlooked is that his theory of the earth was founded primarily on a philosophical rather than a scientific basis. He observed that the continents are gradually being eroded, and concluded that without some means for their renewal then they would erode away completely. Hence, believing in a God who would not leave the earth to a downward spiral of decay but would maintain the earth in a viable state, he reasoned that there must be a way to counteract the effects of erosion, and hypothesized some means for thrusting land upwards. It was generally recognised that volcanoes result from magma erupting to the earth's surface; what Hutton proposed is that sometimes an upwelling of magma does not actually reach the surface but its force nevertheless pushes up the crust overlying it; this provided a possible mechanism for raising the continents and for mountain-building. In support of his theory he pointed out that most land is covered with sedimentary rocks, indicating that they must have been under the sea at some time.

Because it was evident to Hutton that the continent of Europe had scarcely altered within recorded history, he extrapolated that the processes of land formation and erosion must be extremely slow. But he did not merely propose that the earth must, therefore, be millions of years old:

in keeping with his philosophy he proposed that the earth had always been in some sort of equilibrium, in other words that it had always been in existence. To use his oft-quoted phrase, 'The result then of our present enquiry is, that we find no vestige of a beginning, – no prospect of an end', which is strikingly reminiscent of Aristotle's ideas. In that sense, Hutton's views were, in fact, retrograde: so much progress had been made in moving away from the traditional concept of the earth being made as it is, and towards the scenario of the universe and earth evolving; but Hutton was rejecting that sort of development.

Conversely, he accepted that man probably appeared on earth fairly recently – consistent with the few thousand years of recorded history, and he accepted that there had been a major flood in the not too distant past; so, remarkably, he could accept much of biblical chronology. But the important difference was, of course, he proposed not just that there had been a long time before the flood, there had been infinite time – which gave plenty of scope for geological processes to shape the earth.

Hutton's views were not widely known during his lifetime. They were publicized to some extent by John Playfair with his *Illustrations of the Huttonian Theory of the Earth* (1802) in which he tried to present Hutton's views in a more scientific light. Yet Hutton's views might well have been all but forgotten were it not for another Scot, who was born in the year that Hutton died.

Lyell

Sir Charles Lyell (1797–1875) studied law at Oxford and practised for a short time, but then turned to science, especially geology. It is thought that his interest in geology might have been prompted partly through hearing Buckland lecture at Oxford, and at first he followed Buckland's views, but gradually his stance became radically different.

His early work included stratigraphical studies on rocks of the Tertiary period – equivalent to those investigated by Cuvier (whom he met on one of his visits to France) – and he came up with similar observations in terms of a gradation of species through the strata. However, he was not inclined to follow Cuvier in accounting for the discontinuities between the strata in terms of catastrophes; in this he was influenced by the French geologist Prevost who was opposed to Cuvier's ideas.

Lyell went on to study Mount Etna and observed that it had been built up by a series of many eruptions, interspersed with periods of erosion – all of which must have taken a long time. This impression of age was compounded by the fact that the base of the volcano is on rock which was considered to be young by geological standards – so, taken with the time required to build the volcano on top of it, this indicated that the earth must be exceedingly old.

In the light of these observations, Lyell came to reject any idea of the earth having a limited life span, and to think in terms of the earth's processes acting over far longer periods than had been considered previously. This allowed time for natural processes, though generally acting very slowly, nevertheless to accumulate a large overall effect; and dispensed with any need for catastrophes. Lyell also read Bakewell's *Introduction to Geology* (1813) which was strongly in favour of Hutton's views, and he became committed to the principle of uniformitarianism. Lyell presented his theories in *Principles of Geology : Being an Attempt to explain the Former Changes of the Earth's Surface, by Reference to Causes now in Operation*, first published in 1830 and revised many times thereafter, in which he gave credit to Hutton for originating the uniformitarian position.

Before Lyell, the stance adopted by geologists such as Buckland was that natural processes, as we now know them, have acted in recent millennia, typically since the biblical flood, but before that, although the same types of process operated, they probably did so much more rapidly or with greater intensity than now, and some of the protagonists would have included the possibility of supernatural events. In direct opposition to this, Lyell insisted that the earth has been shaped in the past by exactly *the same processes*, and *acting with comparable intensities*, as we see operating in the modern world. For example, mountains had arisen not by violent upheaval but by the successive effects of ordinary earthquakes which elevated the land little by little; and, similarly, valleys were not formed rapidly by a gigantic deluge, but by gradual erosion over thousands of millennia. In other words, violent processes were replaced by vast periods of time as the means of explaining how the geological face of the earth had been formed. So far as Lyell was concerned, this was the only possible scientific view, and had to be adopted if geology was to leave behind mythical ideas and advance as a true science.

However, in his efforts to oust catastrophism with the general principle of uniformitarianism, Lyell overstated his case and adopted a position similar to Hutton's – that the current natural processes have always operated in *exactly* the present manner – a view which almost inevitably leads to the idea of an eternal earth. Clearly this conflicted with the growing sense that the earth had had a beginning and that in its earliest stages it must have been a very different place from what it is now. The mere fact of the earth having been hotter in the past necessarily implies that some geological processes must have operated differently then than they do at present. The obvious conflict with the prevailing cosmogony caused some resistance to Lyell's views, with Buckland and several other geologists plainly refuting the idea of the earth being eternal or in a steady state, and reasserting its initial creation. When uniformitarianism was finally accepted it was more in the sense of the uniform action of the laws of physics and chemistry which underlie geological processes rather than these processes themselves. In fact, this position was more 'scientific' than Lyell's: it was consistent with the Newtonian principle of there being regular scientific laws behind observed phenomena. Nevertheless, aside from this debate about exactly what was meant by uniformitarianism, the particularly important and immediate impact of Lyell's *Principles of Geology* was that geologists began to accept the earth was very old, with an age of tens or even hundreds of millions of years.

So far as the fossiliferous strata were concerned, it is obvious that Lyell did not interpret the discontinuities between strata as evidence of catastrophes; instead, he saw them simply as gaps in the record – arising from a cessation of the conditions which had led to their deposition and/or subsequent removal by erosion of strata that had been laid down. This was, therefore, further evidence of earth's great age because the strata of which we are aware are only a part of earth's history. However, because of Lyell's commitment to the extreme form of uniformitarianism (and despite his early stratigraphical work), he did not see the fossils as demonstrating a progression of life, as he felt that would contradict his concept of an eternal earth which has always been inhabited. He suggested that the discontinuities between strata in a particular location could be caused by migrations in or out of the region concerned.

A New Perspective

So publication of Lyell's *Principles of Geology* may be seen as a turning point. It signified that geology had become a science in its own right. And, with this, the idea of the earth being only a few thousand years old, which had so dominated people's thinking since the beginning of the Christian era, was completely abandoned in mainstream geology and replaced by an age measured in hundreds of millions of years – much longer than we can readily comprehend. In some respects it might be seen as a restoration of the early beliefs in a very ancient earth – but now with an objective basis, and without the mythical trappings or philosophical speculations of periodic time or an eternal earth. Also, just as Newton did so much more than establish the heliocentric view of the solar system, so the revolution in geology had far greater significance than merely revising the accepted age of the earth.

Scientific method

In the first instance, the emergence of geological science demonstrated the application of the scientific approach to a new area of knowledge. Towards the end of the 18th century there had been a sense that a scientific outlook was being employed in many disciplines – such as physics, chemistry and biology – but that somehow geology was lagging behind, still including unnatural or even supernatural events in its explanations. This, of course was the very point that Lyell sought to correct, and to put the discipline of geology on to a modern scientific footing.

Where we left the story of geology, it was recognised that the earth must be very old, but it was unclear as to how old, or even how to calculate how old, and there was much debate on the matter throughout the 19th century. However, the critical step had been taken – an answer would be sought from science and science alone: from then on, the views of traditional authorities would be ignored, and the search was on to find methods that relied on consistent physical or chemical processes which could provide a basis for an objective measure of the age of the earth.

Natural laws

All of which served to reinforce and broaden an understanding of natural laws and how they operate. Newton had demonstrated that physical laws which operate on earth also operate in the solar system and probably throughout the universe. In a similar way, with Lyell, it became accepted that processes which operate on earth (or elsewhere) at present will have operated in much the same way (at least in principle) in the past, and are expected to operate similarly in the future. Indeed, without uniformitarianism in this temporal as well as spatial sense, there can be no science.

Similarly, the perception of the universe as a machine, which had started with Descartes and Newton, gained further weight through the 18th century and into the 19th. Whereas most of the 17th-century scientists had still very much seen God's hand behind the workings of the machine, any such God was taking an increasingly remote position. It was at this time, during the Enlightenment, that the school of thought known as Deism reached its height. Deists accepted evidence for a creator in the design of the universe i.e. that the cosmos had a Maker or First Cause; but maintained that having started things, the creator then took a back seat and had no further dealings with it. The universe now functioned solely by means of secondary laws, which operated and could be investigated without any reference at all to the creator who did not intervene in the affairs of the universe. In fact, extension of the operation of natural laws temporally as well as spatially increasingly outlawed any possibility of supernatural events. Consistent with this, they also rejected any religious claims, especially the claims of revealed religion.

Along with this, even less weight was given to what the Bible may have to say so far as nature is concerned. It had already suffered through the church's stance on astronomy, and was further eroded by what was seen to be an irrational stance adopted by some regarding geology – insisting on a young earth and catastrophic flood despite a weight of contrary evidence. Some, following Burnet's position, maintained that harmony was possible between religion and science. But after Lyell, whilst the likes of Buckland continued to give some weight to what the Bible had to say, geology as a whole was freed from its influence and could proceed along strictly scientific lines.

Preparing the ground for organic evolution

Of course, one of the really important implications arising from the revised age of the earth was that it gave time for things to develop – to evolve. This idea had been sown with that of the earth and cosmos developing from some early simpler form to its present complex state. The evident age of the earth and the gradual formation of its geological structures gave biologists the time for evolution to occur and encouragement to think along those lines. Darwin took a copy of Lyell's *Principles of Geology* with him on the *Beagle* and, although it seems his evolutionary ideas did not crystallize until after his return, no doubt the voyage gave him the opportunity to assimilate Lyell's ideas, not least in freeing him from the traditional time-scale.

The suggestion of an evolutionary process having occurred was reinforced by work done on fossils. By the 1840s all the major geological strata had been identified and characterized by their fossils, with the general pattern of invertebrates—fish—reptiles—mammals indicating an overall progression.

So far as most people were concerned, biology remained the ultimate evidence of design in the world and hence of a Creator. But the success of seeing the universe as a machine which followed natural laws, and gradual erosion of the miraculous as an explanatory factor in nature, also set the scene to look for a natural explanation for life itself.

4

Fauna and Flora

The theory of evolution is generally attributed to Charles Darwin due to his best-selling *Origin of Species* (1859) in which his ideas were first published. However, it will already be apparent from comments regarding fossils in the preceding chapter that the concept of species changing, even in a progressive evolutionary sense, was not new. Darwin's main contribution was to propose a plausible guiding mechanism for this to occur, which he described as Natural Selection, and which accorded with the growing scientific understanding of nature. Further, the theory of evolution was far from complete with Darwin's work because the rules governing heredity were still a mystery and there was no knowledge whatever of genetic mechanisms. Not until the development of genetics in the first half of the 20th century was an understanding gained of how characteristics[1] are transmitted to offspring; and the advances of biochemistry in the second half-century revealed the mechanisms of genetics at the molecular level. So to gain an adequate appreciation of the theory of evolution we must look at the rise of biology before Darwin, and at how evolutionary ideas have matured since then. Having said that, I do not provide a detailed account of how the theory of evolution developed as there are many comprehensive works available. Rather, especially for the benefit of readers who are less familiar with evolution, the aim is to identify major milestones and highlight the important issues, especially with regard to my subsequent discussion.

BIOLOGY BEFORE THE SCIENTIFIC REVOLUTION

Classical biology

It may seem odd that an account of the relatively recent theory of evolution should start with the classical natural philosophers. However, not only did they undertake the first studies in biology, they also began to ponder related issues such as design, purpose, classification and the origin of life – issues which are still the subject of research and debate today. Further, as I will explain in due course, I think our current theories about biology are still being blinkered by deep-rooted ideas which stem from classical times.

---

[1] Biologists generally use 'character' rather than 'characteristic'; as I am writing for non-biologists too, I use the latter with which more readers may feel familiar.

Ancient evolutionary ideas

Following publication of Darwin's *Origin* some authors have found indications of evolutionary thinking in the ancient philosophers, and see them as pioneers of evolution. But this is probably more with the benefit of hindsight than that they actually contributed to formulation of the theory.

It is true that some of the early philosophers, in their endeavour to break away from mythological ideas of origins and to provide explanations which completely excluded anything supernatural, put forward naturalistic theories for the origin of life. For example, it is not surprising that Thales (*c.*625–546 BC), who believed that the fundamental element was water, should suggest that life itself arose from water. This view was supported by the familiar observations that water is necessary for life and that moulds etc. grow in damp conditions, apparently from nothing. (Spontaneous generation was still being proposed in the middle of the 19th century, and not abandoned completely until about a hundred years ago.) Anaximander (611–547 BC) went further, proposing that life began in the sea, and that land animals, including man, developed from fish; in fact he had an evolutionary understanding of the whole of nature and the cosmos.

Another idea came from Empedocles (493–433 BC) who suggested that existing animals had arisen from the coming together of various organs, limbs etc.; unsuitable combinations perished, whereas viable ones survived and are the creatures we now know. In this scheme we can see some concept of natural selection and survival of the fittest, although not in the sense of adaptation or progress as we would understand it today. His views are significant more because they clearly extended the philosophy of rejecting the supernatural and seeking exclusively naturalistic explanations for the world, such that even living things were derived from random associations rather than having been specifically designed by some superior being.

It should be noted, however, that all of these ideas originated with men who were predominantly philosophers rather than scientists, and they preceded Aristotle who is generally regarded as the first to have observed nature in anything like a scientific manner.

Plato's essentialism

Before discussing Aristotle, we must briefly mention Plato because, as we have seen, what little he did write on natural history affected for a long time how nature was perceived.

His concept of 'Forms' (see Ch. 1) meant that material entities were regarded merely as imperfect reflections of fundamental realities. In particular, his creation account describes how a god fashioned the universe, including forming living things as representatives of corresponding ideal entities. Plato also believed that all possible entities must have a real existence, so there were no gaps in nature. This philosophy conveys two concepts to an understanding of biology:

First, a consequence of invoking a rational maker is to imply that everything is designed and made for a purpose, and following on from this a teleological approach to the understanding of biological structure. This contrasts with the views of many of the pre-Socratic philosophers such as Empedocles.

Second, it implies that species are unchanging. This is because (i) they are designed, so to change would imply imperfection in the initial design; (ii) since material organisms are representative of an essential ideal, this automatically conveys a sense of fixity – the ideal is unchanging, so there is no latitude for living things to develop or evolve; and (iii) nature is complete – there is no opportunity for new kinds to arise because there are no gaps to be filled. It seems unlikely that the actual doctrine of the *fixity of species* was comprehended as clearly as this at the time, but the underlying concept became firmly accepted as part of the scientific or biological outlook, of its understanding and expectation of nature.

These ideas are at the core of what is called an essentialist view of nature, which came to dominate western biology for the next two millennia.

Aristotle's biology

As mentioned previously, although the early Greek philosophers tried to rise above mythological traditions, their theories regarding nature were still largely speculative; and the philosophy of such as Plato even militated against observation rather than encouraging it. Despite being a close student of Plato, Aristotle broke away from this and, especially in the natural sciences, pioneered an empirical approach based on careful observation. Aristotle's major scientific work was to identify and categorize many types of organism: his *History of Animals* describes more than 500 kinds.

Aristotle laid the groundwork for the concept of species – being natural kinds which reproduce true to type – whilst at the same time recognising that the individuals of a species are not exactly alike but exhibit some degree of variation. Although he did not make any formal classification of animals, always seeking order (patterns) behind the facts, he grouped those that had similar physical characteristics (morphology) and/or modes of life. This is similar in idea to modern genera, but his groups were much broader, for example embracing all fish or all birds; and he sometimes grouped animals in different ways, so a given species did not necessarily have a unique allocation.

He also placed animals into a hierarchy: the various kinds forming a ladder of nature (*scala naturae*) or Chain of Being from 'simple' organisms such as worms and flies at the bottom to complex animals, notably humans, at the top. In *History of Animals* (Book VIII) he wrote that 'Nature proceeds little by little from things lifeless to animal life in such a way that it is impossible to determine the exact line of demarcation, nor on which side thereof an intermediate form should lie.' So he saw this hierarchy as a gradation of complexity and 'perfection', with each organism closely related to those either side of it, and all together providing a complete order of nature. But it seems he did not see it as evolutionary, with progression possible up the scale; as the gradual change from one generation to another which progression requires would have contradicted his understanding that species produce like for like.

His attempt to classify organisms was consistent with his four causes (Ch. 1), with the *formal cause* being the type of organism i.e. the species to which it belonged. Then, the *material cause* was the substance which makes up the tissues and organs of the organism, the *efficient cause* was its parents, and its *final cause* would be a mature adult.

In recognising that everything needs an *efficient cause* (which most closely compares with our modern concept of 'cause') Aristotle provided a significant stimulus to scientific enquiry: It prompted him to ask questions of 'how?', and we still ask the same sort of questions, continually probing deeper into how things work. For example, we would now be completely dissatisfied with identifying the *efficient cause* of an organism as being its parents: we wanted to know how the characteristics of the parents were combined in the offspring (hereditary mechanisms), and how the information for a new individual is stored (DNA), and we are now working on how that information is decoded, e.g. to transform a single fertilized egg into a complete individual comprising perhaps trillions of cells of many very different types. This series of questions illustrates the application of Newton's science too – seeking explanations at progressively more fundamental levels.

Aristotle held very strongly to a teleological view of nature (his *final cause*) and of biology in particular. It was evident to him that the various structures of animals – their limbs and organs – are exceptionally well-suited to their function (see especially his *Parts of Animals*). So, for Aristotle, just as human implements are well-suited for their intended use because they have been specifically designed and made, it seemed certain that organisms must have been designed and made in an analogous way. One of the widely accepted claims of the theory of evolution is that it provides an alternative non-teleological view of nature – a different sort of explanation for how organisms seem so well-suited (so well-adapted, to use the evolutionary term) to their environment and/or way of life, other than having been specifically designed that way. Aristotle was convinced that teleology was the key to a proper study of organisms; the modern view is that a scientific study of biology can be achieved only once we abandon teleological explanation.

Roman biology

After George Sarton's scathing remark about Roman science, which I quoted in Chapter 1, it may come as something of a surprise to find that some useful progress was made in biology under the Romans. But there were two individuals, both of Greek descent but by this time living within the Roman Empire, who made such exemplary contributions to biology that I felt they should be mentioned even in such a brief account as this.

It was arising from the Romans' practical approach that they had an interest in plants for their medicinal properties. Dioscorides (*c.*40–*c.*90) served as a physician in the Roman armies, and in the course of his work collated a vast amount of information on plants which he presented in his *On Medical Matters*. This was a creditable scientific work, accurately describing a wide range of plants and their uses in medicine, and substantially free of superstitious ideas. Botanical illustration was developed by the Romans, and Dioscorides included high quality drawings which enabled accurate identification of the plants.

Galen (129–199) was born in Asia Minor (Turkey), trained in medicine at Smyrna, and then lived in Rome for most of his life, for some of that time working as physician for the Emperor. Through the dissection of many different animals he acquired a detailed knowledge of anatomy, including comparative anatomy, noting the similarity between monkeys and humans. He carried out a number of experiments to determine the function of organs, notably how muscles are controlled from the spinal cord. He also demonstrated that arteries carry blood rather than air as had been thought previously, but wrongly believed that blood passed from one side of the heart to the other through pores in the dividing wall between them (the septum). Galen followed Aristotle's conviction that, in view of the exquisite structure and function of organs, as well as the body as a whole, they must have been designed and made for their purpose, and wrote *On the usefulness of parts* which is a treatise on teleology, praising a creator. He was influential through his lectures and prolific writings, which included philosophy as well as medicine; and, although he was not a Christian, the church commended him because of his teleological views.

Similar views had been expounded by the statesman and writer Cicero in the first century BC, who argued that just as it is evident a statue could not have arisen naturally but must have had a sculptor, so living things cannot have arisen by chance but must have had a designer. In contrast to this, at about the same time Lucretius wrote his epic poem *On the Nature of Things* in which he expounded the Epicurean philosophy which originated with the Greek philosophers Democritus and Epicurus, extending the ideas of Empedocles whom I mentioned earlier. Their view was that the world is exclusively material, that everything is but a fortuitous aggregation of atoms[2]: there is no designer or maker, no divine being, or at least none that has any relevance to human life or destiny. This materialistic view of the world extended to all biology, all forms of life, including whatever the human soul may be if it exists.

---

[2] By 'atom' they simply meant the smallest, indivisible part of matter that could exist, not our modern concept.

The Middle Ages

From what has been said in previous chapters, it will come as no surprise to learn that, with the general decline of science during the Middle Ages, biology too was very much in the doldrums. Hardly any original work was carried out – whether in terms of observation or theorizing about nature. As with other areas of knowledge, so with biology, the scholars of the Middle Ages were primarily concerned with preserving what they could of the ancient works, especially of the Greek natural philosophers.

Arising from the Christian emphasis on spiritual values, nature was no longer studied for its own sake, but rather as a means for glorifying God; and the prime interest in nature became allegorical – for illustrating truths of morality and religion rather than for an objective understanding.

Typical of medieval works were the bestiaries which described various creatures, but included many weird fictional beasts along with actual ones; and the prime purpose of such books was not so much to instruct about animals, but to use them symbolically – as a vehicle for moralising – somewhat like Aesop's fables. To appreciate the sorry state of medieval biology, it should be realised that these mystical interpretations of nature were not merely an embellishment on what would otherwise have been an objective study of nature. Rather, identifying such meanings and deriving 'spiritual' applications was the main purpose of studying nature. Herbals, the medieval books about plants, were freer of such mysticism, but still focused on the use of plants to man, especially their medicinal value, rather than their role in nature.

Also during this period there was widespread acceptance of the teleological view of nature. The church, of course, saw this as part of its belief in a creator; but it should be noted that non-Christians such as Galen and Cicero believed nature must have been designed, and so did Pliny whose *Natural History* also advanced this view. Because of the church's dominant position, those who held a contrary view, such as the Epicureans, were progressively marginalized. And, tied up with this teleological outlook, the fixity of species became an established doctrine.

Revival of learning

When the classical works were rediscovered in the West, after the initial reaction against Aristotle his works were received enthusiastically. They reinforced the teleological frame of mind which was already prevalent, and Aristotle's *scala naturae*, with its sense of man being at the top of nature's hierarchy, also suited the anthropocentric teaching of the church that man was made in the image of God and has dominion over the rest of creation. The scholars of the time were immediately impressed with Aristotle's acute observation and knowledge of nature, and readily adopted him as an authority on the natural world.

Albertus Magnus stands head and shoulders above the rest, not just for his promotion of Aristotle's works, but for his interest in and study of nature. The breadth and quality of Albert's observations were outstanding, especially for his time; and the books on nature which he wrote, chiefly *On animals* and *On vegetables and plants*, were probably the finest works on natural science to emerge during the Middle Ages. Of particular note, from his researches Albert realised there were many more plants and animals than those known to the classical philosophers; but this fact was largely overlooked by his immediate successors.

An important stimulus to the observation of nature in the Renaissance period was the revival of art. Artists were increasingly concerned to depict their subjects correctly, and consequently studied the form and structure of plants and animals. Undoubtedly the pre-eminent representative of the Renaissance artist–naturalists was Leonardo da Vinci, notably with his detailed studies of anatomy which he used in order to represent bodily forms and positions accurately. Leonardo's work went well beyond his artistic requirements, for example in demonstrating the skeletal equivalence of man and horse; indeed Leonardo is often regarded as a scientist and engineer as much as he was an artist.

However, individuals such as Albertus Magnus and Leonardo da Vinci were exceptional. For most, such was their sense that the ancients had known everything that, even with the reawakening of interest in the natural world, for a long time the focus of activity was more on recovering as much as possible of the classical works and studying those rather than looking at nature for themselves. However, by the 15th century it was becoming apparent that most of these works had been recovered and there was a gradual increase in direct observation of nature, with the first books resulting from this fresh approach appearing early in the next century. Yet still it took a long time to accept that the ancients had not known everything and the Renaissance authors tried to equate the plants and animals they found with those described in the classics. For example the German botanist Otto Brunfels (1489–1534), who produced what is believed to be the first modern book on plants – based on observation rather than literary sources – nevertheless tried to identify the plants he knew in Germany with those described by Dioscorides which came mainly from around the Mediterranean. Such was the deference shown to the classical authorities that it was quite common in works from this period for the authors to apologise if they described species that had not been mentioned by them!

However, slowly but surely there was a realisation that the ancients had not known everything or always been right. The Belgian anatomist Andreas Vesalius (1514–64) illustrates the gradual change of attitude that was taking place about this time. He trained in medicine at the Universities of Paris and Padua, and no doubt his studies were based largely on the works of Galen. But he went on to pursue his own investigations, including many dissections of cadavers, and used his observations to produce *On the Structure of the Human Body* (1543) which is regarded as the foundation of modern anatomy. In this Vesalius follows Galen in having a strongly teleological approach to the structure and function of the body; but his investigations convinced him that Galen had been wrong on some matters of fact, notably whether blood passes directly from one side of the heart to the other, and he wrote as follows:

> Not long ago I would not have dared to turn aside even a hair's breadth from Galen. But it seems to me that the septum of the heart is as thick, dense and compact as the rest of the heart. I do not see, therefore, how even the smallest particle can be transferred from the right to the left ventricle through the septum. [2nd edn, 1555]

It was well into the next century before the renowned English physician William Harvey (1578–1657) published his *Anatomical Essay on the Motion of the Heart and Blood in Animals* (1628) in which he describes how the blood circulates from the left side of the heart to the right through the body's tissues, and proposed the existence of capillaries to convey blood from arteries to veins even though, lacking a microscope, he could not actually see them. His work is notable not just for its correct conclusion, but because in it he carefully and accurately described the experiments which were the basis for his conclusions. Yet, despite the progressive nature of his work, Harvey claimed to be following Aristotle; this might have been little more than paying lip-service, but nevertheless it is an indication of the mood of the times, even as late as the 17th century.

One of the principal causes of the eventual overthrow of the perception that the ancients had comprehensive knowledge, at least so far as biology was concerned, was the expansion of geographical exploration which led to the discovery of creatures which were quite different from those familiar to the medieval Europeans. As early as the 13th century Marco Polo and his father had made their epic journeys to the East, and overland trade continued to increase in the ensuing centuries, with travellers bringing back descriptions of oriental plants and animals. Though some of these stories were the source of continued belief in mythical beasts as well. By the end of the 15th century the voyages of discovery were taking place, sea routes were opened up to the Far East and the New World, and voyagers started to bring to Europe samples of exotic fauna and flora – species which were totally different from those known in Europe, whether from classical or modern times. In particular, it became increasingly evident that many of the new plants and animals bore no relation to those described by the classical natural philosophers, which showed they had not known everything and began to indicate that there was plenty of nature still to be discovered. This helped to break the blinkered, subservient attitude towards the ancients, and biological investigation gradually gained momentum.

In the 15th century only a few hundred plants were known, but by the end of the next century there were several thousands. This huge increase in the number of known species prompted various attempts to revise the medieval encyclopaedias of nature (which had remained fairly static for so long) to assimilate the new knowledge. Probably the best known of these encyclopaedic naturalists was Konrad von Gesner who practised as a physician and had an extraordinarily wide interest in nature. Gesner was known better to his contemporaries as a botanist, but is remembered more now as a zoologist because of his *Historia Animalium* (1551-8) which encompassed practically all known animals, notably including those that had come to light as a result of recent exploration. This work is typical of the times – with one foot in the future but still one in the past: on the one side it was based largely on firsthand information and endeavoured an objective account of the different types of animals, but on the other it resembled the bestiaries with its allegories and moralising, and still included some fabulous beasts. Gesner began a similar work covering plant life, but was not able to complete it before his death.

Typical of medieval works, Gesner arranged animals and plants predominantly in alphabetical order. This had been adequate when the number of species was limited (and when the prime interest was in their use to man), but the 16th-century naturalists recognised that it was no longer satisfactory, and there was an increasingly urgent need to have a more rational, a more 'scientific', organization of the growing number of species. Devising a meaningful classification of the living world was probably the principal achievement of the scientific revolution in biology, and out of the recognition of natural groupings of species came ideas of organic evolution.

BIOLOGY OF THE SCIENTIFIC REVOLUTION

Classification

So, as the 16th century drew to its close, at about the same time that the revolution in astronomy was taking place and man's view of the universe was expanding, biologists were beginning to realise that the world of nature was vastly more diverse than their predecessors had imagined could be possible. One wonders how Plato or Aristotle, with their belief that there are no gaps in nature even though they knew of only a few hundred types of plants and animals, would have come to terms with the many thousands of species known by the end of the 16th century. For the naturalists it was an increasing headache trying to manage the growing body of data (without PCs!), and it was becoming essential to organize it. Although it was the sheer volume of material that posed the immediate problem, it is important to recognise that the aim of the 17th- and 18th-century naturalists was not just to find an effective way of cataloguing or categorizing thousands of species: they saw the development of a classification scheme as a means of understanding nature – of identifying the fundamental relationships within nature, and its underlying principles. They perceived their work as no less revolutionary for biology than did the physical scientists for their mechanics and astronomy.

Progress was made initially with plants, and Gesner was one of the first to attempt something like a scientific classification, based on the structure of their flowers. More influential was the work of his contemporary, Andrea Cesalpino (1519–1603), who acquired an interest in botany from his work as a physician. However, unlike so many of his predecessors, he did not look upon plants solely from a medicinal perspective, but began to group them according to similarities of structure of their fruits and seeds. In 1583 he produced *De Plantis* in which he set out a quite comprehensive classification for plants. Early in the next century Kaspar Bauhin (1560–1624) introduced the concept of a genus to describe a group of closely related species, and began to use, albeit intermittently, a binomial system of nomenclature in which each species is assigned one name to identify its genus and another to identify the particular species within that genus, much as is the case today.

Ray

Although the son of a blacksmith, by means of a scholarship John Ray was able to study at Cambridge, and after graduating in 1648 he obtained a fellowship to enable him to carry on his biological investigations. He lost this official income in 1662 because of his Puritan sympathies, and thereafter was supported financially by friends, in particular by one of his former students, Francis Willughby.

Ray's first interest was in botany, and in 1660 he produced a comprehensive catalogue of plants in the Cambridge area. There had been earlier attempts at describing the natural history of an area, but his was the first to be so complete, describing over 600 plants, and to do so in such a systematic and scientific manner; it was not superseded for nearly 200 years. He followed this with similar catalogues for the British Isles, then Europe, culminating in his greatest botanical work *Historia generalis plantarum*, which issued in three volumes from 1686 to 1704. This was a wide-ranging encyclopaedia of plants, including a description of all known European species and many others world-wide, totalling more than 18,000.

What was so pioneering about Ray's work was not so much its comprehensiveness, but the advances he made in classification. Although there had been some idea of what a species is for a long time, even going back to Aristotle, the existence of variations had always posed the problem of whether two dissimilar specimens were of the same or different species. Ray is credited with providing a clear modern definition of a species in terms of it being a group of organisms which are capable of interbreeding (and having fertile offspring). When it came to classification, whereas most of his predecessors had tended to focus on just one or two characteristics, Ray sought to take cognisance of the organism as a whole, and to identify key distinguishing features which were associated with substantial overall differences. In this way he hoped to determine what are the important groupings and divisions in nature, and achieve a more objective or 'natural' classification rather than the subjective categories of the early classifiers. Many of his natural groupings are still in use today; for example, he divided plants first into flowering (angiosperms) and non-flowering (gymnosperms), and the former into monocotyledons and dicotyledons, which are terms Ray introduced, based on their number of seed-leaves.

Willughby and Ray had set out to produce a systematic description of the whole of nature, with Ray focusing on plants and Willughby on animals. Unfortunately Willughby died young and left Ray to complete the work he had started on birds. In this *Ornithology* they made a clear break from the medieval tradition of the bestiaries, rejecting any human-centred approach and explicitly stating that 'we have wholly omitted what we find in other authors concerning ... emblems, morals, fables, presage or aught else appertaining to divinity, ethics, grammar or any sort of human learning'. It is interesting to note that, although Ray was a committed Christian and saw nature as God's handiwork, he expressly rejected reading spiritual meanings into nature, or any other sort of allegory or symbolism. So, as with his works on plants, *Ornithology* set a new standard in the classification of animals, by grouping birds according to their intrinsic structural features – such as beak, foot or body size – not merely by their names or utility to man, or whatever. This new approach was even more important in Ray's major later work on animals *Synopsis methodica animalium quadrupedum et serpentini generis* (1693) which was the first truly systematic classification of animals. The principal distinguishing features he used for these were the teeth and toes, e.g. making a major distinction between animals with nails from those with hoofs (ungulates), which is still followed today.

Ray established a definition of species which lasted for at least two hundred years and is still the basis of the modern 'biological' definition (more on this later). Along with this, he held staunchly to the view that species are fixed – none has been lost since creation (hence his problem with fossils mentioned in Ch. 3) and no new ones ever arise, ever can arise. 'The number of species in nature is fixed and limited, and as we may reasonably believe, constant and unchangeable from the first creation to the present day', he wrote. And the standing of Ray as a biologist served to reinforce belief in the fixity of species. However, even in Ray's experience there were one or two observations which did not quite fit in with this; for example he learned of a seed merchant who sold 'cauliflower' seeds that grew into cabbages. Ray came to terms with this sort of thing by accepting that at least some variation or degeneration could occur.

Linnaeus

I think for most people today classification does not seem a very exciting pursuit and the subject is more likely to conjure up images of fusty libraries with short-sighted academics thumbing through dusty tomes or scrutinizing antique display cases. However, the Swede, Carl von Linne (1707–78), better known by his Latin name of Linnaeus, the chief architect of modern taxonomy, does not fit this stereotype. He was, it is true, a compulsive classifier, and, for example, developed a classification of minerals and diseases as well as his comprehensive work on plants and animals for which he is famous. Yet in his twenties he embarked on a solo scientific exploration of nearly 8000 kilometres across the wilderness of Lapland, and throughout his life had a keen interest in everything to do with the natural world. He spent most of his working life as Professor of Botany at the University of Uppsala, where he was an exceptionally popular teacher, and despatched many of his students to explore the world in search of new plants and animals.

By the beginning of the 18th century, many biologists – notably John Ray, but also the likes of Cesalpino and Bauhin – had made substantial progress towards a systematic classification of living things. Linnaeus brought all of this earlier work together into a consistent and comprehensive taxonomic system which in due course he applied to all the known organic world. As well as recognising species and genera, he constructed a hierarchy in which genera were grouped into Orders, and Orders into Classes, based on ideas of the French botanist Joseph Pitton de Tournefort (1656–1708). Many of his higher groupings are still followed today. Linnaeus produced the first edition of his *Systema naturae* which presented his new taxonomic arrangement for the animal, plant, and mineral kingdoms in 1735, and added to it progressively in successive editions. Over the years he allotted a place in his system, and assigned a Latin binomial scientific name, for every known plant and animal species. This, of course, included man – Linnaeus was the first to identify man as *Homo sapiens*, a species within the genus *Homo*. This was of note because his *Homo* genus included the orang utang, so it fostered the idea of seeing man as alongside some of the great apes, rather than completely separate from the rest of the animal kingdom.

Linnaeus also believed in the fixity of species. In fact the sort of classification system he implemented, and his practice of keeping a collection of archetypal specimens, all but presumes the fixity of species – otherwise it would be necessary to keep revising the definitions of species. So it is to be expected that Linnaeus, like Ray, should hold firmly to the view that species are immutable. But, again like Ray, in the course of his work he encountered instances which challenged that position: he came across a toadflax which, instead of the usual snapdragon type of flower, had a symmetrical one and, although he assumed it to be a hybrid, it was completely fertile. Also, he recognised that some species within a genus could be very difficult to differentiate. So by the end of his career he had modified his position, and adopted the view that all the species classed within a genus might have initially constituted a single species, and the present species had arisen from that by degradation and/or crossbreeding.

There is another important point arising from Linnaeus' work. By the time of Linnaeus, biologists had generally adopted John Ray's definition of species – that it was a group of organisms capable of interbreeding. However, in carrying out his work, with his emphasis on categorizing organisms, Linnaeus focused on identifying differences in the forms of the organisms such as size, shape, colour or number of parts – what are called morphological features. In effect, whilst recognising that most natural species exhibit some morphological variation, where two samples could be distinguished morphologically, especially if they came from different geographical locations, they would generally be classed as different species – without considering whether or not they were capable of interbreeding. Even today, with some notable exceptions such as the dog, the principal consideration as to whether a group is a distinct species is whether it is morphologically distinct, even though, in theory, we still cite something more like Ray's definition.

Because of Linnaeus' important role in developing taxonomy it is easy to overlook his other work, arising from his wider appreciation of nature. However, he did not see individual species merely as isolated units, but recognised that they have a wider role in nature – they interact with other species and the environment – and there is an overall balance between them. These ideas are presented in his *Economy of Nature* (1749), in which we can see the beginnings of an understanding of ecology.

Linnaeus set in place a comprehensive and workable classification system. There were, of course, further developments in classification by his immediate successors, and changes continue to be made; but the basic approach, and much of the present framework, was established by Linnaeus. Of interest to us here is that Lamarck (see below) made the fundamental distinction between vertebrates and invertebrates and arranged the animal kingdom into 14 classes, in a continuous scale from the 'Infusoria' (protozoa) at the bottom to mammals at the top – which followed his belief in the Chain of Being. In contrast to this, Cuvier divided the animal kingdom into four completely separate groups ('embranchments') based on their distinct overall body plans. However, by this time we have reached the early decades of the 19th century and increasingly the focus of interest was not so much on pigeon-holing fixed species, but on vigorous debate as to what extent they are fixed. Once Darwin had published his theory of evolution, this was seen to provide a fundamental, 'scientific', explanation for groupings; and thereafter the basis of classification shifted to that of inferred common descent, albeit based on morphological features, rather than primarily on morphological similarities.[3] However, before we embark on the development of evolutionary ideas we should pause to reflect on the

---

[3] In recent years there has been a shift back on the part of at least some biologists to an emphasis on morphological groupings without regard to evolutionary implications, but we need not be concerned with that here.

outlook of the naturalists in the light of their recent discoveries.

Natural theology

At the end of the preceding two chapters I have drawn attention to how advances in the physical sciences led to an increasingly mechanistic view of the universe – away from the medieval concept of God being intimately concerned with the affairs of mankind, towards seeing the world as a product of exclusively natural laws. This emphasis on the operation of cause and effect eroded the previously well established view that the world had been designed, because it now seemed it could be explained solely in terms of the outworking of natural laws from some primordial form. Along with these developments of scientific ideas, the philosophers of the Enlightenment increasingly challenged any form of 'authority', especially of revealed religion, and their views came to prominence through the 18th century.

However, at the same time that this was happening, biologists of the 17th and 18th centuries were beginning to appreciate more of the richness, diversity and complexity of the natural world, and for many of them it reinforced the traditional view that the world of nature must have had a designer and maker. Although many of the philosophers and physical scientists were rejecting final causes, most naturalists saw the importance of purpose in the structure and function of various parts of organisms, and continued to have a teleological outlook.

Perhaps the best example of this is John Ray. Towards the end of his distinguished career as a biologist he wrote *The Wisdom of God manifested in the Works of the Creation* (1691) in which he explicitly rejected Descartes' mechanistic view of the world and, in the light of the vast array of organisms, concluded, 'What can we infer from all this? If the number of creatures be so exceeding great, how immense must be the power and wisdom of him who formed them all!' And it was not just the richness of nature that impressed Ray, but also the exquisite form and function of living things. For example, with respect to the well-known example of the eye, he concluded that it is so admirably fitted and adapted for vision that 'it must needs be highly absurd and unreasonable to affirm, either that it was not designed at all for this use, or that it is impossible for man to know whether it was or not.'

Whilst John Ray was a sincerely religious man and ordained, we cannot dismiss his views as merely the half-baked conclusions of some cleric who had a passing interest in nature. Rather, this was the considered verdict of a scientist – who happened to be ordained as well, as were many academics of his day – based on his scientific observations of nature. And, we should note, that neither was this a hangover from the medieval religious outlook on the world: Ray held his teleological views at the same time as he rejected a symbolic, allegorical or man-centred approach to nature. Ray's views carried much weight because of his evident ability and standing as a naturalist.

We have already met Robert Hooke who was contemporary with Ray and made valuable contributions to biology through his microscopical investigations. He saw the complexity and fine detailed structure of many biological things such as feathers and leaves, various insects and moulds, and, noting especially the exquisite construction of insect legs in relation to their various functions, concluded that they could not possibly have arisen by chance but must have been fashioned by a Creator (*Micrographia*, Observation XXXVII). Also, as he discovered the intricacy of even 'simple' living things, he began to question the reality of spontaneous generation.

Linnaeus, too, perceived God's hand in nature and believed that the organic world must have been designed. He saw his work on classification not merely as identifying 'natural' groupings, but even as uncovering a divine plan of creation. Similarly in his work on the ecological interactions between species he saw God's hand in the overall harmony of nature.

Unlike John Ray, the Rev. William Paley (1743–1805) was not a scientist but primarily a theologian and philosopher. He should be mentioned here because he was a strong advocate of the view that nature is evidence of a Maker; in fact he believed that the wonders of nature are not merely evidence of God, but prove his existence – what is called the argument from design for there being a deity. His *Natural Theology, or Evidences of the Existence and Attributes of the Deity collected from the Appearances of Nature* (1802) became a standard university text, especially at Cambridge where in due course it was studied by Darwin. Like Ray, and Aristotle and Galen long before him, he points to the complexity and appropriateness of biological structure and function as unequivocal evidence of a Creator.

In an age which increasingly saw the universe in mechanistic terms, Paley used the mechanical example of a watch to illustrate his case. He pointed out that if someone came upon a watch, even well away from civilisation, 'crossing a heath', no-one would imagine that it had arisen naturally, but everyone would know it had been specifically designed and made, that it must have had a maker, that there must have been a watchmaker. Paley argued that the exquisite structure and function of living things are analogous to the detailed, purposeful construction, assembly and operation of all the parts of a watch; and that, like a watch, their existence can be explained only in terms of there being an intentional and competent maker. He then discussed the form and function of various organs of the body, especially of the eye, to illustrate how well they are made and how suited for their function, stressing their similarity to man-made machines.

EVOLUTION

The rise of evolutionary ideas

Although there were many naturalists who saw in the expanding knowledge of nature increasing evidence of God's handiwork, at the same time others began to question whether this traditional interpretation could really stand. Doubts came from various directions. First was simply the vast number of species: whilst the likes of Ray and Linnaeus saw this as evidence of a creator's ability and benevolence, others doubted that any creator would individually have made quite so many different species. Even the work of the classifiers, because it identified and highlighted the relationships between species, began to raise doubts about fixity and prompt questions as to whether morphologically similar species might have been derived in some way from a common source. Finally, the ideas of special creation and the fixity of species were associated with traditional religious views; so the growing anti-authoritarian and especially anti-religious attitudes of the Enlightenment provided the right atmosphere in which deeply-rooted beliefs could be challenged and radical new ideas might arise.

Buffon

The first person from this time clearly to reject a rigid fixity of species and propose significant changes in their forms which might be regarded as evolutionary, or at least the first to do so openly, was Buffon whom we met in Chapter 3 because of his cosmogonical theories in the *Epochs of Nature*. He was born in the same year as Linnaeus, and these two were the leading natural scientists of the mid 18th century. But there the similarity ends. They had diametrically opposite philosophies of nature which resulted in open rivalry and considerable ill-will between them. Whereas Linnaeus viewed nature as God's handiwork and saw his own work as discovering the divine plan for the world, Buffon actively sought to rid natural science of its religious connotations. We have already seen this in his rejection of the biblical account of creation, and he went on to challenge the idea that species are specially created or in any way fixed to some sort of Platonic ideal.

Buffon adopted a view of nature along the lines of Aristotle's *scala naturae*, with an emphasis on gradation rather than demarcation, so he maintained that any classification system is nothing more than an invention of the classifier, and that it has no real meaning so far as nature is concerned. So he despised the core of Linnaeus' life's work with its focus on, as it seemed to him, seeking meaningless minutiae, with which to make ever smaller (and less significant) categorizations of nature. In his *Histoire Naturelle* he said that 'we shall make no use of families, orders or classes, any more than nature makes use of them.'

Instead of highlighting morphological differences, Buffon focused on the idea of a species including all those individuals capable of inter-breeding, considering that organisms close together in the Chain of Being would be able to interbreed even if they looked somewhat different. That is, animal types which are usually classed as quite separate species, such as the European ox and American bison, or various large cats from the Old and New Worlds (e.g. lion and puma, or leopard and jaguar), he grouped together. Buffon considered that these different types had arisen from a common stock, and attributed their divergence in appearance to the fact that they had lived in different environments for a long time, 'their present differences have proceeded only from the long influence of their new situation' – a kind of adaptation. Further, he wrote:

> if it were proved ... that a single species was ever produced by the degen-eration of another, that the ass, for instance, was only a degenerate horse, no bounds could be fixed on the power of Nature: She might, with equal reason, be supposed to have been able, in the course of time, to produce, from a single individual, all the organized bodies in the universe.

> But this is by no means a proper representation of Nature. We are assured by the authority of revelation, that all animals have participated equally of the favours of creation; that the two first of each species were formed by the hands of the Almighty [*Natural History*, Vol. 3]

So it is clear that Buffon did speculate about substantial evolutionary changes, but only in a degenerative sense, rather than in the modern pro-gressive sense. We also see here something of Buffon's struggle to express unconventional views, whilst at the same time making sure that what he wrote was acceptable to his masters, the Sorbonne.

Buffon was one of the leading writers of the French Enlightenment, and represents one of the principal attempts at the time to move away from traditional religious explanations of the world and towards naturalis-tic ones. Although his ideas were based on little science and consequently rather speculative, his *Histoire Naturelle* was well written and widely read across Europe, and started to open up people's minds to the possibility of organic change.

Erasmus Darwin

Erasmus Darwin (1731–1802) was highly regarded as a practising physi-cian and a man of science. He founded various scientific societies, and his enthusiasm for biology is shown in a poem he wrote entitled *The Botanic Garden*. Of particular relevance here is his book, *Zoonomia* (Laws of Or-ganic Life), which was intended to 'unravel the theory of diseases' but is better known for its evolutionary ideas. Thanks to his reputation, it be-came a popular work and many took note of his views, although scientists criticised it for being much too speculative and short on facts. Because of the evolutionary views it expressed it was placed on the Papal Index of proscribed books. In *Zoonomia* Erasmus presented three important ideas. First, he gave reasons why he believed species are not fixed but susceptible to change. These were (i) metamorphosis e.g. a tadpole developing into a frog, or a caterpillar into a butterfly; (ii) hybridization which results in offspring differing from their parents; and (iii) progressive changes in the form of an organism over successive generations due to breeding. Second, he suggested that all organisms might have a common origin, supporting this with (i) the basic similarity in structure (homology) of vertebrate animals, and (ii) the existence of vestigial organs. And, thirdly, like Buffon, he believed that species changed primarily as a result of external influences, and that these changes were then transmitted to their offspring. In this he clearly introduced the concept of the inheritance of acquired characteristics which was to be influential for over a century, and is still maintained by a few even today.

Despite his reputation at the time, if he had not been Charles' grandfather then Erasmus Darwin would almost certainly not be remembered as well as he is and would be given less credit for his ideas.

Lamarck

Jean Baptiste de Monet (1744–1829), Chevalier de Lamarck, was a scientist with wide interests and abilities, making useful contributions to meteorology, chemistry, geology and palaeontology as well as the biological work for which he is better known. After a short military career he studied medicine for four years and then produced his first work on botany, *Plants of France*, which was published in 1779 with the help of Buffon. Partly due to this he became Keeper of the Jardin du Roi where he pursued his interest in botany. In 1793, following the French Revolution, the 'King's Garden' was reorganized as the 'Museum of Natural History' and Lamarck lost his botanical position. Despite his lack of relevant experience he was appointed professor of the section for insects and worms, and in this post carried out valuable work on the classification of invertebrates. It was Lamarck who coined the word biology to refer to both botany and zoology.

Lamarck is best known for being the first to propose an explicit and coherent theory of evolution, which was presented in his *Philosophie zoologique* (1809), and elaborated in subsequent publications. He believed that life arose readily by spontaneous generation, and that living things had an innate drive to develop. Hence, starting with the simplest forms of life, with successive generations progressively more complex organisms emerged which gradually ascended the ladder of life by virtue of their inner force. Thus Lamarck accepted the idea of a continuous Chain of Being, but presumed there were two – one each for the plant and animal kingdoms.

Whilst the primary force behind progressive evolution was this innate drive, Lamarck also believed that creatures were moulded by and adapted to their environment; and an important feature of this process was use and disuse of organs or limbs, combined with the inheritance of acquired characteristics. The giraffe is the classic example used by Lamarck to illustrate his theory: He proposed that the modern giraffe arose from an ancestral species which had only a short neck, but this species habitually stretched for high branches, and this stretching caused slight extension of the individual's neck (an acquired characteristic due to use); any elongation achieved by a parent was passed on to its offspring (inheritance of acquired characteristic), so that over many generations the species developed its extra long neck. His ideas of use and disuse and inheritance of acquired characteristics are the best known features of his theory, but, as Lamarck presented it, they were subordinate to the inner drive to develop.

It is evident that Lamarck's theory was similar to the ideas of Erasmus Darwin and, whilst he developed it further, it was still excessively speculative, and his style of writing did not have the popular appeal of Buffon's. Also, his notion of an inner drive for progress, which was central to his theory, was not popular at the time because it was perceived as encouraging a desire for progress by the lower classes in society and promoting social instability. For these reasons Lamarck's theories were not well received and were largely ignored until after Charles Darwin's *Origin of Species* when there was renewed interest in the mechanisms of inheritance. A further major stumbling block was the staunch anti-evolutionary stance of his contemporary at the Paris Museum, Cuvier, who had a much more scientific approach and far greater influence in both scientific circles and in the eyes of the public. In particular, he criticised Lamarck's idea of an inner drive as unscientific – a reversion to mystical medieval notions.

Cuvier

In Chapter 3 we saw how Cuvier worked primarily on comparative anatomy. For him, the evident interrelation of structure and function reinforced the view that animals are designed. Further, he recognised four basic body types within the animal kingdom, which he identified as the vertebrates (animals with backbones), molluscs, articulates (insects, spiders and crustaceans), and radiates (grouping together anything else, and mainly displaying a radial symmetry, such as the echinoderms, e.g. sea urchin). So far as Cuvier was concerned, all of these body plans are equally successful and it would be quite inappropriate, for example, to place the vertebrates at the top and radiates at the bottom, so he rejected the idea of a hierarchy or Chain of Being. These four embranchments, as he called them, can be compared with the modern concept of a phylum; and the main differences from a modern classification is that his articulates and radiates are no longer regarded as single, but each is now divided into several phyla. Because he saw these groups as completely distinct, Cuvier believed they supported the view that organisms are substantially fixed, and in particular that all the different forms could not have evolved from a common ancestor.

The early nineteenth century debate about evolution

By the early decades of the 19th century there was much debate about
whether or not species had arisen by special creation or by some sort of
evolutionary process. At the time it was in terms of whether or not species
were immutable, or referred to simply as the 'species question' – could
new species arise, and if so, how? And it is worth recalling that this was
the same time that the debate was taking place in geology between catastrophism and uniformitarianism.

On one side were the conservatives, typified by Paley, representing traditional religious views that organic beings must have been created more
or less as they are, and arguing that they are clear evidence of a creator.
Some plainly condemned evolutionary ideas as atheistic. Others wanted to
hold on to conservative ideas for more political rather than religious reasons, to protect the social status quo. Their scientific voice came primarily
from Cuvier and then, especially in England, from Richard Owen (1804–
92) who was the leading biologist there, becoming Director of the Natural
History Museum, carried out palaeontological work of equal importance to
Cuvier, and was just as determined in his anti-evolutionary stance.

On the other side, establishment of the stratigraphical column in geology convinced most that there had been a progression of species in the
past and, whilst some maintained that each change could be the result of a
special act of creation, others felt there must be a more natural explanation. Equally, the evident natural groupings of some species clearly suggested that they had arisen from a common source, and in the not too
distant past. And some wanted to find a naturalistic explanation for life
simply in order to oust outdated religious dogma. However, in the main,
evolutionary ideas were based on those of Lamarck, and we have seen that
those were generally regarded as unsatisfactory.

A book that helped to pave the way for a wider acceptance of evolution
was *Vestiges of the Natural History of Creation*, published in 1844 by Robert
Chambers (1802–71). Chambers was a writer and publisher rather than a
scientist, but was well-read and had a good grasp of the prevailing scientific issues. In *Vestiges* he brought together current ideas about the condensation of the solar system from a nebulous cloud of dust, formation of
the early earth, and the progressive emergence of life forms on the earth.
All of this he portrayed as by the operation of natural laws: he proposed
that, just as the law of gravitation controls the physical world, so there was
an inner law of development which controlled the organic world. And, he
argued, these laws had been set in place by God, such that the whole was
a working out of God's plan for the World, i.e. God was acting through
secondary causes. In this way he hoped to avoid criticism of atheism, and
help people to see that evolution need not be an anti-religious concept.
Nevertheless, for fear of adverse reaction, he went to great lengths to publish the book anonymously, and some thought Darwin might have written
it. It received a mixed reception. Many scientists were scornful of this

'popular' science, especially as he was somewhat credulous and included some fanciful ideas; but it was widely read and brought the subject of evolution to the public attention and more open debate.

But scientists were no longer content just to describe, they wanted to explain. So it was inadequate merely to describe the changes undergone by species, as evidenced in the fossil record, as Chambers had done. To be satisfied that evolution had in fact occurred required a convincing scientific theory of *how* organisms might change in a progressive way, and it had to be something more 'scientific' than Lamarck's mystical inner force.

Natural selection

The idea of evolution through the operation of natural selection was proposed by the naturalist Alfred Russel Wallace (1823–1913) as well as by Darwin, and both arrived at the idea through reading a work by the Rev. (Thomas) Robert Malthus (1766–1834). Although a parish curate for a short time, Malthus was primarily an economist and spent the latter half of his life as professor of political economy and modern history at the college of the East India Company. He is best known for writing *An Essay on the Principle of Population* (1798) in which he challenged the prevailing view, arising from the industrial revolution, that increased population would lead to prosperity. He argued that the human population has the capacity to increase faster than the expansion that can be achieved in food production, so food supply will always limit growth of population, either directly through famine or indirectly through war over the world's resources. Malthus made a useful contribution to economics and demographic studies; but important for us here is the application of his ideas by Darwin and Wallace in the development of their theories of evolution.

Darwin

The life of Charles Darwin (1809–82) is so well documented, with his autobiography and many subsequent biographies, looking at his life and work from various angles, that, despite his importance, there is no need here to give anything more than the briefest of sketches. He was born into a wealthy family, with sufficient means that throughout his life he had a private income and never needed to earn a living, which freed him to pursue his interests. It is well known that his academic studies were undistinguished – leaving medical training at Edinburgh after only one year and then gaining a mediocre degree from Cambridge where, by his own admission, he wasted much of his time. However, it was at Cambridge that he developed an interest in natural history and geology, which was fostered by John Henslow who was professor of botany; and it was through Henslow that, after graduating, Darwin obtained the post of naturalist on HMS *Beagle*.

The voyage of the *Beagle* lasted five years, most of that time being spent surveying South America, and then visiting the Galapagos Islands before returning to Britain via Australia and the Cape of Good Hope to complete a circumnavigation of the globe. At every opportunity Darwin went ashore and carried out investigations of the local geology, flora and fauna, collecting numerous specimens, and despatching many back to Britain. On his return, he abandoned his earlier intentions of entering the church, and instead embarked upon a career in science by writing up the observations from his voyage. It was these writings which earned him a scientific reputation and meant people were eager to hear what he had to say when the time came to publish his theory of evolution. After working for about five years in London, during which time he married, he withdrew to Down House in Kent where he continued with his researches, and lived out the rest of his life.

Arising from his many experiences on the voyage, there were several findings which made a particularly significant impression. So far as geology was concerned, he carried Lyell's *Principles of Geology* with him and carried out many geological investigations, including observations in South America relating to the slow uplift of the Andes. This reinforced in Darwin's mind that the earth had a very long history, much more than the traditional few thousand years, and certainly plenty of time for species to accumulate many small modifications, resulting in substantial overall change.

Regarding natural history, he was impressed by the enormous abundance and variety of nature, the exquisite adaptation of many organisms, and the way in which similar ecological niches are occupied by different species depending on the wider geography. He was also struck by the way in which there were subtle differences between very similar species across a region, and especially between islands such as the Galapagos. It was drawn to his attention that the native islanders could easily identify from the appearance of a giant tortoise which island it had originated from. Similarly, on his return to London he learned there were three distinct species of mocking birds from the islands, and several of the well-known Galapagos finches which are similar to each other but evidently adapted – especially in the size and shape of their beaks – to different feeding habits. It seemed incredible to Darwin that any creator, no matter how 'bountiful', should make so many separate species, and with such little difference between many of them.

So, on his return, as well as writing up his experiences and observations from the voyage, he began to address the species question. He started to amass data to demonstrate that species do change and, perhaps more importantly, to consider *how* species change and *how* new ones might arise – to come up with a plausible mechanism that could operate in nature and lead to the transmutation of species.

It was obvious for him to compare the different but similar species in nature with the varieties of the same species that result from domestic breeding, especially of pigeons which he bred. He was well aware that substantial morphological changes can be and have been achieved by breeding, and wondered whether analogous changes might be possible in nature; but at first he could not see a mechanism which could operate in nature that was comparable to the breeder's action of selecting which plants or animals to breed from. Then in 1838 he came across Malthus' *Essay*, and on reading this Darwin realised that, in a similar way to the human competition for resources, there would also be competition in nature between species in their struggle for life, and Darwin saw how this competition could provide a means of selection: Whereas in domestic breeding man chooses which individuals to use for breeding the next generation – selecting on the basis of individuals having the features which he wants to maintain or develop, nature simply selects those that live and reproduce best. That is, those individuals having certain characteristics which mean they thrive (especially to reproduce) in general will breed preferentially compared with those that lack those characteristics and fare less well. In this way, the advantageous characteristics tend to be passed on to the next generations, from which further advantageous variations can develop. In time, the progressive occurrence and selection of small variations could lead to substantial morphological change, i.e. evolution occurred.

Thus, Darwin had formulated the essence of his theory of evolution as early as 1839, and wrote short accounts of it soon after, but did not rush to publish. He was well aware of the scientific and public hostility towards evolutionary ideas, and opted not to publish until he had collated overwhelming evidence to support his case. Eventually, his hand was forced by Wallace coming up with essentially the same idea, though with much less supporting evidence. Wallace carried out extensive work on the geographical distribution of species and, after an expedition to the Amazon, went to the Far East specifically seeking data to shed light on the species question. It was there that he, too, in 1858, read Malthus' book and saw how the concept of competition could apply in nature and form the basis for a progressive transformation of species. Unlike Darwin, he sought to publish immediately and sent a manuscript of his paper entitled *On the Tendency of Varieties to Depart Indefinitely from the Original Type* to Darwin for comment. The outcome was a joint presentation of the two men's theories to the Linnaean Society in July 1858, and Darwin hurriedly wrote a fuller account which emerged in November of the following year as *The Origin of Species by means of Natural Selection, or the Preservation of Favoured Races in the Struggle for Life.*

The Origin of Species

In the *Origin* Darwin points out that many domestic varieties are so different morphologically that if they were found in the wild they would be considered separate species. He then goes on to say how, in nature, there is a continuous gradation from the minor morphological differences between varieties of the same species, to the larger differences between species, and to the even greater differences between genera and higher taxa. Arising from this, he saw varieties as the early stages of divergence which may eventually lead to separate species, and, as the separate species diversify further, they become increasingly different from each other and in time have their own daughter species. He argued that diversification occurs because species exhibit a range of variations, different variations may be advantageous to different groups of individuals within the species depending on their particular circumstances, so different variations will be selected in those different subpopulations. That is, just as breeders, by selecting for different traits, can produce very different varieties from the same stock, so nature, by selecting different preferred characteristics in different parts of a species' range, results in diversification to provide two or more species from one. Convinced of the efficacy of this process, Darwin could see no reason why it should not account for the production of all species from a common simple ancestor. He saw the whole of nature as a branching tree, with existing species at the extremities, closely related species stemming from the same branch, and very different species having their common source near the trunk.

This mechanism of natural selection can also account for adaptation: Adaptation results from variations being selected which are progressively more suited to the species' way of life, just as breeders can take a characteristic to extremes by persistently selecting individuals which exhibit a particular desired trait. In the main, he says that, for both natural selection and domestic breeding, they do not cause variation, they merely act on variations which are already present; but at times he is not too sure of this because he does not know what is the source or cause of variations. He also admits to the laws of heredity being a mystery, but in general follows the ideas of use and disuse and inheritance of acquired characteristics. We shall look at heredity more closely in the next chapter.

In support of his theory he cited the homologous anatomical structures already noted by his grandfather: if all vertebrates were descended from a common source then one would expect to see similarities in their structure – a basic pattern which has been modified in various ways to suit differing roles. Even though phyla, such as Cuvier's four embranchments, appear completely distinct, Darwin argued that their lack of similarity is only because their common ancestor is a very long way back. Needless to say, he accepted the developing geological view that the earth was very much older than traditionally accepted – at least many millions of years – time enough for the evolutionary process to take place. Darwin also felt that fossils lent support, whilst recognising that they did not provide the intermediates he would like to have seen, but put these omissions down to the scanty knowledge of the fossil record, and was confident that in time the missing links would turn up. He also pointed out that, because common ancestors are in the past, we should not necessarily expect to see extant intermediates between present-day species.

Apart from the lack of intermediates in the fossil record, probably the difficulty he was most keenly aware of was the development of what he called highly specialised organs, notably the eye. He speculated as to how an eye might have come about through a series of small developments from a rudimentary light-sensitive structure, each development offering a minor improvement on its predecessor. Darwin argued that this sort of progressive process could account for all those organisms and organs which previously had been cited as clear evidence of design. In other words, variations arise spontaneously; these may be good, bad or indifferent, but only those that offer an improvement are assimilated by natural selection. The gradual accumulation of slight desirable variations eventually yields complexity and/or adaptation to such a high degree that it gives the appearance of design, though in fact each step had been but the product of natural processes. Indeed, from the viewpoint of the growing explanatory power of science, this was the real achievement of Darwin's theory – it provided a mechanism by which the apparent design of biology could be accounted for simply in terms of the operation of natural laws, and was seen to put a final end to the question of teleology. In his autobiography Darwin put it like this:

> The old argument of design..., as given by Paley, which formerly seemed to me so conclusive, fails, now that the law of Natural Selection has been discovered. There seems to be no more design in the variability of organic beings and in the action of natural selection, than in the course which the wind blows. Everything in nature is the result of fixed laws. [1959]

Evolution as a scientific theory

So far in this chapter I have presented the emergence of the theory of evolution primarily in its biological context. But the first half of the 19th century saw important developments in the philosophy of science – developments which affected how Darwin presented his theory and how it was received. So before proceeding it is worth pausing to comment on these.

Following Aristotle's lead, the general view was that scientific laws were determined by a process of induction from the facts; and over the centuries many scholars such as Roger and Francis Bacon had delineated rules for carrying out this process. William Whewell (1794–1866), an influential scientist and philosopher, added to this by portraying science as a pyramid: a broad base of facts upon which were built progressively higher tiers of scientific laws and explanatory theories. For example, elliptical orbits explained the observations of planetary movements, then Newton's laws of gravity and motion explained why the orbits are elliptical, and many other things as well.

However, contemporary with Whewell, John Herschel (1792–1871) argued that how scientists came upon a hypothesis was not really important – it did not matter whether it arose by long rigorous analysis of the data (e.g. using the rules for induction) or by a flash of inspiration like Archimedes' 'eureka!'. What really mattered was testing the hypothesis – by deduction and further experiment. In other words he distinguished between what he called the context of *discovery* of a scientific theory and the context of its *justification*. Hence, he argued, contrary to the general emphasis on induction, it was not necessary to show that the idea followed logically from the data, but it was necessary to show that deductions from the hypothesis were consistent with the facts.

So when it comes to Darwin's theory of evolution, first it is interesting to look at how he came across the idea. On the one hand, he had examined thousands of observations relating to the natural world, consciously looking for explanations of diversity and adaptation, and mulling over possible ideas. But it seems the final insight came quite suddenly, prompted from an unexpected source – that of reading Malthus just 'for amusement' is how Darwin put it in his autobiography.

Then, it should be emphasized that Darwin saw natural selection as a scientific law of biology, to be compared with Newton's law of gravity for the physical sciences, with comparable explanatory power for biology as the laws of motion have for physics; and one of his prime aims was to present natural selection in this light. In particular, in view of the reputation of Whewell, who was Master of Trinity College and with whom he had become friends while at Cambridge, Darwin wanted to convince him of the scientific standing of his theory. So in writing the *Origin* he presented the facts of variation, the struggle for life, and the analogy of domestic breeding as the basis for his proposed Law of Natural Selection.

However, following Herschel, what mattered was justification of his hypothesis. It was all very well to postulate natural selection, but what was the evidence that it really happened? The obvious deduction from his theory is that species will be seen to evolve from one to another. But Darwin knew he would be unable to demonstrate this, if for no other reason than the limited time scale of human experience compared with the geological eras over which he believed evolution had occurred. So, instead, he presented a range of alternative supporting arguments: the morphological changes achieved by artificial selection, the fossil record, homology and comparative embryology, each supported with many examples. He recognised that these did not *prove* evolution by natural selection, and that there were substantial objections to his theory; but Darwin was convinced of its truth because it seemed to him to be a unifying explanation for a host of facts from these diverse disciplines.

Reception of Darwin's theory

There were some scientists who, right from the start, supported Darwin and his theory. Notable among them was the biologist Thomas Huxley (1825–95) who, on reading the *Origin* declared how stupid not to have thought of that before, and eagerly championed Darwin's cause. It was

Huxley, rather than Darwin, who engaged in public debate with opponents of evolution, famously in his debate with Bishop Wilberforce at the Oxford Union, for which he acquired the nickname of Darwin's bulldog. (Due to his generally poor health, Darwin confined his exchanges to correspondence.) Other keen supporters were Joseph Hooker who had assisted Darwin with the initial presentation of his theory, and the American botanist Asa Gray. All of these were satisfied that in natural selection Darwin had found the key to evolution, finally providing a natural explanation for the adaptation and diversification of earth's living things. Indeed, for Huxley and Darwin, the principal attraction of evolution was that it provided a credible naturalistic alternative to supernatural forces and teleological explanation in biology.

Limitations to evolution by natural selection

Many others were not so sure. As I have already outlined, by mid 19th century there was much speculation that some sort of evolutionary development had taken place, so there was a general willingness to see Darwin's natural selection as having a part in that process, but many were unconvinced that it could achieve all that Darwin claimed. They were well aware that the scope for effecting morphological change by domestic breeding was limited, and doubted that nature was free of similar limitations; so they questioned Darwin's unlimited extrapolation based on nature's superiority and extended time scale.

Many felt that natural selection might be effective for fine-tuning, but could not account for the formation of completely new complicated structures such as eyes and wings. A common argument was that part-formed structures would be a disadvantage rather than an advantage – so natural selection would work against them rather than for them. So several scientists proposed that, at least from time to time, evolution would require the occurrence of gross mutations to effect large changes such as these. How such gross novelties might arise was still a mystery; but the occasional appearance of monstrosities was seen as a possible indication.

There were also doubts about the inheritance of variations, notably by Fleeming Jenkin. These arose mainly from the almost complete ignorance at the time of hereditary mechanisms, and we will see in the next chapter how these were unravelled.

Geological objections

One of the key areas for dispute was the fossil record. Several eminent geologists, notably Adam Sedgwick and Richard Owen, strongly contested Darwin's interpretation of the fossils, especially pointing to the abrupt rather than gradual appearance of many distinct forms. Darwin had accepted that the intermediates his theory predicted were generally absent and appealed that the fossil record was far less complete than was generally believed. But this interpretation was contested, typically by François

Pictet who wrote:

> Why don't we find these gradations in the fossil record, and why, instead of collecting thousands of identical individuals, do we not find more intermediary forms? To this Mr. Darwin replies that we have ... only a few incomplete pages in the great book of nature and the transitions have been in the pages which we lack. By when then and by what peculiar rules of probability does it happen that the species which we find most frequently and most abundantly in all the newly discovered beds are in the immense majority of the cases species which we already have in our collections? [Hull]

However, not long after, in 1861, there was the discovery of *Archaeopteryx* which is a fully-fledged bird but with some reptilian features, especially in having teeth which are absent from modern birds, and was an obvious candidate for an intermediate form. Later in the 1860s and 1870s various fossils were found in Europe and America which were pieced together to show the development of the modern horse through the Tertiary period from a dog-sized ancestor in the early Eocene, about 50 million years ago (more on this in Ch. 10). Such discoveries were seen to support Darwin's belief that the missing pages would be found.

Philosophical objections

Another common criticism of Darwin's theory was that it was not sufficiently inductive. Darwin recognised that his theory of natural selection was a hypothesis to explain various observations about the natural world and that he could not prove it. But, despite Newton's pioneering work long before, most scientists of Darwin's day still saw science primarily as the derivation by induction of generalised laws from sets of facts; so they felt Darwin's approach was not really scientific. Darwin was especially disappointed by the response of the philosophers of science. Even Herschel who had realised that science sometimes advanced by bold hypotheses rather than rigorous induction referred to natural selection as 'the law of higgledy-piggledy'. It seems they were so accustomed to the rigorous, quantitative laws of the physical sciences that they found Darwin's qualitative description of natural selection for biology as too woolly to be truly scientific.

In addition, both Whewell and Herschel, though not biologists, believed that teleology was an important aspect of biology and that Darwin was wrong to reject it. It is interesting that even some of Darwin' supporters, such as Wallace, Lyell and Gray urged him to see God's hand behind evolution, i.e. that evolution occurred through the action of secondary laws.

Others were unhappy with the amorality of natural selection: the idea of nature advancing 'red in tooth and claw' (Tennyson, *In Memoriam*) did not sit comfortably with many Victorians, whether for religious or other reasons.

Acceptance of evolution

However, within a decade of the *Origin*, probably most scientists and a significant proportion of the public were inclined to accept that evolution had occurred. There was, of course, much debate at first as to whether and to what extent man had evolved; but with Lyell's *Antiquity of Man* (1863) and Darwin's *Descent of Man* (1871) an increasing number, including many theologians, were persuaded that the biological *Homo sapiens* had arisen by an evolutionary route, though many reserved judgement regarding his 'spiritual' nature.

In the latter half of the 19th century much scientific work focused on determining evolutionary histories (phylogenies) of various groups of organisms, and there was a shift towards phylogenies being the basis for classification rather than strictly morphology. At about the same time, mainly through writings of the German biologist Ernst Haeckel (1834–1919), the idea that evolutionary history can be traced in the development of the embryo, encapsulated in the phrase that 'ontogeny recapitulates phylogeny', came to be seen as further evidence for evolution.

Even before the *Origin*, the social philosopher Herbert Spencer (1820–1903) had proposed that society advances in an evolutionary manner, in the context of which he coined the phrase 'survival of the fittest'. And as the theory of biological evolution gained wider acceptance there was increasing application of the evolutionary concept to many aspects of human life.

As the 19th century drew to its close, the outstanding problem with the theory of evolution was the lack of understanding of how the all-important variations originated and are propagated. The answers to these questions came early in the next.

5

Parents and Offspring

It will be apparent from the preceding account that the theory of evolution arose largely from naturalists trying to make sense of the vast diversity of life that had been discovered since the Renaissance. Classification had been the first step, and this, in highlighting the similarities between various groups of organisms, taken with the increased awareness of variation and adaptation, led many to question the traditional belief in the fixity of species, and hence came the concept of evolution. Darwin provided the guiding principle of natural selection which made the whole process seem possible, even probable.

However, despite all the evidence for evolution on the large scale – in terms of comparative anatomy (homology), adaptation and fossils – what was lacking, to give the theory a sound footing, was a scientific understanding of heredity. How do offspring resemble their parents? Why are they not identical to their parents? Or why are siblings not identical? It is evident that, above all, any theory of inheritance must explain reproducibility – how a species produces more of essentially the same; but, especially important for the theory of evolution, it also needs to explain the origin and inheritance of variations. Clearly, without the production and reliable transmission of advantageous variations there can be no long-term progress. So any hereditary mechanism, to be compatible with evolution, had to satisfy the conflicting demands of both constancy and change; or, as Darwin put it, 'descent with modification as the means of production of new species.'

Darwin on inheritance

Right from the outset, Darwin recognised that the production and propagation of beneficial variations were crucial to his theory, but that he lacked a clear understanding of either. Acutely conscious of this deficiency, in the years following publication of the *Origin* he amassed information on the subject and eventually presented his ideas on inheritance in *Variation of Plants and Animals under Domestication* (1868). In this he grapples with what he found confusing and all too often seemingly contradictory observations.

Origin of variations

So far as the source or causes of variations are concerned, Darwin was influenced by the ideas of Buffon and especially Lamarck, that variations arose due to an organism's mode of life, and on the effects of use and disuse. Common observations supported this view, for example that physical work or training leads to muscular development, and sons often follow their father's career, e.g. in sport or manual labour, or in more academic pursuits where intellectual rather than physical abilities are prominent. Also, Darwin saw the breeding of domestic varieties of plants and animals as due to their improved way of life when cultivated, and similar reasoning could explain why many reverted when released into the wild. Of course, this idea for the origin of variations suited his theory of evolution because it made sense that an organism acquired characteristics appropriate to its way of life, indeed, which further adapted it to that way of life. However, whilst he was convinced that variations arose from changes in conditions of life, he could not discern any clear cause and effect, and he was well aware of exceptions to these general observations. For example, why was it that of a large number of individuals living under the same conditions only some produced variations, or why did the same variations arise under totally different conditions? He concluded that 'Our ignorance of the laws of variation is profound.'

Inheritance

When it came to inheritance, his basic problem was that he could see this only in terms of *blending* characteristics from the parents, which was the prevailing view among biologists at the time. Whilst this interpretation fitted many common observations, especially where offspring were intermediate the characteristics of their parents, such as in height or coloration, there are others where it does not fit at all. For example, there are many instances where a particular characteristic is possessed by only one parent, yet is passed on fully-developed to its offspring, i.e. not partly-formed, as would be expected if the offspring were a blend of the parents' characteristics. Much more difficult to reconcile with blending were cases where both parents had a fully-developed characteristic but it was completely absent from their offspring; and, even more confusing, where offspring had a distinctive characteristic which was absent in both parents, but might have been present in a grandparent or even further back.

For instance, the common snapdragon has the well-known asymmetric flower, but there is a variety having an abnormal symmetrical (peloric) flower. Darwin crossed the two varieties and found that all the first generation resembled the common form; and on interbreeding these he obtained 88 asymmetrical and 37 peloric (with two intermediate in appearance). Darwin just could not understand how, when he crossed a mixture of types, only one form resulted, but then when he crossed plants which all looked the same (the first generation offspring), he got a mixture of progeny. We can now see that these results compare reasonably well with the Mendelian understanding of inheritance described below, but Darwin could explain them only in terms of the peloric form being 'latent' and then 'to gain strength by the intermission of a generation'.

Pangenesis

Eventually Darwin settled on the theory of pangenesis, which he elaborated from ideas of the classical Greek physician, Hippocrates, even though Aristotle had rejected it because he recognised that it did not account for instances where children resemble more distant ancestors than their immediate parents. Many, including some of his contemporaries, have criticised Darwin for adopting this view, seeing it as an 'unfortunate anomaly', but he clearly presented it only as a hypothesis in the absence of anything better.

The principal assumption of pangenesis was that all parts of the body release minute particles (gemmules) which permeate the whole body and in particular the genitalia. They pack into the sex cells, unit on fertilization, and then direct development of the embryo to produce organs and tissues corresponding to the originating organs and tissues. The gemmules were not considered to produce cells, but penetrate new ones to direct their development. In this scheme, if, due to an organism's mode of life, a particular organ was much used or developed, then it produced more and/or stronger gemmules, which enhanced development of that organ in the offspring; conversely, a little-used organ produced fewer or weaker gemmules so that the offspring's organ was less well developed.

Pangenesis was, of course, a blending scheme, with gemmules from both parents being mixed and affecting the form of the progeny. However, as pointed out at the time, most prominently by the engineer Fleeming Jenkin, there are difficulties with any blending mechanism of inheritance for the theory of evolution: If offspring contain a blend of their parents' characteristics, then if one of the parents has a beneficial variation, because its hereditary substance is blended with that from the other 'normal' parent, it will only be half as distinct in its offspring, a quarter in the next generation, and so on. In other words, beneficial changes would be in danger of continual dilution – ultimately of being washed out – rather than accruing.

As it happened, it was soon after Darwin published his ideas on pangenesis that real progress began to be made towards an understanding of heredity. Population genetics followed in the succeeding decades, leading to Neo-Darwinism or the Synthetic Theory of Evolution (because it was a *synthesis* of Darwin's natural selection and Mendel's genetics), which had been formulated by the middle of the 20th century. Knowledge of the biochemical basis of heredity has arisen only since then, with perhaps the major breakthroughs being made mainly in the 1950s to 1970s.

This chapter outlines the elucidation of hereditary mechanisms and the consequent development of Neo-Darwinism. It includes some technical points that are necessary for an adequate understanding of the theory of evolution and discussions in subsequent chapters. I have dealt with these in subsidiary boxes to make the main text less ponderous; hopefully this also means they can easily be skipped by those readers who are already familiar with them, and more conveniently available for reference later if needed. Fuller accounts are readily available from any reasonably modern standard textbook on genetics.

HEREDITARY MECHANISMS

Early cytology

Cells had first been observed as long ago as the 17th century by Robert Hooke who, you may recall, carried out important microscopical work: in his *Micrographia* he described cork as being composed of discrete cells, arranged rather like a honeycomb. At about the same time another pioneering microscopist, Antoni van Leeuwenhoek (1632–1723), included cell walls in his drawings of plant and animal tissues, but probably without realising their structural significance. Leeuwenhoek also examined a range of single-celled entities such as blood cells, sperm, protozoa and, amazingly, given the 'simple' single-lens microscope which he used, he observed individual bacteria, and well enough to distinguish different types! Unfortunately little further progress was made in cytology for about a hundred years.

The cellular nature of biological organisms was established in the first half of the 19th century. As early as 1805, Lorenz Oken (1779–1851) in his *Generation* stated that 'all organic beings originate from and consist of vesicles or cells', but his ideas did not carry much weight at the time. Credit for recognising the role of cells is usually given to Matthias Schleiden (1804–81) who was Professor of Botany at Jena where Oken had worked, and wrote *On Phytogenesis* (1838) in which he describes plants as being composed of cells. The cellular structure of animals is generally much less apparent than of plants, and Schleiden implies that animals are not composed of cells in the same sort of way; but Theodor Schwann (1810–82) put this point right the following year in his *Microscopical researches on the similarity in structure and growth of animals and plants*. In this he gives detailed accounts of the structures of many animal organs, clearly demonstrating their cellular structure, and stresses that all animal tissues are derived from cells, even those such as teeth and feathers for which it might have seemed unlikely.

Cell division: mitosis

Schwann also recognised that eggs are single cells, and observed that newly-fertilized eggs develop by dividing progressively into two, four etc. cells. This led to acceptance of the view that cells can arise only from another cell, and not by, for example, crystallization of protoplasm which had been an earlier idea. It was soon suspected that nuclei are important for cell division, and this was confirmed in the 1870s when staining techniques were developed which highlighted the nuclear material and al-

Box 5.1   Mitosis

In a normal, non-dividing cell only very fine stained filaments are discernible, if at all, in the nucleus.

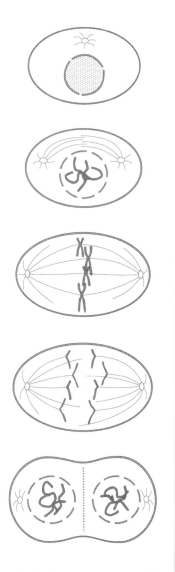

As cell division commences, the filaments become obvious and progressively more thread-like in appearance, then contract and thicken, eventually leading to a number of discrete bodies, the chromosomes. While this is happening, the membrane surrounding the nucleus, which normally separates it from the other cell contents, disintegrates. In addition, a small feature called a centrosome divides, and one migrates to either side of the cell.

The chromosomes align across the equatorial plane, between poles defined by the centrosomes. Fine lines (spindle fibres) radiate from the centrosomes, some towards and contacting the chromosomes.

It is now evident that each chromosome comprises a pair of chromatids, and one of each pair migrates towards each centrosome.

As chromatids aggregate near the centrosomes, a new nuclear membrane surrounds each set to form two new nuclei, and at the same time a cell membrane forms between them, completing cell division.

The genetic material (DNA) duplicates very early in the process, even before cell division is evident: right from when the chromosomes first appear they are already double, containing two chromatids which are separated in the course of mitosis.

lowed the process of cell division, led by division of the cell nucleus, to be observed. The process was called mitosis, and is outlined in Box 5.1. An important discovery was of chromosomes ('coloured bodies', because they took up the stain), the number of which is specific for each species, ranging from two to hundreds: man has 46.

Weismann's germ-plasm

By now, of course, the *Origin of Species* had been published for several years and aroused much interest in the mechanisms of inheritance. And it was against this background of early observations regarding the cellular composition of organisms and increasing evidence for the critical role of the nucleus in the formation of new cells that August Weismann (1834–1914) made a key contribution. He recognised that the essence of heredity is the transmission from parent to offspring of information – information that determines the development of the embryo into a new individual of the same species. In order to fulfil this role, he concluded that the information must be transmitted by means of 'a substance with a definite chemical and, above all, molecular constitution' (i.e. not just an overall chemical composition, but the arrangement of the constituent atoms was important, which was a remarkably advanced concept), which he called germ-plasm. He believed the germ-plasm to be associated with chromosomes, and that the elaborate process of mitosis in which they were clearly seen to be shared equally between daughter cells was to achieve accurate duplication and transmission of the hereditary information.

An important aspect of Weismann's theory was continuity of the germ-plasm. Most of the cells arising from the growth of an embryo become the various organs and tissues of the individual – and ultimately die. However, those which become the germ-cells do not: in due course at least some of them (or, more importantly, at least copies of their hereditary material) are transmitted to the next generation. This is how Weismann concluded that the hereditary material is constantly being propagated.

A significant consequence of his theory of the germ-plasm and its role in directing development of the individual was that it contradicted the long-standing doctrines of the effect of use and disuse, and of the inheritance of acquired characteristics: the hereditary material controlled the form of the organism's body, not vice versa. Weismann concluded that this was inconsistent with these doctrines or pangenesis.

Not only did his work help to put an end to these erroneous ideas, it also shed light on the origin of variations. With the germ-plasm in control of an organism's development, it cast doubt on the idea that an organism's environment or way of life could in some way give rise to appropriate variations; mutations in the tissues would not be heritable anyway. So the clear conclusion from this was that variations of relevance to evolution (which clearly need to be heritable) must arise by mutations of the germ-plasm. It seemed highly unlikely that the environment could in any way direct such mutations, so the clear consequence was that mutations must be substantially random (undirected) events. Weismann envisaged that mutations arose through miscopying of the germ-plasm, and that natural selection prevented the propagation of grossly disadvantageous ones.

Mendelian inheritance

One of the main areas of debate at the time was Darwin's claim that natural selection acting on a series of small variations could eventually yield the substantial morphological changes involved in large scale evolution. Some questioned this and thought that much more dramatic variations due to large mutations would be required. A key figure was Hugo de Vries (1848–1935) who was Professor of Botany in Amsterdam.

De Vries and pangenes

The traditional view of species, especially stemming from work on classification, was that a species was characterized by a fixed combination of features (characteristics). However, from studying plant varieties it seemed to de Vries that for many species there was a range of features which could be selected and combined in different ways, resulting in different varieties, but still the same species. In other words, the characteristics of a species are not inseparable (as had been thought), but consist of discrete attributes which can be combined in a variety of ways. Hence, starting from Weismann's idea of the germ-plasm, de Vries went further and proposed that the germ-plasm comprised individual hereditary units, which he called pangenes, each controlling specific characteristics or features of the organism as a whole. In time his concept of a pangene developed into the modern idea of a gene.

De Vries was particularly influenced by some breeding experiments he carried out with the evening primrose. Due to a peculiar genetic make-up (which was unknown to de Vries) the common form of this plant produces new varieties generation after generation, many of which were quite different from the parent plant and would breed true. To de Vries, these looked like the product of significant mutations, and he concluded that mutations which produce substantial changes were far more important in the production of new species than the gradual accumulation of small changes by progressive natural selection. To develop his ideas further he carried out various experiments crossing plant varieties which had different distinct characteristics; and it was in the course of researching for this work that he came across Mendel's crossbreeding studies.

Mendel

Gregor Mendel (1822–84) was an Augustinian monk at Brünn in Moravia, but he follows more in the tradition of scholars such as Albertus Magnus, especially with his wide-ranging interest in natural history, than our usual image of a monk. He studied physical sciences at the University of Vienna, and returned to the monastery in 1851 where he stayed for the rest of his life as a teacher and then Abbot. He carried out his plant breeding experiments in the monastery gardens (at about the same time as Darwin was writing the *Origin*) and published his results in 1866 and 1869. Unfortunately, although these papers were included in bibliographies at the time and copies sent to one or two leading biologists, they were not known by the general scientific community until publicized by de Vries and others in 1900.

Crossbreeding experiments had been carried out by several others before Mendel, some even looking at the same crosses as he; but their prime interest had been investigation of the species question and they had looked at their results only qualitatively rather than quantitatively. It was because Mendel counted the results of his experiments, and used sufficiently large numbers to mitigate the effects of random variation, that he was able to come up with his radical, and correct, conclusions for an underlying hereditary mechanism.

He selected several strains of pea plant with various contrasting characteristics such as short (dwarf) versus tall stems and green versus yellow peas, ensured that each variety bred true, and then carried out a series of experiments in which he crossed plants having contrasting characteristics. When he crossed plants having yellow peas with those having green peas, all the offspring had yellow peas, rather than any being of intermediate or mixed colour (which might have been expected if inheritance were based on blending characteristics); and when he interbred these offspring, three quarters of their progeny had yellow peas, the remaining quarter having green, again with none being intermediate.

To explain these results he proposed a theory of heredity, which can be summarised as follows:

Each plant carries two factors which determine the colour of the peas; either of these may be for yellow or green.

When a plant reproduces, it passes on a copy of one of its factors to each of its offspring; either factor may be passed to any offspring. Thus each offspring derives one of its two factors from each of its two parents. The combination of factors is random, and it is irrelevant which sex donates which factor.

The factor for yellow peas is dominant: this means that if either factor is for yellow then the plant will have yellow peas; only if both are for green will the plant have green peas, i.e. the factor for green peas is recessive.

How this theory explains his observations is illustrated in Box 5.2. He did similar experiments with peas differing in six other contrasting characteristics and all gave comparable results.

Mendel checked the validity of his theory by back-crossing the hybrid offspring, which he proposed had a combination of yellow and green factors even though they had completely yellow peas, with each type of true-breeding parent. As shown in Box 5.2, back-crossing with the green parent strain resulted in half the offspring having green peas – compared with his first cross between yellow- and green-bearing plants which had resulted in all the offspring having yellow peas. This showed that the recessive green

---

Box 5.2   **Mendel's experiments**

For true-breeding varieties, i.e. the offspring for several generations show no variation, it can be assumed that all the individuals have a consistent set of genes (at least for the characteristics under consideration), and that both copies of the genes (alleles) are identical, i.e. they are homozygous.

On crossing two varieties, one breeding true for yellow peas and one true for green (represented by AA and aa respectively), each offspring receives one factor from each parent, as indicated here (offspring gene combinations are shaded).

		possible genes from parent AA	
		A	A
possible genes	a	Aa	Aa
from parent aa	a	Aa	Aa

This means that all of the offspring are Aa; and, because yellow is dominant, all the offspring have yellow peas – fully yellow, not a blend or mosaic of yellow and green.

Interbreeding these offspring (first generation, F1) produces the following gene combinations:

		possible genes from F1 parent	
		A	a
possible genes	A	AA	aA
from F1 parent	a	Aa	aa

Second generation offspring (F2) with aA or Aa are identical, so the overall result is 25% having AA, 50% Aa and 25% aa; and since both AA and Aa appear fully yellow, the appearance is 75% yellow and 25% green, i.e. in a ratio of 3:1.

To check his theory, Mendel backcrossed F1 plants with each of the initial pure-breeding parents:

		possible genes from AA		possible genes from aa	
		A	A	a	a
possible genes	A	AA	AA	Aa	Aa
from F1 parent	a	Aa	Aa	aa	aa

When F1 (yellow) was crossed with the dominant parent (yellow) all the new offspring were yellow (whether their genotype was AA or Aa); when crossed with the recessive parent (green), 50% were yellow (Aa) and 50% were green (aa).

---

factor was present in the hybrid progeny, but was masked by the presence of the dominant yellow factor, but nevertheless was available for transmission to the next generation.

The particularly important point to arise from Mendel's work, which contrasted with earlier ideas about inheritance, was that inheritance is indirect, via factors or determinants (which we now think of as genes). This means that transmission of a characteristic depends on possession of the relevant gene within the individual, not whether or not the characteristic is evident (i.e. the gene does not have to be expressed in order to be heritable). Whether genes are available for transmission is simply dependent on whether or not they are present in the organism's hereditary material, the factors are not produced by the tissue or organ concerned.

Secondly, and of particular relevance to evolution, Mendel's work showed that inheritance depended on discrete factors which remained intact and were not subject to dilution by blending, so this overcame one of the fundamental problems with Darwin's initial theory. His work also pointed to heritable variations being due to mutation of the genetic material rather than to changes in the tissues or organs themselves.

Following recovery of Mendel's work, two biologists, William Bateson (1861–1926) and Edith Saunders (1865–1945), showed that many studies on inheritance dating from before 1900, such as Darwin's snapdragons, which had previously seemed inconclusive or completely baffling, could now be explained readily in terms of discrete heritable factors, coupled with the idea of some factors being dominant and others recessive.

These two workers also introduced some of the terms which are now commonly used in genetics. They designated the initial parent generation as P, the first (filial) generation as F1, the next as F2, and so on. Each of the possible alternative forms of a heritable factor (gene) they called allelomorphs (now usually shortened to allele); an individual having identical alleles they called a homozygote, and a heterozygote where the alleles are different.

Continuous variation

As soon as Mendel's work was discovered, there was much enthusiasm for his theory, because of the questions it resolved. But the discrete nature of Mendelian genetics posed other problems. Heredity had now been found to entail discrete units controlling characteristics in an 'all or nothing' way, frequently resulting in substantial changes in the organism, rather than the gradual changes predicted by Darwin. So this encouraged some biologists, such as de Vries and Bateson, to think that significant evolutionary change could be achieved only by means of large mutations; that is, they put an emphasis on obtaining the right mutations rather than the cumulative operation of natural selection on small random variations.

Also, quite apart from any evolutionary implications, it was evident that some characteristics do vary gradually or continuously rather than in discrete amounts, such as the height or weight of individuals within a population; so this raised the question as to whether Mendelian particulate inheritance was the whole story. Some suggested that Mendelian heredity controlled major features, but smaller changes were possible by some other mechanism. This problem of the gradation of some properties was addressed in the first decade or so of the 20th century, and finally explained within a framework of Mendelian inheritance.

First it was recognised that any given characteristic may be controlled by many genes, with some genes having a large effect and others small, and even that the effects of the different genes need not be additive but have a complex interaction. It was also realised that some alleles were not clearly dominant or recessive, so heterozygotes could be intermediate in character. As the complexity of genetic interactions began to emerge, this enabled most of the observed variations to be explained, even where a feature appeared to vary continuously rather than in discrete steps.

Then it was recognised that the environment of the organism could also have an effect. Since Weismann had proposed the germ-plasm with his emphasis that this directed the form of the organism rather than vice versa, his views had been so persuasive that biologists had begun to assume that an organism was exclusively the product of its genetic makeup, and completely unaffected by its way of life. And this was reinforced by Mendel's work. However, in 1911 the botanist Wilhelm Johannsen (1857–1927) showed that this was not the whole story. Beans from a true-breeding line nevertheless vary in size, and even when he took subsets of these beans – whether small, medium or large – and grew new plants from them, in each case the resulting beans had exactly the same range of sizes and average value as the initial parent plants. Because he started with a true-breeding line, all of the plants had the same genes, so any differences in the sizes of the beans must have been due to environmental factors such as the local climate or even just position in the pod. So this work showed that some variation can be due to environmental factors, and this could further smooth out discrete variation due to genes alone.

Also arising from this work Johannsen introduced the important distinction between the genetic composition of an individual (its genotype), and the resultant characteristics (its phenotype). Phenotype usually refers to physical or morphological appearance, but can mean less obvious features such as the presence or absence of a particular protein, or even which type of a protein is present. In addition, it highlighted the important point so far as evolution is concerned that selection for favourable characteristics in an organism, whether by natural selection or domestic breeding, will be effective (i.e. assimilated) only if the difference in characteristic has a genetic basis, and is not merely a consequence of appropriate environmental conditions or way of life.

Meiosis

We now come to meiosis which is part of the overall process of inheritance, and is especially important for producing, from limited genetic material, enormous variation – variation which is essential for evolution because it is on this that natural selection can act.

In the 1880s it was observed that the nuclei of eggs and sperm (gametes) contain only half the number of chromosomes of the fertilized egg (zygote). It was evident that each gamete contributes half the chromosomes to the zygote, and that the chromosomes of the latter are paired, one coming from each parent, the members of a pair being called homologous chromosomes. Further observation showed that the final cell division in the formation of gametes was unusual, with two nuclear divisions occurring in quick succession, and the outcome was a halving of the number of chromosomes. This double division was named meiosis which means reduction. Cytological and hereditary studies went hand in hand during the first 30 years of the 20th century to elucidate the mechanism of meiosis, and the main features are as follows:

As meiosis starts, threads of chromatin appear within the nucleus, but only half the number that appear during mitosis. As with mitosis, each chromosome has been duplicated, to yield two chromatids which are intimately associated with each other; but, unlike mitosis, the homologous chromosomes are also associated. This means that at the start of meiosis each visible 'chromosome' comprises four chromatids.

At the first division, the parental pairs (homologous chromosomes) are separated so that each daughter nucleus has half the number of chromosomes, one of each parental type. But this separation is not between the original parents' chromosomes, it is random so that the chromosomes of each daughter nucleus is a mix of maternal and paternal chromosomes. At this stage, each is present as two copies (chromatids). At the second division, the chromatids are separated (analogously to that in mitosis) so that each new nucleus now has only one copy. It will be apparent from this random shuffling that even with only a few chromosomes a very large number of possible combinations of chromosomes can occur. This alone was a remarkable finding, but it turned out to be hardly a drop in the ocean compared with the actual variability that is possible.

Gene shuffling

Mendel had investigated multiple characteristics and found they were distributed independently among offspring. That is, it did not matter whether his parent yellow- or green-seeded plants were dwarf or tall, or whether the yellow-seeded were dwarf and the green-seeded were tall, or vice versa – in every case the characteristics of the F1 and F2 generations were distributed as shown in Box 5.2. However, in 1905, Bateson found that in the sweet pea the apparently unrelated characteristics of purple flowers and long pollen grains were inherited together, so they proposed that such coupled genes were linked physically by being on the same chromosome, such that they were segregated together at meiosis. Further studies showed that each chromosome could carry several genes affecting a whole range of characteristics. But other results were rather confusing because some characteristics were neither wholly independent nor wholly dependent – their inheritance was linked, but not absolutely tied to one another.

As early as 1903 de Vries, recognising the wide variety of characteristics that seemed to conform with Mendelian heredity, had postulated that some exchange of genetic material may occur between paired (homologous) chromosomes in order to provide the random mixing of factors which Mendelian inheritance required. But this idea was largely ignored at the time because chromosomes were thought to be indivisible and unchangeable entities. However, it was the most likely explanation for partial linkage, and was confirmed after many years of research. The process may be summarised as follows:

In the early stage of meiosis, when the chromosomes first appear, each comprising four chromatids, one chromatid of each pair (i.e. one maternal and one paternal) crossover. That is, each chromatid breaks and rejoins to its partner, resulting in part of one chromatid being exchanged for the corresponding part of the other. Crossover is illustrated in Figure 2. Not only may this arise on all the chromosomes, but there can be multiple crossovers on the same chromosome, especially for longer ones (so for two crossover points, only the portion of the chromatids between the crossovers is exchanged, and so on for multiple crossover points). The result of crossing-over is to produce four different versions of each chromosome: two are the same as in the original parents, but two are different from either parent or each other. It seems that crossing over can take place at almost any point on the chromosomes, and which versions of the chromosomes are associated in the daughter nuclei (gametes) seems to be entirely random. In fact meiosis is such an effective mechanism for shuffling or recombining genes that the number of possible gene combinations becomes astronomical.

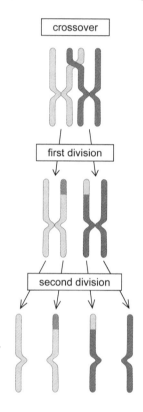

FIGURE 2. Chromosome crossover and distribution between gametes.

Before describing how this capacity for gene shuffling became the basis for population genetics which developed in the 1920s and 1930s, mention should be made of the work on linkage maps which took place alongside the studies used to elucidate meiosis. Notable among them was Thomas Morgan (1866–1945) and his colleagues at Columbia University (New York) with their work on the fruit fly, *Drosophila*. They carried out numerous hereditary studies with many strains of *Drosophila* which differed in recognisable ways as a result of mutations. He crossed various strains and examined the degree of association between the various traits, and gradually built up maps of the *Drosophila* chromosomes (of which there are four pairs) based on the degree of linkage between different genes. Where traits were totally independent it was recognised that the genes responsible were located on different chromosomes; conversely, where genes showed some linkage they were on the same chromosome. For the latter, it was found that the degree of association was constant for any two genes, and, interestingly, the level of linkage (expressed as percent of individuals exhibiting evidence of crossovers) between different pairs of genes was substantially additive. This was a clear indication that the degree of linkage found in hereditary studies reflected the physical separation of genes on the chromosomes: the closer together genes are, the less likely that crossing-over will take place between them, and the closer will be the association between the traits. In this way linkage maps were drawn up which were seen as an actual representation of the physical placement of genes on the chromosomes. I shall pursue the physical nature of genes in the next chapter.

POPULATION GENETICS

In the first couple of decades of the 20th century there was much debate between biologists as to the relative importance of the roles of mutation and natural selection in the process of evolution. On one side were de Vries and Bateson who favoured mutations because of their observations of substantial changes occurring rapidly, and they were followed by Morgan because of his work on mutations in *Drosophila*; and on the other side were Ronald Fisher (1890–1966), J. B. S. Haldane (1892–1964) and Sewall Wright (1889–1988) who emphasized the role of natural selection. The two sides were brought together through the rise of population genetics.

Darwin had seen evolution largely in terms of individuals: he envisaged individual organisms developing preferential variations and, because they survived a little better, overall they produced more offspring and gradually their advantageous variations were propagated. As genetic mechanisms began to be understood, in particular that an individual's genotype was fixed at fertilization, and there was no role for use and disuse or the inheritance of acquired characteristics, it became apparent that individuals do not, indeed cannot, evolve; rather, it is species which are composed of individuals that evolve. This realisation gradually led from a focus on the individual to increased attention on the population of individuals which comprise a species; and so population genetics emerged as a discipline in its own right.

Starting with the basic framework of Mendelian hereditary mechanisms and gene shuffling, various mathematicians showed how these determine the behaviour of mutations within a species' population: how the relative proportions of genes in a population were affected by natural selection, and how segregation of genes could lead to divergence within a species, leading to the formation of new species.

**Nyack College
Eastman Library**

The Hardy–Weinberg equilibrium

Probably the first step towards an understanding of population genetics was by the English mathematician Godfrey Hardy (1877–1947) and German physician Wilhelm Weinberg (1862–1937), working independently, in 1908. Taking on board Mendelian genetics which clearly indicated that at least for some genes there are alternative forms (alleles), they investigated the fate of such alternatives. They showed that if there are two alleles for a particular gene locus (i.e. two options for the same function, such as yellow and green in Mendel's pea plants), and if neither conferred any advantage compared with the other, then their relative proportions within the species' population would remain constant. That is, where there are two alleles, designated here as A and B, the ratio of AA:AB:BB, (representing homozygous for A, heterozygous for A and B, and homozygous for B), is $p^2:2pq:q^2$ where p and q are the relative proportions of the two alleles A and B respectively (so $p + q = 1$). And if for some reason these ratios were disturbed from this equilibrium (for example, by artificial segregation of some of the phenotypes) then the Hardy–Weinberg ratios would be restored within one generation of free mating being reinstituted within the population as a whole. This applies not only for two alternative alleles but also where there are many alleles for a given locus (the formulae for the distributions of alleles in such cases can readily be found in texts on population genetics). Obviously any single individual can have only one or two alleles within its genotype, but within the large number of individuals which make up a species' population, many alleles are possible for each gene locus.

This result is independent of whether either allele is dominant. If in the above example A were dominant to B, it simply means that AB would be phenotypically like AA, but the relative proportions of the three genotypes would be unaffected. Taking the example of the yellow and green pea plants: all of the F1 plants were yellow, and when crossed the F2 generation had 75% yellow and 25% green; but in both generations the overall proportions of both yellow and green alleles was 50%. (And if all the F2 individuals interbred freely, the proportions would remain at these percentages in all subsequent generations.)

The Hardy–Weinberg equilibrium is important in demonstrating that intrinsic variability within a species – where the variations are substantially of comparable survival and/or reproductive value – is maintained indefinitely.

Mutations

An appreciation of the Hardy–Weinberg equilibrium may be the starting point of population genetics, but it is immediately apparent that in describing a steady-state situation, it patently is not evolution which necessarily involves a change in genetic composition. Although variation becomes

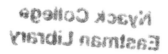
Nyack College
Eastman Library

evident through the shuffling of genes, the ultimate source of variability is thought to be mutations, so it was obvious to the early population geneticists to consider the occurrence of these, and their fate in terms of how readily they are incorporated within a species.

Fate of a new mutation

One of the early conclusions, which was somewhat unexpected, is that the vast majority of new mutations are lost simply due to the vagaries of inheritance. This can be shown by considering the situation where a single mutation arises – i.e. the individual concerned, for the relevant gene, has one normal allele and one mutated allele – and assessing its prospects for being passed on to subsequent generations. To begin with, I will assume that the mutation is neutral, i.e. that it conveys no advantage or disadvantage, but we will see below that the fate of a non-neutral one is not very different.

To illustrate this, first consider the idealised situation of assuming that every individual mates with one other, and the pair produces exactly two mature (reproducing) offspring, i.e. just enough to maintain the population. In this case, when an individual carrying a single mutation mates with a normal member of the species (i.e. one not having the mutation), with random segregation of chromosomes at meiosis and random association of the gametes at fertilization, the mutated gene has exactly the same chance of being passed on to progeny as the normal gene. Hence there is a 1 in 2 chance that the mutation will be passed on to any particular offspring, or, conversely, the probability of not being passed on to a particular offspring is 0.5. So for a family size of two there is a $0.5^2 = 0.25$ probability that the mutation will not be passed on to any of the next generation, or a 75% chance $(1 - 0.25 = 0.75)$ that it will be. These prospects of survival or loss of the mutation will apply to subsequent generations[1]; so the probability of still having the mutation present after two generations is 0.75^2 which is about 56%, after three generations it will be down to about 42% (0.75^3), and so on. Consequently, the most likely outcome for any neutral mutation is that it will be lost from the population within a few generations.

The preceding paragraph describes an overly simple way of looking at the fate of a mutation, but hopefully it will help readers to appreciate why most mutations are lost, without being unduly mathematical. A more rigorous mathematical treatment of the problem was carried out by Ronald Fisher who was primarily a statistician and, by applying statistical techniques to genetics, became one of the pioneers of population genetics. In *The Genetical Theory of Natural Selection* (1930) he considers the more realistic situation that the number of progeny from a given individual can

---

[1] This is true while the mutation is rare so that it is present predominantly in heterozygous form which usually mates with a homozygous 'normal' individual.

Box 5.3   Fate of a mutation

Ronald Fisher, in *The Genetical Theory of Natural Selection*, derived the following relationship to describe the probable survival of a gene from one generation to the next, i.e. from x_i to x_j

$$f(x_j) = e^{c(x_i - 1)}$$

where x is the probability of survival, and c is a measure of the relative advantage conferred by the gene, such that if the gene confers a 2% advantage then $c = 1 + 2/100 = 1.02$.

The following values are based on this equation and a table provided by Fisher.

number of generations	probability of survival	
	no advantage	1% advantage
1	0.6321	0.6358
2	0.4685	0.4738
3	0.3741	0.3803
5	0.2681	0.2754
10	0.1582	0.1667
30	0.0607	0.0705
100	0.0194	0.0307
limit	0.0000	0.0197

Note: In effect, Fisher considered the chance of a new mutation gaining a foothold in a population of infinite size. It was about 50 years later that other population geneticists, notably Kimura who is mentioned in Chapter 7, considered the effect of population size. The main effect of this in relation to the results given here is that the chance of a neutral mutation spreading to the population is not actually zero, but approximately 1/(2N) where N is the number of individuals in the population, which is usually of the order of millions, so the chance of survival, though not zero, is very small.

---

range from zero to many, with each number having an appropriate probability of occurrence, and then considers the probability of the mutation being present for each case, and evaluates the fate of the mutation for an unlimited number of generations in order to assess the overall outcome. The resulting formulation is given in Box 5.3 which also includes some tabulated results. It can be seen from this that the fate of a mutation is even worse in this more realistic analysis than in the simpler situation outlined above – there is about a 37% chance of loss in the first generation compared with 25%. These results clearly show that most neutral mutations are lost, and usually in less than ten generations.

Calculations such as these convinced most biologists that mutation alone was insufficient to produce substantial evolutionary progress. Even with evidence that mutations recurred at a measurable rate (see below), this was found to be insufficient by itself to be a major cause of evolutionary change. It became increasingly evident that natural selection must have a key role to play in the assimilation of mutations, so the next stage was to consider those situations where the alleles are not comparable but confer different levels of advantage or fitness to the individual.

Natural selection

Fate of an advantageous mutation

Along with his analysis of the fate of neutral mutations Ronald Fisher showed the more surprising result that, even when a mutation confers significant advantage (improved fitness), still most will be lost, as may be seen from Box 5.3. The reason for this is that in the early stages, while a mutation is present in relatively few individuals, its fate is determined much more by chance than by any advantage it may confer. But whereas the probability of a neutral mutation spreading through a population is virtually zero, an advantageous gene has a significant chance of success: if the gene confers an advantage of $X\%$ (i.e. on average its bearers produce $X\%$ more viable offspring than their 'normal' counterparts), then its chance of spreading through the population is approximately $2X\%$. Box 5.3 includes Fisher's results for the case where a gene confers a 1% advantage. Clearly this is still a rather low probability of success, and means that a given mutation, even an advantageous one, will generally need to arise many times before we can be reasonably sure that it will be assimilated by the species. For example, in the case of a mutation conferring a 1% advantage, it would take 35 recurrences for the mutation to have a 50% chance of being retained by the population.

Propagation of an advantageous mutation

Having determined that advantageous mutations can, with enough attempts, gain a foothold in a species' population, this leads to the question of how long it takes (in terms of number of generations) for it to spread to a reasonable proportion of the population – so that it can be a stepping stone to further advance. (Note that the table in Box 5.3 relates only to the chance of survival after a certain number of generations, which is quite different from the matter, if it has survived, of how extensive it has become within the population.)

First we should see how an advantageous mutation will fare at each generation; and note that we are now assuming a mutation is present in sufficiently large numbers that we can ignore chance fluctuations. A basic model for representing the effect of selection is given in Box 5.4. The essential idea behind a model of selection is as follows: whereas with no differential advantage or fitness the proportion of the genotypes remains constant from generation to generation (i.e. the Hardy–Weinberg equilibrium), with selection the proportions of the genotypes contributing to each succeeding generation is in proportion to the relative fitness (reproductive

Box 5.4   The effect of selection on gene frequencies

In the following table the letter U represents the overall fitness of an individual having the relevant genotype. It takes account of factors such as survival of the individual, fertility, and viability of gametes and offspring. The values of U are proportional to fitness.

Fitness of the heterozygote (U_y) is commonly the same as one of the homozygotes (U_x or U_z) or something in between, but this is not necessarily the case; some heterozygotes are fitter than either homozygote, a situation referred to as heterozygote advantage.

The calculations assume that the alleles are fairly common in a large population, so that random fluctuations can be ignored.

	AA	AB	BB	total
Fitness	U_x	U_y	U_z	
Initial genotype frequency	p_i^2	$2p_iq_i$	q_i^2	1
Proportionate contribution to (alleles of) next generation	$p_i^2U_x$	$2p_iq_iU_y$	$q_i^2U_z$	$\bar{U}$
Genotype frequency in next generation	p_j^2	$2p_jq_j$	q_j^2	1

$\bar{U}$ denotes the (weighted) mean value of U for the whole population.

Allele frequency after selection

Bearing in mind that U represents overall reproductive fitness, the genotypes reproducing successfully are in proportion to both their initial frequency and their relative fitness. These genotypes then contribute proportionately to the alleles passed on to the next generation, as follows:

frequency of A in next generation, $p_j = (p_i^2U_x+p_iq_iU_y)/\bar{U}$

frequency of B in next generation, $q_j = (p_iq_iU_y+q_i^2U_z)/\bar{U}$

It is then assumed that theses alleles combine randomly, i.e. independently of their source genotype, to form the genotypes of the next generation, which arise in proportions consistent with the new allele frequencies, i.e. AA:AB:BB in the proportion $p_j^2:2p_jq_j:q_j^2$.

An important point to note is that the rate at which the frequencies (p and q) of the alleles change is roughly proportional to the product of the prevailing allele frequencies (p x q). This product is greatest when p = q = 0.5, and tends to zero as p or q tends to zero. This means that p and q change most quickly when they are about equal, and most slowly when one or other allele is present at a very low level – regardless of relative fitness. This is seen clearly in the following table.

continued....

success) of the phenotype having the relevant genotype. In this way the proportion of preferential alleles within a population gradually increases at the expense of unfavourable ones. This does not necessarily imply any change in the total number of individuals; the population size depends on various factors, and may increase or decrease independently of any differential selection of genotype occurring within the species.

It should be borne in mind that selection is not between the alleles

Box 5.4 *continued*

Number of generations to pass from one allele frequency to the next

p	variety is dominant selective advantage			variety is recessive selective advantage			q
	50%	10%	1%	50%	10%	1%	
.999							.001
	6	25	234	1807	9026	90235	
.99							.01
	7	27	252	188	927	9243	
.9							.1
	3	10	96	13	59	582	
.8							.2
	2	8	73	5	23	222	
.7							.3
	2	8	69	3	14	128	
.6							.4
	2	8	75	3	10	91	
.5							.5

Adapted from the Appendix by H T J Norton in *Mimicry in Butterflies* (1915) by R C Purnett.

themselves but between the individuals bearing the alleles, i.e. between phenotypes and not genotypes. Hence the existence and extent of dominance affects the outcome. For example, if B starts at a low frequency within the population, and BB confers a significant advantage over AA, its spread through the population will be much more rapid if B is dominant such that the heterozygote AB shares the advantage of BB than if it were recessive and AB resembled AA (this is especially important in the early stages, because when B is rare it will be present much more as AB than as BB). If there is no dominance, then the heterozygote is likely to have fitness intermediate the two homozygotes.

However, it should be noted that, although both relative fitness and the degree of dominance affect the outcome of selection, there is no necessary relationship between fitness and dominance.[2] For example, a gene which conveys substantial phenotypic advantage when homozygous may be totally recessive with the heterozygote being no fitter than the homozygous dominant, and conversely a dominant variant could be very disadvantageous. But clearly, an allele that is both advantageous and dominant will spread through a population much more rapidly than one that is advantageous but recessive.

Once we have seen how an advantageous mutation changes in frequency from one generation to the next, we can assess how rapidly allele frequencies can change within populations in response to relative fitness. Box 5.4 shows that a major part of the change in frequency over one gen-

---

[2] It is therefore unfortunate and misleading that the situation where the heterozygote is fitter than either homozygote is often termed over-dominance; but there is a growing tendency to refer to this situation as heterozygote advantage, which is much more satisfactory.

eration is described by the expression $p_0q_0U_y/N$, from which it can be seen that the change in frequency is proportional to pq, the product of the frequencies of the two alleles, which is maximal when $p = q = 0.5$, i.e. when the two alleles are equally distributed. The important point to draw from this is that the spread is most rapid when the advantageous variant already occupies 50% of the population, and the rate of increase tails off at both low and high frequencies; until, as we have seen for the fate of a new mutation, at extremely low or high frequencies it is determined primarily by chance rather than relative fitness. This may be seen from the last table in Box 5.4 which gives the calculated number of generations for a favourable allele to increase between given frequencies within a population (up to 50%), including different levels of selective advantage and compares whether the favourable allele is dominant or recessive.

An example of natural selection: industrial melanism

Undoubtedly the best-known example of natural selection at work is that of 'industrial melanism' in moths. The peppered moth *Biston betularia* normally had mottled but predominantly pale grey wings, but from the middle of the 19th century a number of individuals were found with very dark wings, and the frequency of occurrence of these progressively increased until by the end of that century it almost completely replaced the earlier form in some industrial areas, especially around Manchester.

Breeding experiments indicated that the dark (melanic) form was due to a single dominant gene. In the 1920s, with the growing understanding of population genetics, John Haldane (1924) used the approach outlined above, together with field data on the relative changes in occurrence of the pale and melanic forms (from an estimated 1% in mid 19th century to around 99% by 1900), to calculate that the latter had a selective advantage of about 50% over the pale form. The selective advantage was explained in terms of the susceptibility of the different forms to predation by birds and the effect of the industrial revolution: Before industrialisation most tree barks were lichen-covered and consequently relatively light-coloured – against which the mottled pale grey provided a good camouflage; but as industrialisation progressed lichens died off and the tree trunks in industrial areas became increasingly dark such that the melanic moths were better camouflaged against this background than the pale ones.

This was an important example: not only did it substantiate the growing science of population genetics and the scientific understanding of natural selection, it was also seen as a clear demonstration of the evolutionary process as a whole.

Small populations and genetic drift

It will, therefore, be apparent that Fisher's work went a long way towards reconciling the contrasting viewpoints that had prevailed regarding the mechanisms of evolution in the early years of the 20th century. In particular, he provided a scientific or mathematical justification for Darwin's belief that evolution could proceed by the accumulation of small favourable mutations; in other words, he demonstrated that natural selection could be effective even where a mutation offered only a modest improvement (e.g. 1%), so it was not necessary to propose large scale mutations for evolution to take place.

However, set against this, the early work on population genetics had highlighted that on the whole most natural populations tend to be rather stable and resistant to evolution due to their large size. This was seen most clearly in the Hardy–Weinberg equilibrium, the resistance to new mutations becoming established, and the very slow rate at which new mutations could spread through a population unless they conferred a very large advantage such as with the moths – but it was recognised that industrial melanism was exceptional in that respect. Also, whilst the mathematical representation of natural selection had explained how a species may change through time by acquiring new mutations, it had not explained how the number of species increases, i.e. how two or more distinct species can arise from a common ancestral one.

Progress with this issue was made by Sewall Wright who demonstrated the potential importance of small populations, or small subpopulations of a large one. First, recall that the early work on the fate of a new mutation had shown how significant an effect random processes may have within a small group of individuals (those having the mutation). From this it became apparent that an advantageous mutation would have a much better chance of spreading through a small population than a large one; this would also apply to a neutral mutation (see the Note in Box 5.3); in fact, due to random effects, even a moderately *dis*advantageous mutation could spread through a population if the population were small enough. So it was evident that small populations could assimilate mutations more readily, and hence that they had the potential to evolve more rapidly, and may have an important role in the overall process of evolution.

Related to the fact that small populations can more easily assimilate mutations, is that genetic variability is more likely to become evident in a small population. To appreciate this, recall that one of the assumptions for the Hardy–Weinberg equilibrium is that the population be large enough for random fluctuations to be ironed out. The converse of this is that in small populations random effects can have a significant effect, and the genetic constitution can move away from the Hardy–Weinberg equilibrium. For example, consider what would happen if a large and more or less uniform population were to become partitioned into subpopulations – perhaps due to geographical separation – such that there was no longer free breeding between them. Then, simply due to random fluctuation, the genetic constitution of each subpopulation could alter from the Hardy–Weinberg equilibrium, and each subpopulation alter in different ways, resulting in the subpopulations becoming quite distinct from each other.

This is important because it is one of the ways in which the potential but hidden genetic diversity of a large population can become evident; and if the subpopulations continue to diverge, then a single population becomes two quite separate ones, ultimately leading to distinct species.

Changes in gene frequencies due primarily to random events rather than selective pressure are called genetic drift. Wright believed that, because of their greater flexibility, subpopulations are especially important for evolution – whether in terms of assimilating new mutations, greater potential diversity, or because perhaps they could respond more quickly to selection pressures.

NEO-DARWINISM

By the middle of the 20th century there was a widespread sense among biologists that evolution had matured. It had left behind the unsatisfactory 19th-century ideas of use and disuse, and the inheritance of acquired characteristics, and replaced them with a proper scientific understanding of heredity. Science itself had advanced dramatically since Darwin's day, with a much clearer methodology and development of mathematical tools. With this scientific approach applied to biology, the theory of evolution was no longer seen only as a historical process discerned in the fossil record; but in genetic and field studies the mechanisms of evolution could be seen taking place in the present, and the whole could be comprehended with mathematical models which showed how gene frequencies changed through time in response to mutation, selection and random factors. 'Neo-Darwinism' refers to the Darwinian theory of evolution framed in terms of Mendelian genetics.

Innate variability

One of the facts to emerge in this period, the extent of which even surprised the biologists, was that most natural species have far more innate variability than had previously been recognised or even imagined. Of course, even from the time of the 17th- and 18th-century classifiers different varieties had been recognised, variation had been an important feature of Darwin's original theory, and de Vries had concluded that species seemed to contain a wide range of features which could be combined in many ways to produce varieties. But, as genetic studies proceeded through the 20th century, there was an increasing realisation that virtually all natural species exhibit significant variation in almost every identifiable characteristic. Although most individuals in a natural population appear much the same, close examination showed that there are in fact minor differences between them – for most characteristics (such as height, weight, colour) there is a range of values.

In addition to these common variations there were significant deviants arising from substantial mutations. Notable was Morgan's work on *Droso-*

phila in which he found a wide range of naturally occurring mutations, and that mutations induced by irradiation or chemically could affect virtually any feature of the fly's morphology, including such bizarre examples as antennapaedia in which legs grow in the place of the usual antennae. Morgan's work on linkage maps clearly showed that mutations were due to specific mutated genes at precise locations on the fly's chromosomes, and that (contrary to early ideas) most morphological features are determined not by one gene but by many, and often these are scattered throughout the genome (all the chromosomes, genetic material). In fact it became evident that even the lowly fruit fly has many thousands of distinct genes, and there are several times that in higher organisms. Also, for most genes within a given species there are multiple alleles available; so, far from having a uniform genetic composition, most individuals are heterozygous at a substantial proportion of their gene loci – typically 6% for vertebrates and about twice that for invertebrates.

Such innate variability, coupled with the mechanisms of inheritance which shuffle the available genes, provides ample raw material for generating variations. Mathematicians such as Fisher showed that having alternatives at only a few sites could give rise to a huge number of gene combinations; so, given the many variable genes which were being found, it was evident that the number of possibilities far exceeded the individuals in most populations: at any one time, the actual genotypes in a species are only a tiny proportion of those possible from its pool of genes. Much of this innate variability is not normally discerned, but is maintained by the Hardy–Weinberg equilibrium, and is a store of genetic flexibility which a species can draw upon when need arises.

Mutations

One of the principal differences between Darwin's thinking and the modern view relates to the origin of variations. Whereas Darwin had thought that variations arose in some way related to a species' way of life, it is now known that essentially mutations occur at random; and they occur in the genetic material which then causes a change in morphology, rather than directly to the tissue(s) concerned.

Work such as Morgan's on *Drosophila* showed that the rate of mutation for most genes is about to 10^{-5}, which means that on average one gene in 100,000 will mutate per generation. So, for example, in a population of one million, for each gene, mutations will arise in about ten individuals every generation. Though, bear in mind that the vast majority of these, even if advantageous, will rapidly disappear.

However, one of the disconcerting findings was that the vast majority of mutations were definitely detrimental, most of the rest were neutral, and there were scarcely any examples of advantageous mutations. In the light of this it was assumed that most potentially beneficial mutations must have already arisen in the past and been absorbed into present populations, and the vast innate variability was seen as evidence of this. Although it was recognised that the raw material for evolution must be mutations, the extent of variation available within almost all species led some biologists to comment that further evolution did not need new mutations, at least not in the short to medium term. For example, Fisher wrote as follows.

It has often been remarked, and truly, that without mutation evolutionary progress, whatever direction it may take, will ultimately come to a standstill for lack of further possible improvements. It has not so often been realised how very far most existing species must be from such a state of stagnation, or how easily with no more than one hundred factors a species may be modified to a condition considerably outside the range of its previous variation, and this in a large number of different characteristics. [*Genetical Theory*, Ch. 4]

Adaptive topology

Darwin's idea of how species adapted to their environment had also been clarified with the concept of an ecological or adaptive topology (due to Sewall Wright). By environment is meant not just the physical habitat, but a species' interaction with it and with other species; and species can interact in all sorts of ways such as being part of a food chain, and can be cooperative, not necessarily competitive. So a species' mode of life can be visualised as an undulating landscape in which there are peaks and troughs of ecological fortune. Peaks are perceived as favourable, and a species seeks to gain as high a ground as possible; and in general, due to the action of natural selection in the past, most established species are on or close to an adaptive peak.

Because a species expresses variation, to a certain extent its population spreads over a small area around the optimal point rather than being concentrated on it. One of the principal roles of natural selection is conservative – to weed out those individuals which by chance are somewhat less well adapted (i.e. stray too far from the optimum), so that the species as a whole remains well adapted to its ecological niche.

Evolution

However, over time the adaptive topology changes. This can be due to changes in the physical environment and/or the organisms with which the species in question interacts. As a result the species may no longer be on an adaptive peak.

It may be that random variation within the species' population produces some individuals which are slightly better suited to the changed conditions (i.e. have moved to higher ground), though others will fare worse. Natural selection favours the former rather than the latter, and overall the species moves some way from its original position towards the higher ground. Provided the new adaptive peak is not too far away and the change in conditions is not too rapid, and especially if there is a reasonably steady upward route to get to it, then the species has a good chance of producing variations which suit it to the new location. That is, it is possible for the species, over successive generations, progressively to move towards and attain a new adaptive peak; i.e. it adapts and evolution has taken place. Conversely, if the move is too far, especially if there is a substantial adaptive valley between the species' current position and a nearby adaptive peak, then the jump may be too great, and instead of evolution the outcome could be extinction.

Gradual adaptation to changes in a species' mode of life is one of the ways in which Darwin saw an individual species evolving, of how organisms gradually progress, of how they attain greater complexity or enhanced adaptation. But we should note that with Neo-Darwinism the emphasis has moved away from a species evolving primarily by acquiring new mutations, to an emphasis on appropriate variations arising by the shuffling and segregation of genes in its gene pool.

Diversification

In the preceding paragraphs it was assumed that the whole of a species was subject to essentially the same ecological conditions, and hence the same evolutionary pressures, so the whole species evolved in essentially the same way. This may apply to a relatively small, confined species; but many species are large and extensive with their population extending across a range of environments. In this latter situation, natural selection can favour different characteristics in different parts of a species' range. To what extent differences are evident depends on the strength of the selective pressures and the degree of segregation between subpopulations – in effect, how freely interbreeding occurs. Where different forms arise within the same species, usually in different parts of the species' range, it is called polymorphism. Many examples of polymorphism are known; a simple one to mention here would be the moths discussed earlier – the pale and melanic forms are polymorphs of the same species; and note that melanism was much more widespread in industrial areas than rural, because selection for melanism was greater there, and there was sufficient segregation of breeding between those moths and the ones in rural areas.

Whilst the principal factor leading to polymorphism is natural selection for differing conditions, it should not be forgotten that divergence of characteristics can also arise by genetic drift, i.e. without selection, especially where interbreeding is restricted so that definable subpopulations arise.

Speciation

Evolution, of course, is not just the modification of a species, or even the production of new species in a one-for-one sense: it is also the multiplication of species – of producing several distinct species from a common ancestral source, i.e. it is branching rather than just linear.

The partial gene segregation evident in polymorphism is seen as a po-

tential first step towards distinct subspecies arising in nature, in time leading to fully separate species (speciation). For polymorphism to lead to distinct subspecies and then species requires a significant level of breeding
isolation, otherwise the genes in different subpopulations would mix too
much to allow segregation to occur. Genetic isolation can arise in many
ways, of which perhaps the easiest to appreciate is simply geographical
separation, such as may occur at opposite ends of a species range, or due
to some intervening physical barrier such as a river or mountain range.
Biologists have also recognised various other isolating mechanisms, for
example based on differing behaviour patterns, which we need not go into
here. The important point is that, as one or more groups within a population become isolated from the main population and start to inbreed, this
allows for continuing divergence of characteristics, whether by selection
or random processes, and may eventually lead to distinct species.

Darwin, and many biologists after him, was especially interested in island species – in relation to the similarities and differences between fauna
and flora of neighbouring islands and any adjacent mainland. It was, of
course, the similar species of tortoises and birds on the Galapagos Islands
that helped prompt his thinking along evolutionary lines, because it was
evident to him that they must have been derived from a common stock,
but diverged and, at least in the case of the finches, become adapted to
different ways of life. It is worth noting that, so far as island populations
are concerned, it is not only that the populations themselves may be small,
but the separation of a common ancestral species between different islands, with limited breeding between them, allows the segregated populations to develop in different directions – whether due to random variations
and/or in response to selection pressure.

Two biologists stand out in the 20th century for their development of
evolutionary thinking, especially in terms of integrating theoretical population genetics with the observed variation occurring in natural populations, and on speciation by the segregation of genes. The first of these was
Theodosius Dobzhansky (1900–75) who wrote, among many other works,
Genetics and the Origin of Species (1937), and second is the zoologist Ernst
Mayr (1904–) who has emphasized the importance of genetic isolation for
the production of new species.

Definition of a species

The current, Neo-Darwinian, understanding of how species arise through
the segregation of genes has renewed debate over the concept and definition of a species. The early evolutionary view was that where different
groups of individuals within a species diverge, they gradually become
more morphologically different until a point is reached when they can no
longer interbreed, and that is the watershed where new species have
arisen. However, nature is rarely so convenient. There are many cases
where 'species' are so morphologically different that no-one wants to

question their status as separate species – even though they interbreed readily; and, more surprisingly, there are cases where two 'species' are unable to interbreed, but are so similar morphologically that it requires an expert to distinguish between them! The latter are termed 'sibling species', and were identified first in *Drosophila*. It is clear that there are genetic isolating mechanisms which have little if anything to do with morphological differences. A related issue is that even interbreeding is not straightforward: Darwin commented on the odd situation that there are many 'varieties' which are hard to cross but the offspring are fully fertile, whilst others cross easily to produce only poorly fertile offspring – of which the mule is a prime example.

The debate about how best to define species goes on. But one thing is certain: with the widespread acceptance of Neo-Darwinism, the idea of species being fixed has been completely abandoned, and with it any notion of special creation. And with that, the old ideas of design and teleology in nature were finally laid to rest.

Darwinism triumphant

With so much achieved in its early decades, by the middle of the 20th century there was a marked sense of success in biology, especially evolutionary biology, and that evolution explains so much of biology. Dobzhansky commented that 'Nothing in biology makes sense except in the light of evolution.' Julian Huxley (1887–1975), grandson of Darwin's champion, captured the sense that a major step forward had been taken with his review *Evolution: The modern synthesis* (1942)[3] in which he presented Darwin's original concept of evolution by natural selection within the modern framework of population genetics and with the new insights into the processes of speciation. And with his many writings and broadcasts he did much to publicize evolution at this time. In his *History of the Environmental Sciences*, Peter Bowler describes this period as 'Darwinism Triumphant'.

Right from the outset the theory of evolution had relied on the occurrence and propagation of variations, but its serious weakness was that it lacked any convincing explanation for these. And when Mendelian inheritance was discovered, the particulate nature of genetic mechanisms at first challenged evolution further. However, with the development of population genetics, not only had evolution come to terms with the mechanisms of inheritance, but it also shed light on the processes of evolution, and evolution emerged a much more robust theory as a result. In particular, there was a much clearer understanding of the concept and workings of natural selection – conservative, adaptive, and diversifying; and the validity of natural selection had been demonstrated in many laboratory and field ex-

---

[3] 'Modern synthesis' now usually refers to evolution in biochemical terms, as described in the next chapter.

periments, with melanism in moths being an especially prominent example. Natural selection and population genetics also fitted in well with the growing disciplines of environmental science and ecology.

Darwin had first proposed evolution by natural selection as a hypothesis to explain various aspects of biology, especially the occurrence of closely related species and adaptations. Many scientists of his day had complained that it was not inductive enough and demanded further proof. Now, just as the concept of natural selection could explain various features of biology, the understanding of genetics was seen to explain how natural selection worked, i.e. natural selection was explained by more fundamental biological mechanisms, in keeping with the aims of science in general. It also had a more mathematical description which suited the physical scientists and philosophers of science. So far as the latter were concerned, Darwin had 'discovered' evolution by natural selection, and with Neo-Darwinism it had now been (more than adequately) 'justified'.

That is not to say biologists thought all the problems had been resolved, for example they were still well aware that most mutations were detrimental. But they were encouraged by the enormous innate variability of species, and by the substantial progress in biology made in the first half of the 20th century, so there was confident optimism that progress would continue to be made, and that outstanding problems would be resolved. And there were no doubts that the resolution of those problems would be entirely consistent with the essential principles of evolution.

However, all of this was with scant knowledge of the chemical basis for heredity: genes were regarded simply as discrete entities, with next to nothing known of their chemical composition or of how they worked. A great deal of the theory of evolution had been determined and totally accepted before the biochemical basis of heredity and genetics was known. It was assumed that any further knowledge could only lend further support to the theory of evolution – not that further support was needed. But in my view the elucidation of the chemical basis of heredity, and of molecular biochemistry in general, has cast substantial doubt on the validity of the theory of evolution, at least in its widest sense; and that the established theories have not been properly examined in the light of this new knowledge, in fact that there has been a reluctance to do so.

6

Nuts and Bolts

In Chapter 5 we saw how the rise in understanding of hereditary mechanisms supplied what was lacking in Darwin's original theory of evolution through natural selection. It provided a clear mechanism for the reproduction of like for like, while allowing for the generation and propagation of the all-important variations. It also explained many of the observations relating to heredity that had been so confusing to Darwin, and naive Lamarckian ideas such as the effects of use and disuse were finally refuted. We saw that at first the theory was challenged by the particulate or discrete nature of Mendelian genetics, which seemed at odds with the gradual evolution proposed by Darwin. But once a clearer understanding was gained of the difference between genotype and phenotype, and the relationship between them, the original theory emerged far stronger. Importantly, it overcame early criticisms that favourable variations would rapidly become ineffective through blending or dilution within the population as a whole (although replaced by the fact that most mutations are lost, even if advantageous). In addition, his somewhat nebulous concepts of variation and adaptation were given a much more substantial basis in the mutation and interaction of genes. Similarly, major advances were made in the understanding of natural selection by quantifying the relative fitnesses of variations and modelling how these affected the distribution of advantageous genes within a species' population. Population genetics also provided insight into the possible segregation of genes, ultimately leading to the formation of distinct species. So by mid 20th century Neo-Darwinism was firmly established in biology.

At the biochemical level, by the 1940s the hereditary material was known to reside in the chromosomes, and linkage maps resulting from recombination studies had demonstrated that genes were almost certainly arranged linearly. In effect, a chromosome was envisaged as a string of beads, with beads as the genes, and the beads could be exchanged between paired (homologous) chromosomes at meiosis. Alternative alleles at a given gene locus resulted from different options for the beads, and mutations arose from alterations in the structure of the bead. Stemming from Weismann's ideas, the gene was thought to be a specific molecule or aggregate of molecules, which in some way controlled the development of a particular characteristic in the organism. But the actual chemical structure of the gene, in particular what determined its specificity, and how this was worked out or expressed as the phenotype, were quite unknown.

This chapter sketches the elucidation of that molecular basis of genetics – the chemical nature of genetic material, including an outline of how it is replicated, and the basics of how the information is translated into the structure of organisms. On the latter subject, in particular how genetic material directs the development of a multicellular organism with its various differentiated tissues, there is still a great deal not yet known. I have included little of the experimental background as, understandably, it becomes somewhat technical and, fascinating though the story is for those with an interest in the subject, my main purpose here is to provide a basic grasp of the biochemistry, with some idea of the continuing sense of scientific progress, rather than an introduction to biochemical techniques.

With knowledge of the chemical basis for genetics it became possible to specify what is meant by a mutation of the genotype in fairly precise chemical terms, and we can now explain the consequences of such mutations in terms of changes to the phenotype. This provides a more fundamental understanding of the nature of evolution, and the chapter closes with a brief summary of this currently prevailing view. It is worth recalling the point I made at the close of the preceding chapter – that, by the time these discoveries were made, Neo-Darwinism was so firmly held by most biologists that any new knowledge which came to light would be interpreted in an evolutionary perspective, and it was inconceivable that the biochemical understanding of biology could challenge the fundamental evolutionary precepts.

DEOXYRIBONUCLEIC ACID (DNA)

Identification of DNA as the core genetic material

Well before the close of the 19th century, along with increasing awareness of the important role of the nucleus in cell multiplication and heredity, the chief constituent of the cell nucleus was identified as nucleoprotein – a combination of protein with nucleic acid, the latter subsequently identified as deoxyribonucleic acid, DNA. As it became accepted that the chromosomes contain the hereditary material, and given their nucleoprotein composition, it was generally assumed that the essential or core genetic material – Weismann's germ-plasm which was responsible for conveying the hereditary information – was predominantly if not exclusively protein. The reason for this was that proteins appeared to offer much greater diversity and potential for variability than DNA: proteins are derived from 20 different types of amino acid (see Boxes 6.2 and 6.4) and evidently could form a diverse and versatile range of compounds; whereas nucleic acids have just four types of base (Box 6.1) and appeared a somewhat uniform class of compounds, especially because at first it was mistakenly thought that all four bases occur in equal proportions. This view persisted well into the 20th century when evidence began to emerge to challenge it.

It had been known for some time that mutations could be induced, e.g. in bacteria and *Drosophila*, by ultraviolet radiation, and in 1939 it was found that the frequency of this radiation that was most effective (having a wavelength of 265 nm) coincided with the maximum absorption of UV by DNA (rather than by protein).

Others measured the DNA content of the nuclei from various tissues from a range of species. Not surprisingly, they found that it differed between species, but for the same species it was constant for different tissues – except that it was approximately only half as much in the nuclei of the gametes i.e. in sperm and egg cells – consistent with these having only half the genetic material, following meiosis.

The name most associated with the identification of DNA as the hereditary material is Oswald Avery (1877–1955) who built on earlier work by a British microbiologist Fred Griffith (1881–1941). Griffith worked on pneumococcus which was a major medical problem in the days before antibiotics. The virulence of pneumococcus arises from its outer capsule made of polysaccharide, and Griffith obtained a non-virulent strain which lacked this capsule. When he injected either this non-virulent strain or heat-killed virulent bacteria into mice they remained healthy; but when he injected both preparations together the mice developed pneumonia. Also, bacteria obtained from the infected mice now developed the polysaccharide capsule. Griffith concluded that some genetic material from the dead virulent strain had become incorporated into the live non-virulent strain and transformed them into a virulent form. Avery extracted, purified and identified DNA from virulent bacteria and found that this alone, when mixed with the non-virulent strain, would result in the production of virulent bacteria.

Avery's work was reported in 1944 but, despite the strength of his evidence, so entrenched was the view against DNA that it took several more years and lines of evidence before the central role of DNA was accepted. What clinched the argument was determination of the structure of DNA by Francis Crick (1916–) and James Watson (1928–) in 1953. This provided an explanation for how DNA, though consisting of only four types of subunit, could encode the genetic information, and indicated a plausible mechanism for its accurate replication.

Structure of DNA

The double helix structure of DNA is of course now well known. As this description implies, DNA comprises two strands, linked to each other, and the whole is twisted so that each strand adopts a right-handed helical configuration. Each strand has a backbone of alternating sugar (deoxyribose) and phosphate groups, and to each sugar is attached a nitrogenous base which pairs up with a corresponding base on the other strand. The two strands together may be compared with a ladder, the alternating sugars and phosphates comprising the uprights, and the paired bases as the rungs. The combination of base+deoxyribose+phosphate is called a nucleotide which is the basic subunit or monomer of DNA.

Box 6.1   Structure of deoxyribonucleic acid (DNA)

Constituents

The basic constituents of DNA are
- phosphate, $[PO_4]$
- deoxyribose, which is a sugar having five carbon atoms, and
- nitrogenous bases, of which there are four types:
  two purines, adenine and guanine;
  two pyrimidines, cytosine and thymine.

Chemical structure of deoxyribose:
(C, H, O represent atoms of carbon,
hydrogen and oxygen)

the nitrogenous base attaches to carbon number 1'
and the adjacent phosphate groups attach to carbons 3' and 5'

Subunits

In DNA, the constituents are joined into subunits called nucleotides, for example, adenine+deoxyribose+phosphate = adenosine phosphate, with the phosphate being attached to the 5' carbon atom.

Polymer

Each strand of DNA consists of nucleotides joined by linking its phosphate group to the 3' carbon of the preceding nucleotide. This results in a backbone of alternating deoxyribose and phosphate, and the base attached to each sugar protruding to the side. The polymer is directional due to the relative attachments of the up- and downstream phosphate groups. It is built up and 'read' in the 5' ⇨ 3' direction, left to right in this diagram.

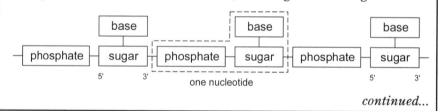

continued...

The helical arrangement of the chain leaves grooves between the backbones. They are of unequal size, identified as major and minor. These grooves are important for the binding of other molecules, especially proteins, in the various mechanisms for duplicating and 'reading' the DNA molecule.

Deoxyribose has five carbon atoms and the neighbouring phosphate groups are attached to the carbon atoms designated 3' and 5' on either side. This gives a sense of direction or orientation to each DNA strand, which is particularly important when it comes to 'reading' the genes, which is done in the 5' ⇨ 3' direction. The strands are orientated in opposite directions, i.e. they are said to be antiparallel, rather than simply parallel if the two strands ran in the same direction.

Each chromosome contains just one chain (double helix) of DNA, typi-

Box 6.1 *continued*

Double helix

Two polymers (strands) associate via the bases: pairs of bases forming rungs of a ladder, held together by hydrogen bonds which are much weaker than the covalent bonds that form the chains. For the hydrogen bonds to form, each base must pair with its appropriate complement: adenine with thymine, cytosine with guanine, forming two and three hydrogen bonds between them, respectively.

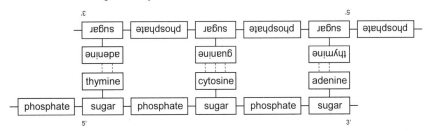

Note that the strands run in opposite directions, and in the diagram dotted lines represent hydrogen bonds.

Base pairing

The overall similarity between the two sets of paired bases, especially in terms of how they fit between the phosphate-deoxyribose backbones, is readily apparent:

adenine – thymine   guanine – cytosine

cally containing many millions of base pairs – the total number in the human genome (i.e. in all 46 chromosomes) is estimated at three billion (3×10^9), which would be about a metre long if it were extended in a straight line. Yet this packs into the nucleus of each cell! This packing is achieved by the double helix coiling upon itself, and these coiled helices themselves coiling – supercoiling – all of this being aided by proteins called histones.

DNA replication

One of the aspects immediately apparent to Watson and Crick regarding their proposed structure for DNA, and indeed one of the features which made their model so appealing and convincing to the scientific world in general, was the opportunity it offered for accurate replication. From the outset it was postulated that the specific sequence of the four bases in some way encode the genetic information; and because of the specific pairing between the bases, it was evident that each strand could act as a template for the other, and so accurately preserve the order of the bases. That is, if the two strands were unwound and separated, a complementary strand could be constructed on each, matching base for base, resulting in two copies of the original paired strands. Various experiments showed that this did indeed occur, i.e. the result of DNA replication is not one old double helix and one new, but each daughter helix contains one old and one new strand. Further details of DNA replication are described in Chapter 8, and illustrated in Figure 6.

GENE EXPRESSION

It is all very well to see that in theory information can be stored by a combination of only four letters (just as information is stored in computers by the two-letter binary system of 0s and 1s), but the important question is to determine how it is done in practice, biologically. So we now turn to the two important and necessarily related issues of how the genetic information is stored, and how it is 'read' and used to direct the workings of the cell.

The link between genes and proteins

In the early years of the 20th century a number of workers, including Bateson, had suggested that at least some genes affected the functioning of enzymes. But the major advance in associating individual genes with particular enzymes came in the 1940s, notably by George Beadle (1903–89) and others who were specifically trying to establish how genes work.

First, they carried out experiments on various naturally-occurring *Drosophila* mutants that had unusual eye colours. They showed that for at least two mutant strains – those called *cinnabar* and *vermilion* (shades of bright red) – it appeared that the eye colours were due to the synthesis of one of the eye pigments being interrupted at two different stages (more on this in Box 9.1).

Then, by irradiating yeast they produced mutant strains which were dependent on various substances (such as thiamine) in their growth medium in order to thrive, and Beadle proposed that the effect of the mutation was to corrupt the gene responsible for a key enzyme in the normal synthesis of the substance. He took this approach further with yeast mutants requiring the amino acid tryptophan. He found that some of the mutant yeasts could grow, not only if tryptophan were included in the growth medium, but if certain structurally related compounds were added instead. To explain this, he proposed that such compounds may be precursors in the synthesis of tryptophan, occurring in the biosynthetic pathway after the step carried out by the affected enzyme. By generating a range of mutant strains, with varying nutritional requirements, he was able to pinpoint the specific metabolic step of the action of different genes. This was key evidence showing that genes appeared to determine specific protein molecules, especially enzymes. Along with this invaluable information, these workers also gradually deciphered many of the metabolic pathways used by organisms for the synthesis of cellular components.

Proteins consist of a linear sequence of amino acids. With acceptance of DNA as the core genetic material and Watson and Crick's model for its structure, it was proposed that the linear sequence of bases in DNA somehow specified the linear sequence of amino acids in proteins, and in this way determined the protein's structure. It was immediately apparent that to code for 20 different types of amino acid with only four types of base requires a minimum of three bases to specify each amino acid (two bases could specify only up to 4 x 4 = 16 options). This is the fundamental concept of the genetic code. Confirmation of this theory, and cracking of the code went hand in hand with elucidation of the mechanism by which proteins are synthesized biologically, and we shall turn to this shortly.

Prokaryotes and eukaryotes

However, before proceeding I need to explain the difference between prokaryotes and eukaryotes.

Almost all of the organisms mentioned so far, in particular all the higher plants and animals, are eukaryotes. The cells of eukaryotes have distinct subcompartments (called organelles), bounded by membranes similar to that around the whole cell, which carry out specific functions. In particular they have a definite nucleus which contains the hereditary material organized into several chromosomes, with each chromosome containing a single linear chain of double helix DNA (but bear in mind that in most cases chromosomes will be present in two copies, one from each parent). In addition, most eukaryotic organisms are multicellular, although some such as yeast are single-celled.

In contrast, prokaryotes do not have a nucleus or other organelles, they have just one chain of double helix DNA which is in the form of a closed circle, and are always single-celled (although several cells may associate). The most important group of prokaryotes are bacteria. Because of their simpler (and more accessible) genetic material and biochemical machinery, most advances in biochemistry have been made first with prokaryotes, and usually with the bacterium E. coli (billions of which innocuously inhabit our intestines, and one or two pathological strains have become well-known in recent years).

One of the striking features of biological systems is the remarkable similarity of genetic material, and of the mechanisms of its expression within cells. The essential processes are substantially consistent, not only in plants and animals, but also prokaryotes and eukaryotes despite the significant differences in the organization of their genetic material. This, of course, is cited as evidence for evolution – of the common origin of all forms of life.

Ribonucleic acid (RNA)

Accordingly, the means of gene expression, of how genetic information directs the structure of proteins, was determined first in prokaryotes; elucidation of the corresponding mechanisms in eukaryotes is more recent and less fully known. The following description is, therefore, based primarily on prokaryotes, but is generally applicable to eukaryotes.

Messenger RNA

The synthesis of proteins is not carried out directly from DNA but from a copy of the relevant gene in the form of ribonucleic acid (RNA). The essential difference between them is that in RNA the sugar is ribose rather than deoxyribose (deoxyribose has one less oxygen atom because a hydroxyl group (OH) of ribose is replaced by a hydrogen atom (H), which makes DNA more stable than RNA). RNA has the same bases as DNA except that one of the pyrimidines, thymine, is usually replaced by another, uracil.

So the first stage of protein synthesis is to make a copy of the appropriate DNA gene as RNA. This is called 'transcription' and is carried out by an enzyme called RNA polymerase, with other supporting proteins and enzymes. It involves identifying the appropriate starting point of the gene, unwinding the DNA and separating the strands, then an RNA strand is synthesized complementary to one of the DNA strands, and the strands of DNA are rejoined. This is analogous to DNA replication except that only one strand is copied, and synthesis proceeds in only one direction (whereas DNA is copied in both directions from a common starting point, see Chapter 8). The end result is the original double helix of DNA and a single molecule of RNA complementary to part of one strand of the DNA. For a long time it was thought that only one strand of a double helix was transcripted, but it is now known that there are occasional exceptions where genes are coded on both strands.

The resulting RNA which carries a copy of the relevant gene is called messenger RNA (mRNA). It is the bases on this that are 'read' in groups of three to determine the order of amino acids to be combined to form a protein. A group of three bases specifying an amino acid is called a 'codon'. The codon in the mRNA is of course the complement of the corresponding codon of the gene in the DNA. For example a codon in DNA of CTG would be represented in the mRNA by GAC, and CAG by GUC (not GTC, because in RNA uracil is used instead of thymine). The genetic code is shown in Box 6.2 and, as indicated above, this code has been found to apply almost universally. It is apparent that there is some degeneracy in the code, i.e. that most amino acids can be represented by more than one codon, some having up to six.

Box 6.2 The Genetic Code

Amino acids incorporated into a protein for the specified codons in mRNA.

1st base	2nd base				3rd base
	U	C	A	G	
U	UUU UUC phenylalanine UUA UUG leucine	UCU UCC UCA UCG sereine	UAU UAC tyrosine UAA UAG 'stop'	UGU UGC cysteine UGA 'stop' UGG tryptophan	U C A G
C	CUU CUG leucine CUA CUG	CCU CCC CCA CCG proline	CAU CAG histidine CAA CAG glutamine	CGU CGC CGA CGG arginine	U C A G
A	AUU AUC isoleucine AUA AUG methionine*	ACU ACC ACA ACG threonine	AAU AAC aspartamine AAA AAG lysine	AGU AGC serine AGA AGG arginine	U C A G
G	GUU GUG valine GUA GUG	GCU GCC GCA GCG alanine	GAU GAG aspartine GAA GAG glutamine	GGU GGC GGA GGG glycine	U C A G

*AUG acts as a 'start' signal as well, see Box 6.3.

Transfer RNA

The group of molecules that implement the genetic code are the transfer ribonucleic acids (tRNAs) in conjunction with their activating enzymes.

Transfer RNAs are single strands of RNA, typically having around 70 to 80 nucleotides. Although single-stranded, there are some nucleotide sequences complementary to each other, allowing some of the tRNA to loop back on itself and form short stretches of double helix. This is illustrated in Figure 3. Also, although there are substantial differences between the nucleotide sequences of the various tRNAs, they all form a similar configuration with four or five loops. The critical feature of a tRNA is that at the end of one loop are three nucleotides, called the anticodon, which are complementary to the three codon bases on the mRNA. For each amino acid there is at least one tRNA, but there can be more if multiple codons are associated with the same amino acid, although there is not necessarily a different tRNA for every codon option.[1]

At the 3' end of every tRNA strand the sequence of bases is -CCA and the amino acid corresponding to the anticodon is attached to the terminal adenine. The tRNA does not of itself 'recognise' the appropriate amino acid; this is done by its activating enzyme which identifies both the amino acid and its corresponding tRNA(s), bringing them together and attaching the amino acid to the tRNA, to form what is called aminoacyl tRNA. For each type of amino acid there is one, and only one, activating enzyme.

---

[1] Due to what is described as 'wobble' some tRNAs can recognise more than one codon.

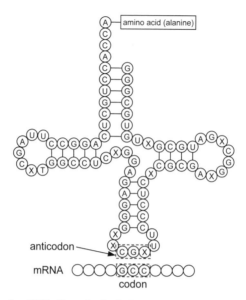

FIGURE 3. Transfer RNA (for alanine), having its anticodon lined up with a codon on mRNA. A, C, G, T, U = nucleotides with standard bases; X = nucleotides with various modified bases.

Transfer RNAs are synthesized as complementary strands from designated stretches of DNA; i.e. these genes code for tRNAs rather than for proteins. One of the notable features of tRNAs is that they contain several unusual bases: the tRNA is first synthesized using only the normal bases and conventional base-pairing, but some of the bases are subsequently modified to produce the unusual ones. It is thought that these unusual bases are required for the tRNA to adopt its necessary 3D configuration and/or enable unambiguous recognition by its activating enzyme.

Ribosomes

The protein synthesizing machinery of cells – where mRNA is 'read' and polypeptides are assembled from the amino acids carried by tRNAs – are the ribosomes. They consist of two subunits, each subunit comprising at least one strand of ribosomal RNA (rRNA) and several protein molecules. In most cells they are the most abundant nucleoprotein complex. Like the tRNAs, ribosomal RNAs are synthesized from dedicated regions of the DNA; unlike the tRNAs, generally they do not have modified bases.

Protein synthesis

Translation of mRNA to produce proteins can conveniently be considered in three phases: initiation, elongation and termination of the polypeptide. These are outlined in Box 6.3, and here I will describe only the process of elongation – the systematic reading of sequential codons and progressive assembly of a polypeptide.

Box 6.3   Protein biosynthesis (translation)

Initiation

Synthesis begins by a methionine tRNA pairing up with the initiating AUG sequence on an mRNA, along with a small ribosomal subunit. A large ribosomal subunit then joins, locating the methionyl tRNA in its P binding site. Both stages of initiation require the presence of protein initiation factors, which are released once synthesis is established.

Elongation

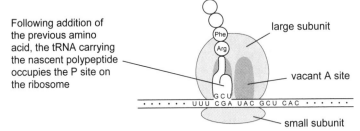

Following addition of the previous amino acid, the tRNA carrying the nascent polypeptide occupies the P site on the ribosome

large subunit

vacant A site

small subunit

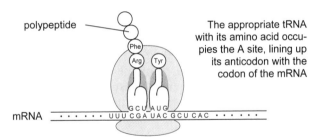

polypeptide

The appropriate tRNA with its amino acid occupies the A site, lining up its anticodon with the codon of the mRNA

mRNA

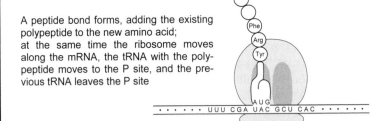

A peptide bond forms, adding the existing polypeptide to the new amino acid; at the same time the ribosome moves along the mRNA, the tRNA with the polypeptide moves to the P site, and the previous tRNA leaves the P site

Termination

Polypeptide synthesis is terminated, not merely by the absence of a suitable tRNA for the 'stop' codon, but a specific protein identifies and binds this codon, causing separation of the ribosome from the mRNA and into its subunits, and release of the polypeptide.

Following synthesis, almost all polypeptides are modified in some way, usually by removal of some amino acids from one or both ends (e.g. the initiating methionine) and/or addition of other chemical groups.

Ribosomes attach to a strand of mRNA with the codons of the mRNA aligned with two binding sites for tRNA. One of the sites is designated P and is occupied by a tRNA carrying the growing polypeptide. The other site is designated A and immediately after completion of a peptide bond this site is empty, with the codon of the mRNA for the next amino acid aligned with it. The first step of adding an amino acid is for an aminoacyl tRNA carrying the appropriate amino acid to occupy the A site, with its anticodon pairing with the codon on the mRNA. It is of course this specific pairing which determines which aminoacyl tRNA engages the site. A peptide bond forms, adding the polypeptide to the new amino acid, resulting in transfer of the polypeptide from one tRNA to the other. At the same time, the ribosome moves along the mRNA so that the tRNA with the polypeptide moves from site A to site P, displacing the previous tRNA from the P site. Polypeptides are biosynthesized in this way at the rate of adding three to five amino acids per second.

PROTEINS

Before summarising how the theory of evolution has been interpreted in the light of the biochemical understanding of genetics, it is appropriate to comment a little on the structure and function of proteins because, to a large extent, they are the direct consequence of the genotype and/or implement it in terms of the visible phenotype. Genes may be the raw material of evolution, but in most cases natural selection is based on the resulting proteins – occasionally due to the proteins themselves, but usually to the products of those proteins that function as enzymes. For example, with the different *Drosophila* eye colours, mutations in the relevant genes resulted in changes to the enzymes which synthesize the eye pigments.

Amino acids

All biological proteins, from those in bacteria to those in higher plants and animals, are built up from the same set of twenty amino acids (although some are modified after their initial synthesis, in an analogous way to the unusual bases of tRNA) – which reinforces the unity of all biological systems. All but one of the amino acids conform to the generalised structure shown in Box 6.4, which also gives the structure of the exception, proline. Any two of the amino acids (including proline) can combine by forming a peptide bond – it is this bond-formation which is carried out on the ribosomes – and several amino acids combined in this way are called a polypeptide.

It is the nature of the side chain, designated R in Box 6.4, which determines the properties of the specific amino acid, and which contributes to the overall character of the protein in which it is incorporated. Obviously, the biological activity of a protein is completely dependent on, in fact dic-

Box 6.4 Amino acids and peptide bond formation

Amino acids

Almost all biologically-occurring amino acids contain an amino group (-NH$_2$) and an acidic group (-COOH), and conform to the general structure shown below, in which 'R' represents the part of the molecule that varies between the different amino acids, and determines its properties. There are 20 different types in biological proteins.

The only exception to this general pattern is the amino acid proline (shown on the right) in which the R incorporates the N of the amino group, as shown. Because of the restriction this causes to the orientation of the peptide bond, proline is often used to introduce a bend in the folding of proteins.

Peptide bonds

Amino acids are joined together by reacting the amino group of one acid with the acidic group of the other (with loss of a molecule of water, H$_2$O), to form what is called a peptide bond.

It is this reaction that takes place on ribosomes. Sequential addition of amino acids results in the formation of polypeptides and proteins. The ends of the polypeptide are designated the C- or N- terminus depending whether it has a free -COOH or -NH$_2$.

tated by, those of its constituent amino acids, and it should be noted that the specific sequence of amino acids is important, not just the overall composition.

This is illustrated in the important matter of protein folding, which requires an appropriate mix and distribution of hydrophilic and hydrophobic amino acids. The range of biologically available amino acids includes those which, by virtue of their side chain, are hydrophilic (readily associate with water) and those that are hydrophobic ('oily'). In general, because the interior of a cell is an aqueous environment, it is favourable to place hydrophilic amino acids on the outside of the folded protein where they will associate with water, and place hydrophobic ones on the inside where they can associate with each other. If a polypeptide can be folded so as to arrange its amino acids in this way then it is likely to adopt that configuration: but it is important to note that the fit of the amino acids, especially in the interior of the folded molecule, must be very good for this to work, rather like a 3D jigsaw. I have more to say on this in the next chapter.

The hierarchy of protein structure

The *primary* structure of a polypeptide or protein refers to its linear sequence of amino acids. Although a protein may be envisaged as a string of amino acids, in order to function biologically proteins fold up on themselves, as just mentioned, to adopt a specific 3D conformation which is called its *tertiary* structure. The term 'tertiary' is used rather than '*secondary*' because the latter refers to subsidiary structural features found in most proteins where some of its amino acids are arranged into configurations which are called α-helices or β-sheets; but I shall make little mention of secondary structure, generally regarding these simply as a part of a protein's tertiary structure. Most proteins fold into their appropriate 3D conformation spontaneously, i.e. without assistance from other molecules and without expenditure of energy, but there are a few exceptions.

Finally, most proteins do not act as single molecules, but need to associate with others if they are to function properly, or even to function at all in many cases. They may form aggregates of the same type of polypeptide and/or combine with other types of polypeptides. The assemblage of multiple polypeptides in this way is called the *quaternary* structure.

Protein functions

The structures and roles of proteins in living things are amazingly diverse. At the small end of the scale are short polypeptides (not usually regarded as proteins because they are so short and do not form a definite tertiary structure) consisting of only a few amino acids, such as the endorphins which are used as chemical signals in the brain. Larger are the peptide hormones, typically with several tens of amino acids, such as insulin. Between 100 and 200 amino acids in length are several of the proteins used for carrying oxygen, notably haemoglobin, or involved in related oxidation-reduction reactions, in particular cytochrome c which we will consider in the next chapter. An important role for proteins is, of course, as enzymes to carry out the many thousands of biochemical reactions which take place within cells – harnessing the energy of food, and synthesizing the wide range of compounds required by the cell or organism as a whole, of which processing the genetic information is an important part. Such enzymes generally have a few hundred amino acids. Also of about this size are proteins known as transcription factors which have a key role in controlling the expression of genes.

Some proteins have important structural roles. Even here there is a wide range of function – from intracellular such as stabilising cell membranes or forming pores through them, to extracellular, especially the various forms of collagen (typically having a thousand or more amino acids) which are the principle components of animal connective tissues, and the keratins which form hair, nails, scales and feathers. Silk, too, is composed of proteins, and includes the silk used by spiders for their webs, which is far stronger than steel or even of any synthetic polymer to date, such as Kevlar®. The long fibres of collagen and silk are formed by individual molecules interacting with each other – a kind of quaternary structure. And last to mention here is titin which, with 25–30,000 amino acids is the largest known protein molecule; although it is a component of muscle, it does not provide any contractile force, but is elastic and protects the muscle from becoming over-stretched (see Fig. 9).

Enzymes

Whilst it is relatively straightforward to envisage how a protein molecule can act structurally – for example that a network of interlinked collagen molecules can form a tendon – I think the functioning of an enzyme requires a little more explanation, especially for the benefit of readers with minimal knowledge of chemistry.

An enzyme is a catalyst. This means that it facilitates a chemical reaction: it takes part in the reaction but is released unchanged at the end of it, so is available to facilitate a subsequent reaction. A single enzyme molecule may catalyse tens or even hundreds of reactions every second.

Enzymes offer several advantages over what can generally be achieved in terms of chemical synthesis in a laboratory. First, whereas for many reactions a chemist has to employ extreme reaction conditions – perhaps strong reagents at high temperature, and/or with an unusual solvent – in almost all cases enzymes carry out their reactions in water, under neutral conditions and at ambient temperature and pressure. Second, even though enzymes normally operate under such mild conditions, they enable remarkable speeds of reaction, often millions or even billions of times faster than would otherwise take place. Third, most enzymes are very specific in terms of which reactants they will interact with and what reactions they will facilitate; whereas a chemist often needs to take elaborate steps to minimise unwanted side reactions.

Enzymes achieve these feats due to three features of their structure: First of all, they have binding sites for the reactants – i.e. part of the enzyme's surface is complementary, in terms of its shape and chemical nature – for each of the compounds involved in the reaction. Secondly, the binding sites are arranged on the enzyme so that the reactants are held in the correct orientation and separation with respect to each other to enable the reaction to proceed. In contrast, in the laboratory, a chemist usually has to rely on the reactants meeting at random – in terms of when and how they collide. And thirdly, where the reaction takes place – at what is called the enzyme's 'active site' – the amino acids have reactive groups (part of their side chain, R) which interact with appropriate chemical groups of the reactants. It is this last feature in particular which greatly facilitates the reaction, and which enables it to proceed under such mild reaction conditions.

This book is not the place to give an adequate account of how enzymes function, readers are referred to modern textbooks on molecular biochemistry for that, but it is important to recognise the specificity of proteins, especially of enzymes: Based solely on their amino acid sequence, they are able to fold up in such a way as to have highly-selective binding sites on their exterior surface, the sites being in appropriate relation to each other and to the active site which has the right reactive groups to enable the reaction to take place.

EVOLUTION: THE MODERN SYNTHESIS

It is appropriate at this juncture to summarise the essential features of the theory of evolution as it is currently understood, taking on board the biochemical understanding of genetics which emerged in the latter half of the 20th century.

Genetic material

DNA is the core genetic material for all living organisms. The only exceptions to this are some viruses that use RNA instead – but viruses are not generally considered to be true living organisms anyway because they have to infect a host cell in order to reproduce. The early idea of genes being like beads on a string has given way to recognising that there is no chemical distinction between the string and the beads: there is just a continuous chain of DNA, in which genes are identified simply because they have specific sequences of bases which are recognised as the starting and finishing points of the gene.

Probably the largest class of genes comprises those which code for proteins. For these genes the sequence of bases prescribes the sequence of amino acids in the resultant protein. Such genes are generally termed structural genes. An important distinction between prokaryotes and eukaryotes is that in prokaryotes nearly all structural genes are continuous so far as the sequence of bases which code for the protein is concerned, whereas in eukaryotes most structural genes are interrupted by non-coding stretches of DNA: the lengths of DNA coding for amino acids are called exons, and the intervening non-coding sections are called introns. The mRNA made at transcription includes copies of both exons and introns, but the latter are excised before the mRNA is used for protein synthesis. Associated with the structural genes of both prokaryotes and eukaryotes are stretches of DNA which are involved in the regulation of the structural gene, i.e. whether and to what extent the gene is copied into RNA for subsequent protein synthesis.

The other important groups of genes are those which code for the transfer and ribosomal RNAs. In both prokaryotes and eukaryotes these do not usually include introns. For many of these genes there are multiple copies, sometimes many thousands, reflecting their importance and high usage within the cell.

In addition to these identifiable genes there are large stretches of DNA with no known function. In the higher plants and animals this 'silent' DNA predominates, even exceeds 90% of the total, and generally comprises multiple copies of base sequences with the individual sequences ranging in length from as little as four bases up to many thousands. Suggested roles for this DNA include to facilitate packing within the nucleus, a wider role in gene regulation, or no use at all.

Mutations

Fundamentally, all mutations of any evolutionary significance must constitute a change to the sequence of bases in the organism's DNA, in germline cells. It is thought that most mutations arise from miscopying when DNA is duplicated in the course of producing gametes; some arise due to the action of mutagens such as chemicals and radiation, and some occur spontaneously. Also, it is thought that the quantity of genetic material can be increased by a mechanism called unequal cross-over which may take place during meiosis and can result in the duplication of a stretch of DNA.

The simplest mutation to consider is where a single base is changed into another. For a structural gene, such a mutation is likely to result in one amino acid being substituted for another, or no change at all if both the old and new codon refer to the same amino acid. More dramatic changes can arise from the insertion or deletion of one or more bases: it will be evident that, because of the way in which bases in structural genes are read in threes to determine the amino acid sequence, inserting or deleting just one or two bases can completely change the resulting sequence, at least for those downstream (in terms of the direction in which the DNA is read) of the mutation and before the next control sequence. It should be recognised that important mutations are not only those to the structure of a protein: equally important can be those which affect the regulation mechanisms, i.e. those that control the expression of the structural gene.

As a result of mutation (which has occurred recently and in the more distant past) a species carries within its population several versions of most of its genes. Much of this variability is not evident, perhaps because the variant in question is recessive to the prevailing type. However, this genetic heterogeneity is the raw material from which variations of genotype and hence phenotype are produced.

I have already outlined at the end of Chapter 5 how this variability can provide the phenotypic variations on which natural selection can act, the different roles for natural selection, and the way in which new species can arise; so there is no need to repeat these here. However, it is important to note that, unlike Darwin's original concept, evolution does not proceed by a simple stepwise accumulation of mutations; rather, mutations provide the basis for genotype variability, recombination provides the variations of phenotype, and natural selection favours advantageous gene combinations.

So succinctly and tidily expressed, it is all too easy to accept this prevailing view of how evolution occurs. But what must be emphasized is that it is completely dependent on random mutations (by whatever mechanisms) for producing meaningful genes, e.g. a base sequence which codes for a protein that can adopt a tertiary structure and then mediate a useful biochemical reaction, or for a tRNA, and not forgetting the control sequences which are essential for the recognition and use of a gene. Great weight was given to Darwin's theory as a way of explaining biological diversity and adaptation, and to population genetics for explaining the operation of evolution in terms of genes, and molecular biochemistry now explains how genes work. The question must now be asked: Does our knowledge of chemistry and biochemistry explain the origin of those genes? Such is the acceptance of evolution as the underlying explanatory thesis of biology that it is assumed that it must; but it is my contention that it does not – and it is to this that I now turn.

7

Chance and Necessity

We have seen how the theory of evolution developed from the revival of biology at the time of the scientific revolution through to modern molecular biochemistry. As noted at the beginning of Chapter 5, the theory was based primarily on morphological observations: in one stroke it provided an explanation for both the diversity and natural groupings of organisms, and for adaptation. Darwin was particularly impressed by the enormous variety of natural species, including the gradation and adaptation of structure, for example in the various beaks of the different species of finches on the Galapagos Islands. He also saw the wide range of varieties within the same species of many domestic breeds such as farm animals and pigeons. And he reasoned that, if significant changes could arise in the course of the limited period over which domestic breeding has taken place, then much larger changes should be possible over geological time. For example, if in the course of a few hundred generations of domestic breeding the fantail pigeon had developed its longer and more numerous tail feathers than the rock pigeon from which it was derived, then it seemed reasonable that feathers themselves could have evolved in a few thousands of generations.

Darwin could make this sort of extrapolation because of the limitations of what was known at the time: in the middle of the 19th century knowledge of biological structure at the cellular level was only just beginning, and there was none at all of their biochemical composition, or of the genetic basis for differences in structure. Consequently, because he was totally unaware of the biochemical implications of doing so, there seemed no reason why he should not extrapolate from the significant observed variations which were known to occur, and propose that much larger changes could arise by progressive variations – changes that constitute the evolution of species. As Buffon had said previously, 'if it were proved ... that a single species was ever produced by the degeneration of another ... no bounds could be fixed on the power of Nature.' In other words, so strong had been the hold of the doctrine of the fixity of species that, once it could be demonstrated a small change in the morphology of a species is possible, it seemed there need be no limit to the changes that could be made. But they were completely unaware of the complexity of biological molecules and structures, or of the elaborate link between genetic material and end structure, or of the innate genetic diversity of species which enables substantial changes of morphology to be achieved simply by shuffling and selecting subsets of the available genes. And – a point I will develop in

Chapter 11 – I think modern biologists, even though they now know of biochemical complexity and of the genetic mechanisms which can produce so much morphological change, still follow this 19th-century attitude to biological variation.

Even though Darwin knew only of morphology and nothing of molecular biology, he recognised that if life had evolved little by little, by a trial and error process, then it would have required a long time. Clearly the traditional western view of the earth being only a few thousand years old was inconsistent with this, but the rise of geology indicated that the earth was several orders of magnitude older, and further support came from developing ideas of cosmogony. The general view, at least in the scientific community, was that the solar system and earth in particular was very old – at least many millions of years, if not hundreds of millions. Darwin considered this was ample time for the evolutionary process to take place:

It cannot be objected that there has not been time sufficient for any amount of organic change; for the lapse of time has been so great as to be utterly inappreciable by the human intellect. [*Origin*, Ch. 14]

In the early decades of the 20th century the theory of evolution gained considerable momentum with the rise in understanding of heredity and genetics. In particular, the theoretical exercises of population genetics, experimental studies such as those relating to the peppered moth which demonstrated natural selection in action, and continuing palaeontological discoveries, firmly established evolution as the underlying explanatory thesis of biology.

However, as late as the 1940s genes were still known merely as nominal entities – there was scarcely any knowledge of their chemical structure or of how they functioned. Mutations and variations were evident, but there was no knowledge of what caused them, what they constituted at the molecular level, or of their implications in terms of biochemistry. There was still virtually no knowledge of the chemical structure of organisms or their tissues, or of cells. These came only from the middle of the 20th century onwards. As outlined at the close of Chapter 6, it is only since then that we have the evolutionary model in biochemical terms, and only in the last couple of decades we have begun to unravel the complexities of gene regulation. But by the time this sort of detail became known, the belief that evolution had occurred was so firmly entrenched that any new knowledge had to be fitted into this paradigm.

Also, the present view is that the earth is considerably older than had been thought in Darwin's time, close to 4.5 billion years, and the universe up to 15 billion years (although there is current debate on this latter figure). So, even though we now know life is much more complex than Darwin's generation anticipated, the common perception is that nevertheless there has been plenty of time for life to evolve. Moreover, we now also have a larger view of the universe, with growing evidence that there are other planetary systems; and clearly life did not have to evolve on earth, but might have arisen on any planet where suitable conditions occurred. Taking into account the aeons of time and vast extent of the universe, surely this can only support Darwin's original assertion. Lawrence Mettler and Thomas Gregg summed up this view at the end of their *Population Genetics and Evolution* (1969):

> If anyone doubts the ability of such a system to generate the diversity of life in the world today, let him reflect on the enormity of time during which this process has been at work, while keeping in mind that, given enough time, events that have a low probability of occurrence at any given moment become certainties.

This is the prevailing view. But my contention is that we have not faced up to what has been discovered of the complexity of biology at the subcellular level, especially in terms of the structure of biological macromolecules – nucleic acids and proteins. My aim in this chapter is to present this bio-chemical challenge to the theory of evolution.

Evolution and biochemistry

The basic issue is this: although the theory of evolution was formulated on morphological variations, and appears to explain many observations regarding adaptation and speciation at the morphological level, now that we know some of the biochemical implications, we must examine whether or not the theory is an adequate explanation at this level too. We have dis-covered that, at the molecular level, biology entails many complex interac-tions between sophisticated macromolecules, and we cannot set this knowledge to one side when we consider evolution. The main reason Darwin gave as to why he was convinced of the truth of evolution was that it explained so many disparate facts.

> The present action of natural selection may seem more or less probable; but I believe in the truth of the theory, because it collects, under one point of view, and gives a rational explanation of, many apparently independent classes of facts. [*Domestication*, Introduction]

We now know new facts – facts which neither Darwin nor even the Neo-Darwinists of the early 20th century could have imagined – and we must examine the theory of evolution in the light of these facts.

The point is, it is no longer sufficient to talk in terms of variation or ad-aptation on the large scale – whether it be something as 'simple' as adapt-ing to a change in temperature of the environment or of being able to utilise a new foodstuff, or of much more dramatic changes such as the de-velopment of a bone or a scale or a wing or an eye. It is no longer suffi-cient to talk about a giraffe stretching its neck, or of a mouse that jumps from a tree developing membranes between its digits and limbs to become a bat, or of a fish that develops limbs from its fins. In all of these cases, and many more like them, we need to account for the changes at the bio-chemical level, in terms of changes to the relevant biological macromolecules, in particular to the genes, the DNA.

It is not only these large changes, of which we have been aware for a long time; in the last half-century we have learned a great deal of the workings of the cell. I have already mentioned some of the mechanisms that are involved in reproducing genetic material and in translating it into a new individual with its range of specialised cells. In addition there is the complex network of biochemical pathways, catalysed and controlled by a large number of proteins, which constitute and sustain basic cellular metabolism. Each and every protein involved in these processes must be accounted for – ultimately in terms of changes to DNA leading to appropriate changes in protein structure. Nor may we overlook the ribonucleic acids, such as the rRNAs and tRNAs, intimately involved in protein synthesis: these too must be accounted for by this kind of process.

We may feel that billions of years should be enough for virtually any improbability to become possible; but now that we have information about biological systems at the molecular level, we must try to look at the implications objectively. Now that we are in a position to test the theory, we should do so. Occasional comments have been made along these lines, since at least as early as 1966 (Moorhead and Kaplan), but on the whole I do not think this manner of critical evaluation has been adequately presented. Although some recent evolutionary texts make some sort of acknowledgement of the improbability of biological macromolecules, all too often it is watered down to the point of misrepresentation. I think the general perception is, on the one hand not to appreciate sufficiently the complexity and specificity of biological macromolecules, and on the other to have an overoptimistic view of the resources available, especially in terms of time. So I will start by trying to redress this misperception.

THE IMPROBABILITY OF BIOLOGICAL MACROMOLECULES

To begin with, we should note that a fundamental postulate of evolutionary theory is that the basic raw materials of evolution are genes which are generated essentially at random. By 'random' all we mean is that the processes which result in the formation of or changes to a gene (or nucleotide sequence) – whether by DNA miscopying, radiation, recombination, or whatever – are undirected: a DNA sequence has no way of causing the changes that would be necessary for it to become a useful or even meaningful gene, nor a gene of causing changes that may bring about an improvement to its function; indeed, of course, they do not even have any way of 'knowing' what may be appropriate changes. Any utility or improvement can be determined only retrospectively, typically by natural selection. That mutations are essentially random is generally understood to be the case by the vast majority of present-day biologists.

In Chapter 6 we began to see something of the complexity of biological macromolecules in terms of their size and specificity in order to function, and I give more detailed examples in Chapter 8. The point I want to make in this chapter is that we cannot assume such macromolecules will arise readily; on the contrary, I will demonstrate what astonishingly improbable structures they are, and that currently we have no satisfactory explanation for their origin.

I should also make clear from the outset that I am not dealing here with the problems of getting life started in the first place. There are substantial difficulties in the production of even the simplest of biological macromolecules in the so-called primeval soup, but I will consider these in Chapter 13. My purpose here is only to look at the probability of obtaining a useful macromolecule even when we have mechanisms in place for generating them efficiently, i.e. through mutations of existing genetic material, the effects being translated into e.g. proteins by the processes of transcription and translation more or less as they are now. Doing this obviates argument over what reactions might have been possible, and the rates of those reactions, in a prebiotic world.

Although, in this context, the actual mutations are of nucleotides in nucleic acids, which must then be transcribed and translated into proteins, for the purpose of the following example I shall consider instead the random ordering of amino acids to form proteins. The main reason for doing this is that it is much simpler to formulate (and to follow, by those not familiar with the biochemistry) as we do not need to consider that most amino acids are represented by more than one codon, i.e. that some mutations will have no effect. Doing this certainly errs on the side of overestimating the probability of obtaining a useful polypeptide by chance. Also, it ignores any problem of making sure that the DNA in question retains its identity as a gene. I make further comment on these points later.

I shall illustrate the improbability of biological macromolecules with cytochrome c. This is a small protein, usually of 104 amino acids, which is present in virtually all cells (notably in both prokaryotic and eukaryotic organisms) where it has a very similar structure and function connected with the production of adenosine triphosphate (ATP), see Box 8.1. It is, therefore, assumed to have evolved very early, probably before any divergence had occurred between different groups of organisms. But just how likely is it that a protein of this size could arise by random ordering of amino acids? To answer this we must calculate how many possible options there are for a protein of this size, and then compare this with how many sequences might have been tried out in the course of evolution.

For a protein 104 amino acids long, and given the 20 different types of amino acid used in biological systems, the number of possible protein molecules is 20^{104} (twenty multiplied by itself 104 times) which evaluates to approximately 2×10^{135}. Written so concisely (even noting that it is '2' followed by 135 noughts) it is difficult to have any idea at all of just how big this number is. A little diversion may help: a chessboard has 64 squares; if you put one grain of rice on the first, two on the second, four on the third ... i.e. keep doubling the number of grains on each square, then the number placed on the last square is 2^{63} which is about 10^{20}. Note that we are multiplying by only 2 each time, not by 20, and have done so only 63 times, not 104 times. But 10^{20} grains of rice exceeds the total worldwide annual production of all cereals – by about a thousand times! More to the point, our current estimate of the size of the universe – say a billion galaxies each with around a thousand billion stars – is reckoned to contain in the order of 10^{80} atoms. This means that the number of possibilities for a protein just 100 amino acids long (and this is small, a typical enzyme has a few hundreds) exceeds the number of atoms in the universe – by a factor of 10^{50} which itself is much too big to comprehend (e.g. consider the grains of rice which were 'only' 10^{20}).

Having gained some idea of how many options there are, now let's try a very generous estimate of how many might have been tried out. Most of the material in the universe is thought to consist of hydrogen and helium; but just suppose that instead there were as much carbon, hydrogen, oxygen, nitrogen etc. that we needed such that the whole mass of the universe consisted of amino acids, and we could employ these to produce proteins. A protein 100 amino acids long will contain at least 1000 atoms, so at any one time we could generate something like $10^{80}/10^3 = 10^{77}$ such proteins.

We have seen that biological systems can synthesize polypeptide bonds at the rate of three to five per second, so to make a protein with 100 amino acids would take perhaps half a minute. However, if we are set simply on trying out alternatives, it is not necessary to synthesize each new polypeptide from scratch, a new amino acid sequence could be generated simply by changing just one bond or maybe two (e.g. by substituting one amino acid for another), so it may be reasonable to suggest that we could try a whole new batch of 10^{77} proteins every second. (Note that this totally ignores the more realistic delay in transcribing and translating a gene.)

So how long have we got? The aeons of geological time are so emphasized in evolutionary literature that it may come as a surprise to realise that the age of the universe (say 15 billion years) is less than a million million million (10^{18}) seconds. But taking this number as an upper limit, the total number of proteins that could be generated in this time would be about $10^{77} \times 10^{18} = 10^{95}$.

This means that even if we could employ the total material resources of the universe, and cycle them as fast as is reasonably conceivable since the universe began, the proportion of the possible combinations that might be produced would have been $10^{95}/10^{135} = 10^{-40}$ or $1/10^{40}$. That is, if we were trying to produce a specific 104 amino acid-long protein randomly, our chance of success could be no better than 1 in 10^{40}. To try to illustrate this, forget about looking for a needle in a haystack, because that doesn't come anywhere close; it is much, much less likely than trying to pick out a *specific* grain from all the cereal crops that have ever been produced – and to do so blindfold!

This may be a somewhat simplistic calculation, but what it demonstrates quite unequivocally is that we cannot rely on random mutations to produce specific proteins, whether directly or of the corresponding nucleic acid genes – quite simply, the resources to do this are not available. It is no longer tenable to hide behind millions or even billions of years – trying to argue that even the improbable becomes probable given time – nor even behind the argument that life did not have to evolve on earth but could have arisen on any one of an astronomical number of possible planets. The conclusion is plain and simple: the universe is not big enough or old enough, not by a factor of trillions of trillions of..., for the complexities of life to have arisen by random associations of simple organic molecules or of random mutations to proteins or nucleic acids. Fred Hoyle and N. C. Wickramasinghe put it like this:

> But in that dawn of certainty, in what might have been a moment of satisfaction, we hit a difficulty that knocked the stuffing out of us. No matter how large an environment one considers, life cannot have had a random beginning. [*Evolution from Space*, Conclusion]

At this point no doubt many readers will be protesting 'Yes, but ...' and put forward various scenarios to circumvent the problem. However, there appear to be only two realistic possibilities demanding proper examination, as follows:

1. In my illustrative calculation I was careful to refer to a *specific* protein, i.e. to obtain a specified amino acid sequence and no other. But it is apparent that some latitude is permitted in the amino acid sequence: First, there is polymorphism within the same species, i.e. where there are slightly different versions of the same protein. Second, equivalent proteins may be present in a wide range of species, carrying out essentially the same role and having similar but different amino acid sequences; as we shall see, cytochrome *c* is a prominent example of this.

2. Some textbooks on evolution, appearing in the last ten to twenty years have accepted that the probability of obtaining useful macromolecules by chance is very low. The usual explanation is to state that it was not necessary to obtain the correct amino acid sequence all at once, but the sequence could develop. In other words, just as species have evolved through natural selection – selecting preferential combinations of genes – so proteins evolved by selecting preferential combinations of amino acids (or, genes evolved by gradual refinement of their nucleotide sequence). In this way, crude short proteins with some biological activity were gradually modified to produce the effective enzymes etc. that we know today.

These two considerations are the basis of current theories of how biological macromolecules have arisen, and I shall consider them in some detail in this chapter. However, for the sake of completeness, here I mention two others:

3. Some suggest that the appropriate sequences for biological molecules might in some way have been directed by the physical or chemical

properties of inorganic molecules such as crystals or clays. Clearly such mechanisms could only be applicable in the prebiotic period. Once life based on proteins and nucleic acids was under way, new genes would have to arise by mutation.

4. Finally, there are still a few Lamarckians around who believe that the environment can influence organisms in the sense of affecting genetic material directly to produce appropriate genes (i.e. apart from indirectly through natural selection); in effect the mutations are no longer random but directed by the environment. There is no known mechanism for exerting this influence, and the general view is that Lamarckism was finished once the one-way street of DNA ⇨ proteins (or, more generally, genotype ⇨ phenotype) was established. The discovery of retrotranscriptases, enzymes which copy a length of RNA into DNA (rather than the usual DNA ⇨ RNA), does not change this view. So I shall not discuss this suggestion any further.[1]

CURRENT THEORIES OF PROTEIN EVOLUTION

So what is the current view of protein evolution, and how has it unfolded over the last forty to fifty years as progressively more of the structure and function of biological macromolecules has become known?

Comparative amino acid sequences

In 1951 Fred Sanger developed a technique for determining the amino acid sequences of proteins and the following decade saw the beginnings of systematic amino acid sequencing, in parallel with elucidation of the mechanisms of DNA replication, transcription and translation, as recounted in the preceding chapter. Understandably, the work started with relatively small and readily accessible proteins – insulin and then cytochrome *c* and the globin family which includes haemoglobin. By the end of the 1970s the amino acid sequences were known for many different proteins, and for equivalent proteins (i.e. those carrying out essentially the same function and having related but not necessarily identical amino acids sequences) from a wide range of species.

Cytochrome c

I explain a little more about the role of cytochrome *c* in Box 8.1; for now it is sufficient to know that the oxidation of food compounds involves the transfer of electrons from them to oxygen, and cytochrome *c* is one of the enzymes taking part in this multi-step transfer. To do this, although the bulk of the molecule is a polypeptide, it also has a non-protein component called haem which contains an iron atom. This iron is alternately reduced and oxidized (between ferrous and ferric) as the cytochrome *c* acquires

---

[1] John Maynard Smith discusses this more fully, with the same conclusion, in his well known book *The Theory of Evolution*.

and then discharges an electron.

Because of its widespread occurrence, small size and being soluble (not bound to membranes), cytochrome c was a convenient choice for early work on protein sequencing, and sequences were determined for equivalent proteins in a range of species. It was found that there were clear similarities between all of the sequences obtained; and, especially interesting, was that in general the closer species are in a morphological or classification sense, the more similar are their amino acid sequences.

On the basis of these similarities it was possible to construct a molecular phylogenetic tree showing how present-day sequences could have developed from a common ancestral one, i.e. by inferring how sequences might have gradually changed, which bore a good resemblance to the accepted phylogenies that had been constructed for the evolution of organisms based on morphology and the fossil record. In other words, just as it had been proposed that organisms have evolved by mutations occurring in genes and natural selection preserved beneficial ones (and/or beneficial combinations of genes), the amino acid sequences were now seen as demonstrating that process at the molecular level. Mutations occur at random in the various genes; it was accepted that many would be detrimental and consequently lost by conservative natural selection, but maintained that some improve the performance of the protein and are thus retained; so, progressively, over hundreds of millions of years, proteins themselves evolve.

There are various ways of constructing these trees. The most popular approach, called maximum parsimony, is to determine the tree which minimises the overall number of substitutions that are presumed to have taken place. This approach is applied not just in terms of minimising the number of changes in amino acid sequence, but in the number of changes of bases in the corresponding genes. An early example of a parsimonious tree for cytochrome c is given in Figure 4, which shows clear similarities to traditional morphology-based phylogenies. This parallel between molecular and morphological phylogenies was seen as strong supporting evidence of the evolutionary process. Not only did these early results strengthen acceptance of Neo-Darwinism, it also indicated that the new findings from molecular biochemistry would be entirely consistent with it.

The globins

At about the same time as amino acid sequences for cytochrome c were being determined, similar work was under way on the globin group of proteins. This includes the well known haemoglobin which carries oxygen in the blood (red cells) from the lungs to all parts of the body, and the less familiar myoglobin which occurs in muscles and stores oxygen ready for use (it is oxygenated myoglobin that gives red meat its colour). Haemoglobin is a tetramer – an association of four polypeptides: two each of two polypeptides designated α (alpha) and β (beta) chains having (in man) 141

FIGURE 4. Phylogeny of cytochrome *c*. Most parsimonious tree, showing inferred number of nucleotide changes for each branch. From Baba *et al*, 1981.

and 146 amino acids respectively; and human myoglobin is a single poly-peptide of 153 amino acids. In the functioning proteins, all three polypep-tides incorporate an iron-containing haem group, and it is this iron atom that binds a molecule of oxygen (but, unlike with cytochrome *c*, the iron is not actually oxidized, it remains in the ferrous state); so each myoglobin molecule can store one molecule of oxygen, and each tetrameric complex of haemoglobin can carry four molecules of oxygen. Further details of the physiological function and complementary roles of haemoglobin and my-oglobin are given in Box 7.1.

The overall amino acid compositions of myoglobin and the haemoglo-bins are not particularly similar and did not suggest a common origin. However, just as their amino acid sequences were becoming available, these proteins were the first to be the subject of X-ray crystallographic studies which revealed the 3D folding of their polypeptide chains.[2] Crystal-lography showed that each of the chains in haemoglobin is folded in a very similar way to the single chain of myoglobin, such that haemoglobin re-sembles four myoglobin molecules packed together. This prompted a care-ful comparison of their amino acid sequences (rather than simply amino acid composition) which showed that exactly the same amino acid is used in several positions of all three polypeptides (myoglobin and both types of haemoglobin chains). A comparison of the human polypeptides, for which there are 26 positions where the same amino acid is used, is shown in Box 7.2. (It can be seen that as the chains are of different lengths they need to be aligned with a few gaps to maximise this similarity.)

It seemed highly improbable that three or even two molecules could have arisen independently having such similar amino acid sequences and 3D folding, so it was suggested that all three polypeptides had derived from a common ancestral molecule. That is, it was proposed that an early gene coding for an oxygen-binding polypeptide at some stage was dupli-cated; the two copies then evolved separately, e.g. one retaining the activ-ity of the original polypeptide, perhaps with some modification, while the other adopted another function. This divergence provided myoglobin and a haemoglobin, and then the latter subsequently duplicated and diverged again to produce the α and β chains.

This view was reinforced as more amino acid sequences of different globins from various species were determined. For a given type of poly-peptide (e.g. myoglobin or one of the haemoglobins) a similar pattern was found to that for cytochrome *c*, i.e. for each type of globin the number of amino acid differences is greater for more widely diverse species, so for each globin a phylogenetic tree could be constructed. In addition, for the globins as a whole, a tree could be constructed which shows how the range

---

[2] Explaining how X-ray diffraction is used to determine the 3D structure of proteins is well outside the scope of this book. A useful account can be found in Jack Kyte's *Struc-ture in Protein Chemistry* which may also be found of interest on some of the other sub-jects discussed in this chapter.

Box 7.1   Physiology of haemoglobin and myoglobin

Requirements

1) The oxygen carrier needs to have a high affinity for oxygen at the lungs so that it readily takes up oxygen from the air.

2) The storage protein (in the tissues) needs to have a higher affinity for oxygen than the carrier does there, so that oxygen is transferred to it.

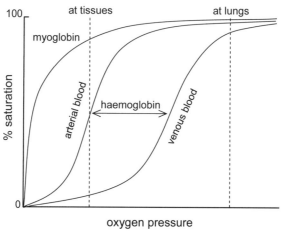

Properties of haemoglobin

1) Interaction between the haemoglobin chains means that when one binds oxygen it encourages the others to do so. This cooperativity of the subunits gives the sigmoid shape of the haemoglobin oxygen affinity curve, which satisfies the basic requirements for the carrier.

2) Performance is further improved by shifting the saturation curve to the left in arterial blood (which increases its affinity for oxygen), and to the right in venous blood (which makes it give up the oxygen more readily). This is achieved by several mechanisms:

a) Bohr effect: in acidic conditions haemoglobin binds H^+ and shifts the saturation curve to the right; venous blood is more acidic due to the presence of carbon dioxide and, during prolonged exercise, to a build up of lactic acid.

b) Carbon dioxide binds directly to haemoglobin, and this also shifts the curve to the right.

c) In some animals, especially most primates, red blood cells contain diphosphoglycerate (DPG) as well as haemoglobin; when the DPG binds to haemoglobin it shifts the curve further to the right. Birds have a similar mechanism, but using inositol pentaphosphate (IPP).

Foetal haemoglobin

In the foetus, normal beta chains are replaced by gamma chains; the resulting haemoglobin has a higher affinity for oxygen than normal (is closer to the myoglobin curve), and enables the foetus to extract oxygen from maternal blood.

Box 7.2 Comparison of amino acid sequences for human myoglobin, α- and β-haemoglobin

```
               1                          10                   20                      30                      40                      50
myoglobin      G    L S D G E W Q L   V L N V W G K V E A D I P G H G Q E V L I R L F K G H P E T L E K F D K F K H L
α-haemoglogin  V    L S P A D K T N V K A A W G K V G A H A G E Y G A E A L E R M F L S F P T T K T Y F P H F D L
β-haemoglobin  V H L  T P E E K S A V T A L W G K V N V D E V G G E A L G R L L V V Y P W T Q R F F E S F G D L

                                        60                   70                      80                      90                      100
myoglobin      K S E D E M K A S E D L K K H G A T V L T A L G G I L K K K G H H E A E I K P L A Q S H A T K H K I
α-haemoglogin  S H       G S A Q V K G H G K K V A D A L T N A V A H V D D M P N A L S A L S D L H A H K L R V
β-haemoglobin  S T P D A V M G N P K V K A H G K K V L G A F S D G L N H L D N L K G T F A T L S E L H C D K L H V

                                        110                  120                     130                     140                     150
myoglobin      P V K Y L E F I S E C I I Q V L Q S K H P G D F G A D A Q G A M N K A L E L F R K D M A S N Y K E L G F Q G
α-haemoglogin  D P V N F K L L S H C L L V T L A A H L P A E F T P A V H A S L D K F L A S V S T V L T S K Y R
β-haemoglobin  D P E N F R L L G N V L V C V L A H H F G K E F T P P V Q A A Y Q K V V A G V A N A L A H K Y H
```

Key

A	alanine	F	phenylalanine	K	lysine	P	proline	T	threonine
C	cysteine	G	glycine	L	leucine	Q	glutamine	V	valine
D	aspartic acid	H	histidine	M	methionine	R	arginine	W	tryptophan
E	glutamic acid	I	isoleucine	N	asparagine	S	serine	Y	tyrosine

of present-day globins might have arisen from a common early polypeptide. An example of such a tree is given in Figure 5 which includes other globin molecules as well: the γ (gamma) chain which is present instead of the β chain in a mammalian foetus, and the ζ (zeta) and ε (epsilon) chains which are produced instead of the α and γ chains at the early embryonic stages in some primates, including man. There is also a δ (delta) chain which performs a similar role to the β chain in adults but usually is present in only small amounts. Other globins are those of the invertebrates, which form a somewhat diverse group, and the leghaemoglobins which are found in leguminous plants such as lupins, peas or clover.[3] Divergence between the plant and animal globins is believed to have occurred very early (perhaps more than a billion years ago), and the divergence to produce the separate myoglobin and haemoglobins is placed at about 500 million years ago, at the time of the early vertebrates (see Chapter 10).

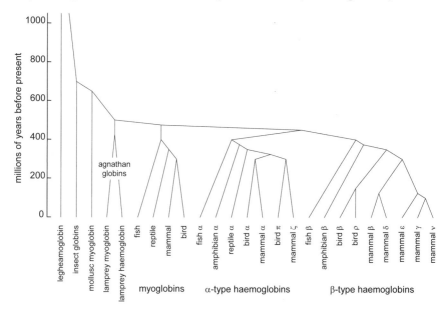

FIGURE 5. Phylogeny of the globins: showing proposed relationships of the different types of globin. Based on various sources, including Dickerson and Geis, and Hardison.

Molecular clock and conserved amino acid sequences

One of the inferences arising from a study of differences in amino acid sequences and comparing them with traditional phylogenies based on fossil records was the rate of molecular evolution – the rate at which one amino acid is thought to have been substituted for another in a given poly-

---

[3] The role of leghaemoglobins is to maintain a very low concentration of oxygen in the vicinity of nitrogen-fixing bacteria which live in root nodules of these plants.

peptide sequence. It appeared that for a particular protein the rate at which amino acid substitutions occurred within its sequence was more or less constant. This seemed to hold even when the protein was evolving in quite diverse lineages in terms of the species from which it was obtained, e.g. in plants and vertebrates for cytochrome c. On the other hand different proteins showed very different rates: for example, from very slow for histone H4, to very fast for the fibrinopeptides which exhibit the greatest variation in amino acid sequence, with the globins being about average. When the fairly constant rates of protein evolution were first noted it was seen as further supporting evidence for the theory of evolution as a whole.

A phenomenon related to the molecular clock is the relative conservatism of proteins. For those proteins that have changed rapidly, amino acid substitutions seem to occur throughout most of the amino acid sequence. Conversely, those that changed slowly seem to have large parts of the sequence which are 'conserved' in the sense that they are substantially if not exactly the same in many if not all species. For most proteins, there are specific sites or even stretches of the amino acid sequence which are conserved in this way.

Within the Neo-Darwinian framework the concept developed that for a given protein there are some amino acids that are essential to its key function – these are consistent across most or all species. Then there are species-specific amino acids that differ between species because in some way they modify or adapt the protein's activity to suit the particular species; for example, avian haemoglobin has binding sites for IPP rather than for DPG in mammals (see Box 7.1).

Domains and exons

As more information became available relating to the 3D structure of proteins it became clear that some of the larger proteins are notionally if not actually resolvable into definite regions that may carry out specific tasks such as binding one of the reactants. Often the regions are constructed from discrete stretches of the primary amino acid sequence. This led to the concept of domains – of proteins being built up from lengths of polypeptide which have definite functions. The combination of different domains is seen as an important aspect of the evolution of genes – enabling large multi-functional proteins to be built up from smaller subunits.

This sort of idea has been extended to the exon structure of eukaryotic genes. It is suggested that the exons code for areas of the final polypeptide that have discrete functions within its overall structure, and that proteins could be built up by combining different exons.

The Neutral Theory of protein evolution

So the predominant picture emerging by the 1970s was that proteins had evolved (and continue to evolve) alongside the more familiar large-scale evolution: Just as natural selection favours advantageous combinations of genes, it was considered that a similar process operated at the molecular level to select advantageous genes; in fact, evolution at the two levels were seen as acting in concert, as two sides of the same coin.

However it was at this time that a theoretical population geneticist, Motoo Kimura, upset this comfortable view by highlighting some difficulties with the accepted theory of molecular evolution. Although Kimura presented his case very mathematically, in the following paragraphs I deliberately omit any mathematical formulations.[4]

The cost of selection

Haldane (1957) had demonstrated that when an advantageous mutation spreads through a species' population, the species has to pay a price for this in terms of the death of some of its members. Selection, which by definition is preferential survival by those individuals possessing an advantageous mutation, necessarily implies relatively more deaths on the part of those that do not. 'Hard' selection is where the deaths due to selection are in addition to the deaths suffered by the species as a whole arising from its environment; 'soft' selection is where the deaths due to selection are mostly included in those that the species would suffer anyway. In nature selection can be anywhere between these two extremes depending on the particular circumstances. Whatever the case, the fitter individuals must be capable of producing sufficient additional surviving offspring to compensate for the deaths due to selection; if not, then the species will dwindle and face extinction. Haldane recognised that the deaths associated with the replacement of one allele by another must limit the rate of evolution – it is sometimes referred to as Haldane's dilemma. He had assumed selection was substantially 'hard' and concluded that a typical replacement rate could not exceed about 1 per 300 generations.[5]

This may not sound very limiting, but Kimura pointed out that, based on the rate of amino acid substitutions deduced from the comparative amino acid sequences and multiplying by the estimated size of e.g. a typical mammalian genome, would require a gene replacement approximately every two to three years. No species should be able to survive such a rapid evolutionary rate because of the aggressive selection associated with it. This was one of the first indications to Kimura that most amino acid substitutions might be neutral in terms of fitness – because no differential fitness would mean no selection, and without selection there would be no cost of substitution.[6]

---

[4] Readers interested in pursuing a more quantitative understanding could consult a fairly recent text on evolutionary population genetics, or Kimura's own account in *The Neutral Theory of Molecular Evolution* (1983) which is very readable.

[5] Note that this relates to the rate at which successive mutations may be assimilated; the number of generations required for an individual mutation to be assimilated is usually much greater than this (see Box 5.4).

[6] For the sake of clarity, what Kimura meant by 'neutral' was that two alleles (typically the established gene and a new variant) confer equivalent fitness, not no fitness; and it is

Polymorphism

Biologists in the first half of the 20th century knew that favourable mutations are rare, and the traditional Neo-Darwinist view of evolution was that in the main genes are stable but occasionally a favourable mutation arises which spreads fairly rapidly through the population. This view was supported by the observed increase of melanism in the peppered moth from practically nil to more than 90% (in some industrial areas) in only 50 years. However, a corollary of this is that most genes should have one predominant allele and that multiple alleles (polymorphism) within a population should be uncommon. Where stable polymorphism did occur this was generally attributed to heterozygote advantage.

However, by the 1960s it was becoming apparent that polymorphism was common (recall the extensive innate variability mentioned at the end of Chapter 5), much more frequent than expected. This was difficult to accommodate within the accepted theory because, if it was due to many gene loci being in transition between one allele and a more advantageous one, then it implied a more rapid evolution than seemed compatible with Haldane's dilemma, as discussed above. But there is also a cost in maintaining heterozygote advantage because this means that individuals having both alleles are fitter than individuals homozygous for either. That is, just as there is a cost due to selection in substituting a fitter allele for another, so there is a cost due to selection in maintaining heterozygotes which are fitter than either homozygote. The other difficulty with heterozygote advantage is that, apart from the well-known case of sickle cell anaemia (see Ch. 9), very few examples have been found. Hence, although when protein polymorphism was first identified it was generally attributed to heterozygote advantage, as evidence for the extent of polymorphism accumulated this explanation was looking increasingly thin. Kimura argued that if most amino acid substitutions are neutral then this could lead to many loci having multiple alleles which are essentially equivalent and could readily coexist.

Pattern of amino acid substitutions

The most direct evidence which Kimura cited as indicating that most amino acid substitutions were neutral was the pattern of the substitutions which was becoming apparent as progressively more amino acid sequences were determined. It had been recognised that the rates of amino acid substitutions were different for different proteins, and that substitutions did not occur evenly along the amino acid sequence but tended to be in clusters. With the increasing use of X-ray crystallography (and other techniques) to determine the 3D structure of proteins, it became possible

---

not necessary that the two alleles should have exactly the same fitness, just so similar that their relative fates are determined more by chance or genetic drift than by differential selection.

to identify the positions and roles of specific amino acids within a sequence. Hence it could be determined not only where substitutions had occurred in terms of their position within the amino acid sequence, but also in terms of their location within the folded protein.

Kimura showed that substitutions of amino acids on the surface of haemoglobin molecules were more frequent than in the interior, and ten times more frequent than the amino acids surrounding the haem group. So he argued that the extent of amino acid variation depended on functional constraints: Mutations arise by chance, most are detrimental because they disrupt the molecule, but a few at some locations are permissible because the substituting amino acid is not too dissimilar in size and functional groups from the amino acid it is replacing and/or the characteristics of the amino acid at that location are not critical. In other words, which substitutions, if any, are permissible depends on the criticality of that part of the molecule and the relative similarity of the substituting amino acid. With this view it is easy to understand why amino acid positions on the outside of the folded protein are likely to be more accommodating of substitution than those in the interior where, typically, hydrophobicity and physical size can be critical. It had been recognised that many alternatives found at a specific site in an amino acid sequence were similar in terms of, for example, their hydrophobicity or hydrophilicity, or polarity, or just size. Or, conversely, the smallest amino acid, glycine, is often not amenable to replacement (i.e. is conserved) because all other amino acids are appreciably larger; and proline which, because of its cyclic structure, forces a change of direction of the peptide backbone also tends to be conserved.

A clear example of the frequency of amino acid substitutions being related to criticality of function came from insulin. Insulin comprises two short polypeptides of 21 and 30 amino acids (designated A and B) which are covalently joined by two disulphide bridges. But insulin is formed from one 84 amino acid long precursor polypeptide, proinsulin, which folds up, the disulphide bonds are formed to stabilise the tertiary structure, and then 33 amino acid are removed from the middle of the sequence. The extent of variation (and hence inferred rate of amino acid substitution) in the excised portion (polypeptide C) which is discarded is about six times greater than in the A and B chains which remain to form the active hormone.

Nucleotide substitutions

When Kimura proposed the neutral theory, it was based entirely on amino acid sequence data as very few nucleotide sequences were known. However, he predicted that there would be similar patterns of nucleotide substitution as that found for amino acids, in particular that there would be greater variation in less critical parts of the nucleotide sequences. The neutral theory gained considerable credibility when these predictions were found to be correct.

You may recall that the genetic code is at least partly redundant in the sense that there are about three times as many codons which specify amino acids as there are amino acids, such that quite often a change in the third base of a codon does not change the specified amino acid. Thus some mutations can arise in a gene – in some bases of the DNA – which do not give rise to any change in the protein that the gene specifies: such mutations are called synonymous or 'silent' mutations. It was found that the rate of synonymous nucleotide mutations was substantially the same for most proteins, including those showing a wide range of amino acid substitution rates. A prime example of this is histone H4 which is a highly conserved protein in terms of its amino acid sequence, but for which the synonymous mutations are just as frequent as for more variable proteins.

Also, in a situation somewhat analogous to the polypeptide C of insulin, it has been found that the nucleotide sequences of introns (which, you may recall from the end of the previous chapter, are excised from transcribed mRNA before it is translated) are far more variable than of exons which are translated.

Last to mention here are so-called 'pseudogenes'. These are stretches of DNA which bear a close resemblance to genes with a known function, but these pseudogenes do not appear to be transcribed for some reason or other; it is generally thought that they are copies of functional genes that have become corrupted. Nucleotide substitutions in pseudogenes are among the highest known. In terms of the neutral theory, it is explained that because pseudogenes are now functionless there is little or no constraint on the nucleotide sequence, so (m)any mutations that arise are accepted – few or even none are rejected by natural selection.

Conversely, the genes coding for the rRNAs and tRNAs are found to be highly conserved, and this is explained in terms of their critical role in protein synthesis, especially in translation of the genetic code.

Selectionist versus neutralist debate

With his neutral theory, Kimura sparked off a debate in molecular evolution that has persisted for more than two decades and has been the focus of many recent texts by opposing protagonists. On one side are the neutralists, championing the case just outlined above, and on the other are the selectionists who advocate a much more prominent role for selection in amino acid substitution. It is not necessary to enter into this debate here, only to give a brief summary of the current situation.

Probably most evolutionary geneticists would now accept that many if not all synonymous substitutions are neutral. But for substitutions which have a phenotypic effect (usually an amino acid substitution) they point to some of Fishers' work which showed that a mutation has a greater chance of being accepted if its phenotypic effect is small rather than large, i.e. it may be not very far from neutral. In other words, whilst the neutralist view is that at the molecular level most substitutions are neutral or nearly so, selectionists would maintain that most have a positive effect but the advantage may be very small. Conversely, even the most ardent advocates of the neutral theory for molecular evolution would accept that natural selection operates in a traditional sense, selecting advantageous gene combinations, at the morphological level.

FAILINGS OF CURRENT THEORIES

The preceding section has given a fair representation, albeit necessarily brief and without mathematics, of current thinking regarding evolution at the molecular level, as may be portrayed in current textbooks on the subject. The variation in amino acid sequence observed in equivalent proteins is taken as clear evidence that unique amino acid sequences are not necessary for protein function, but that substantial latitude is permissible, hence mitigating the prohibitive improbability of a *specific* amino acid sequence. Then, molecular phylogenetic trees which show how protein sequences are believed to have changed over time are seen as supporting the idea of the progressive evolution of proteins – that it is not necessary to obtain a fully-formed protein in one go, but the optimal structure could have developed from an early inferior amino acid sequence which nevertheless has some activity. Additional evidence for the evolution of proteins is seen in the similar primary and tertiary structures of proteins that have related though different functions – notably the globin family – the similarities being seen as clearly indicative of a common origin, i.e. as homologies (see Ch. 12). Further, although the main focus of my discussion is proteins, as the nucleotide sequences of various rRNAs and tRNAs became available, it was found that these could be arranged into family trees in the same sort of way as proteins. So these, too, are seen as further evidence of evolution and, like the protein family trees, are used to construct phylogenies.

The facts of amino acid (and nucleotide) sequences are not in question. However, much inference from these facts, such as the points just outlined above, is at best a biased interpretation and in many respects is clearly flawed, completely overlooking major objections to this perceived evolutionary scenario. It is these failings of the evolutionary interpretation of the comparative amino acid sequences that I discuss here. Given the strength of Neo-Darwinism, it is not surprising that the early results were interpreted from an evolutionary perspective; but, as more information has become available, instead of confirming the early interpretations, they cast increasing doubt on them. However, it is also increasingly evident that the theory of evolution has become so entrenched that many scientists are blinkered – seeing the facts only in an evolutionary context and reluctant to give due weight to the anomalies that arise, convinced there must be satisfactory evolutionary explanations for them, which will emerge one day. They are caught up in the evolutionary paradigm – an issue I will discuss more fully in a later chapter.

Specificity of amino acid sequence

The first issue to consider is the extent of possible variation in the amino acid sequences of proteins.

To begin with, we should note that, before we can accept that differences in amino acid sequences between polypeptides from different species represent permissible variations in the amino acid sequence for the relevant protein, we must ensure that the polypeptides in question do actually represent viable alternatives of what is essentially the same, i.e. equivalent, protein. For example, in the early days of amino acid sequencing, any cytochrome *c* was considered to be equivalent to any other; but as more information emerged it became evident that, especially in bacteria, there are many types of cytochrome *c*. The bacterial cytochromes *c* have a range of different functions (many not connected with ATP production), reflecting their diverse environments and modes of life, so they represent many different proteins, not variations of the same one.

However, at least so far as the mitochondrial cytochromes *c* in all eukaryotes (see Box 8.1) are concerned, it has been demonstrated that many can operate in a foreign species, i.e. a mitochondrial cytochrome *c* from one species will work with mitochondrial components (principally the enzymes cytochrome oxidase and cytochrome reductase) from another, although usually not so well as in its native environment; but this is good evidence that we are looking at variations on a common theme.[7] Within the mitochondrial cytochromes *c* there is variation of many of the amino acids, and this is seen as clear evidence that the amino acid sequence for proteins need not be specific. – But that is looking at only one side of the coin.

The other side of the coin is, of course, the many amino acids which are substantially if not exactly the same for mitochondrial cytochrome *c* from all eukaryotic species. For this protein, the amino acid sequence is now known for about a hundred different species. On the basis of these and the 3D structure of this protein it has been possible to determine which are the essential amino acids for a mitochondrial cytochrome *c*. It should be noted that whereas essential amino acids were recognised first solely on the basis of their persistence in amino acid sequences, i.e. as 'highly conserved' amino acids (and it might have been argued that these was merely chance recurrences), as detailed investigations into the structure and mode of action of proteins have progressed, it has become possible to determine the role of these amino acids within the protein, and

---

[7] The proteins designated cytochrome c_2 from bacteria perform a similar role, though with somewhat different characteristics (e.g. having a higher redox potential) reflecting the fact that these operate in the bacterial cell wall rather than mitochondria.

hence confirm that they are essential for the protein's function. So far as mitochondrial cytochrome c is concerned, first to mention should be the cysteines at positions 14 and 17 which covalently bind the haem, and the other amino acids which interact with the haem either directly (his 18, met 80, arg 38, tyr 48, tryp 59, ser 49, asp 50, thr 78) or indirectly (pro 30, tyr 67). In addition there are many highly conserved hydrophobic amino acids surrounding the haem, notably lysines at positions 8, 13, 25, 27, 72, 73, 79, 86 and 87. Then, essential parts of the overall 3D structure of the protein are α-helices at each end of the polypeptide, which interact with each other, resulting in the conserved amino acids gly 1, asp 2, gly 6, and an aromatic amino acid at position 10 at one end, and five successive hydrophobic amino acids at positions 94–98 of which 97 must be aromatic at the other. In total there are 27 positions where the amino acid is invariant and 20 others where only two options are possible. Also, of course, it should not be assumed that the remaining amino acid positions are completely non-specific; on the contrary, in order to maintain the overall 3D shape of the molecule there are considerable limitations on what is acceptable. In several cases a change of amino acid is permissible provided it is accompanied by a compensating change in another position – a so-called tandem replacement; so, whilst the individual amino acids concerned do not appear to be highly conserved, there is in fact substantial constraint on the viable alternatives.

On the basis of these amino acids which are invariant or nearly so, we could estimate what may be considered a more realistic number of possible amino acid sequences that would provide a functional cytochrome c than the 20^{104} I used previously. We could then try to make a sensible estimate of how many options might reasonably be produced and tried out with the available resources, taking a more realistic view than the gross overestimate I made earlier, and taking into account other considerations mentioned towards the end of this chapter, in order to re-assess the probability of obtaining a useful polypeptide.

However, whilst I invite readers who are so minded to make an attempt at this, such an exercise is quite irrelevant to the basic problem of the improbability of proteins because there are many others that pose a much greater challenge than does cytochrome c. I used cytochrome c because it was one of the earliest proteins to be sequenced, it was a major source of interest in the idea of protein evolution, and because it remains one of the prime examples given in evolutionary texts. It also provides a good example of how further investigations revealed details about the proteins' structure and function, and illustrates some of constraints on an amino acid sequence. But since the 1960s we have obtained comparative amino acid sequences for many other proteins and some of them are much more specific than cytochrome c. Notable among them are the eukaryotic proteins histone H4 and ubiquitin.

Ubiquitin has just 76 amino acids, but 69 of these are totally invariant, and there are only three differences between the sequences found in yeast and humans. Although many readers may not have heard of ubiquitin, they should not think that this is some obscure protein with a minor role. Quite the opposite: as its name implies, it is found almost everywhere in eukaryotic organisms (there does not seem to be a prokaryotic equivalent); it is a key component in the mechanisms regulating the degradation of proteins, and has an important role in cellular functions as diverse as DNA repair, cell differentiation, and the immune response.

Histone H4 is one of the proteins which is intimately associated with DNA, involved with its packing and possibly having a role in transcription. It has 105 amino acids and usually there are no more than two differences between most higher plants and most higher animals, though with greater differences in some lower organisms.

Further, as a general rule, as investigations have moved on from the relatively small proteins of the early work to larger and more complex ones, especially those which interact with other macromolecules, it has been observed that such proteins exhibit relatively less variable amino acid sequences, and the total number of highly conserved amino acids can far exceed the 100 or so on which I based the initial illustrative calculation. For example, the protein actin which has as structural role in all eukaryotic cells, and is their most abundant protein, has a sequence of typically 375 amino acids, 80% of which is the same in all animals from amoebas to humans.

Of especial note is the plant protein rubisco (ribulose 1,5-bisphosphate carboxylase) which is the enzyme that fixes carbon dioxide in the process of photosynthesis. Although it is usually associated with a small subunit, the core part of rubisco which includes the active site is a large unit comprising 476 amino acids. In a study (Kellog and Julian) of 499 plant species, it was found that 105 amino acids of this subunit are totally invariant, and in a further 110 positions only one alternative is possible. Although some of these highly conserved amino acids could be clearly associated with the active site or with regions of interaction with other subunits, many could not – again indicating the importance of specific amino acids or combinations of amino acids for the internal structure of a protein. This is reinforced by the fact that in a further 189 positions the choice of amino acid was limited to only three or four possibilities.

Hence, whilst accepting that the variations evident in the amino acid sequences of equivalent proteins demonstrate that some latitude or lack of specificity is permissible, and that possibly no protein requires 100% specificity, it is equally evident from the same comparative amino acid sequences that for most proteins the bounds of variability are strictly limited. Importantly for the present discussion, the limitations are such that in no way does the lack of specificity for some amino acid positions overcome the prohibitive improbability of biologically useful macromolecules outlined at the beginning of the chapter. On the contrary, there are many proteins with many invariant amino acids (and as investigations proceed their number increases rather than decreases), such that even if we consider only those amino acids that have been demonstrated to be essential for their function, any one of these proteins presents a colossal mountain of improbability.

Molecular phylogenies

This leads us to consider the suggestion that for each protein the mountain can be scaled gradually, and that the phylogenetic trees based on comparative amino acid sequences illustrate this process. But this interpretation no longer stands up to what we have learned about the structure and function of proteins.

Significance

As I have just discussed, for all enzymes there is a core of 'highly conserved' if not invariant amino acids – those that are critical for its active site(s) and/or interaction with other molecules, or for its overall 3D conformation – which are consistent across a wide range of species. Conversely, where variations of sequence occur, these are usually between amino acids which are similar in terms of size and/or chemical properties, and/or are on the periphery of the folded protein. The nature of these differences indicates that they are generally of minor importance in terms of the protein's function, and this is supported by the various arguments outlined above regarding the neutral theory. In fact, crystallographic studies have shown that the 3D structures of eukaryotic cytochromes c are scarcely distinguishable from each other, including those from higher plants to higher animals even though they are presumed to be separated by many hundreds of millions of years of evolution. Further support that the proteins from different species have comparable function and activity are the many instances, mentioned above, where enzymes from one species have been found to function reasonably well in a very different one. Such differences in structure and/or function as there are tend to be in terms of fine tuning to the other macromolecules with which it interacts, e.g. yeast cytochrome c tends to work best with cytochrome oxidase and cytochrome reductase from yeast, and human cytochrome c with the corresponding enzymes from human mitochondria, but the overall performance of the human cytochrome c system is substantially the same as that from yeast. In other words, although differences in amino acid sequence do arise for a given protein such as cytochrome c, from both a structural and functional point of view, these differences are of little significance.

Bearing this in mind, we now turn to the phylogenetic trees that are constructed on the basis of the variations in amino acid sequence of present day proteins. The rationale behind the construction of the phylogenies is to show how the present differences between species have arisen progressively from a common ancestral sequence. But what needs to be recognised is that, even if the phylogenetic trees were valid (and there are reservations about them, noted below), they relate only to the occurrence of what are substantially neutral substitutions – the highly conserved amino acids remain unchanged (it could hardly be otherwise for *conserved* amino acids). So, at best, they can show only how relatively inconsequential changes might have taken place, and they certainly do not show how efficient modern proteins have developed from crude early ones.

It is appropriate to comment here on an aspect of the selectionist:neutralist debate. Readers will recall from earlier this chapter that a selectionist response to the neutralist argument is that Fisher had shown that mutations having a small phenotypic effect were much more likely to be accepted than those having a large phenotypic effect. In this way selectionists could come very close to the neutralist position (because the advantage could be *very* small) whilst remaining true to their selectionist credentials. What is interesting about this response is that Fisher's argument is in the context of a species which is already close to an adaptive peak: the rationale being that a small mutation has a 50:50 chance of taking it higher up the adaptive peak and so improving its fitness, whereas a large mutation could throw it off the peak altogether. (I discuss this a little more towards the end of this chapter.) This is noteworthy in the present discussion because it means that whether one adopts a neutralist or selectionist stance regarding interpretation of the comparative amino acid sequences, it implies that the proposed early proteins had an activity which was similar to that of modern ones.

That is, the evidence of the comparative amino acid sequences is that when a protein first appears on the scene (in so far as this is inferred from the phylogenetic trees) it is already in a fully functioning form, already more or less within the same bounds in terms of its amino acid sequence and function that we know today. In other words, although the comparative amino acid sequences, and the phylogenies constructed from them, are cited as demonstrating protein evolution, in fact any evolution they demonstrate is of a very limited nature. In particular, they cannot be used to support any sort of progression from some simpler protein structure to a substantially more complex or refined or improved version; and certainly they are of no assistance whatever in explaining how a functional protein arose in the first place. This view is clearly supported by the following comment from Jack Kyte:

> It is also quite clear that the evolutionary divergence that produced most of the proteins that are universally distributed among present living organisms, e.g. the metabolic enzymes, occurred before divergence of the organisms themselves. This follows from the observation that the proteins from all living organisms responsible for one particular biological function are usually superposable [i.e. have the same 3D shape], but proteins from the same organism responsible for different functions are usually impossible to relate to each other. Thus the lineages of these fundamental proteins may

have remained almost unbranched since the evolution of the earliest organisms, and the radiation producing these lineages must have occurred
before that time. [*Structure in Protein Chemistry*, Ch. 7]

The idea that molecular phylogenetic trees demonstrate constructive protein evolution arose on the basis of the early amino acid sequences – before there was an appreciation that most substitutions are more or less
neutral, and when (given the background of Neo-Darwinism) it was obvious to see the comparative amino acid sequences in this evolutionary light.
Although we now understand the insignificance of most amino acid substitutions, the idea persists because it suits most biologists to retain it as
support for the evolutionary scenario.

Of course, it can still be proposed that effective proteins developed in
an earlier period, before the earliest point on the inferred phylogenies. But
that is only conjecture – we are left with the daunting improbability of biologically active macromolecules, and no indication of how it was overcome.

Consistency

In the present context of highlighting the extreme improbability of biological macromolecules and considering how they might have arisen, the
main point I want to emphasize about the molecular phylogenies is their
insignificance, i.e. that they do not demonstrate constructive protein evolution. However, no doubt many will still argue that they at least provide
circumstantial support for evolution in view of the parallels which can be
drawn between molecular and morphological phylogenies. But when we
take a closer look at the molecular phylogenies we find that the more we
learn the less convincing they become. The early phylogenies were compiled with a rush of enthusiasm emanating from the dramatic progress of
the 1950s and '60s when scientists were beginning to open up the molecular aspects of genetics and biochemistry. On finding that the proteins from
morphologically related species had similar amino acid sequences it was
natural to presume that evolution at the morphological level was being
paralleled at the molecular level; and with the limited number of sequences available it was relatively easy to construct self-consistent phylogenetic trees showing how amino acid sequences might have
progressively developed. The initial impression of a constant rate of molecular change further encouraged this view. But as further data came to
light it became apparent that the straightforward early picture could not
be maintained.

First to go were the steady rates of amino acid substitution, i.e. it was
found that not only were the rates different for different proteins, but different rates occurred for the same (equivalent) protein in different lineages. It is of course recognised that, being subject to random mutations,
evolution is bound to be variable; but the differences in rate between various lines is significantly greater than would have been expected even al-

lowing for this. For example, to account for the various sequences of cytochrome c, it was concluded that in fish and amphibians this molecule had evolved a few times faster than it had in reptiles, birds and mammals (Baba *et al.*). It was also found that the lack of consistency is not only between the higher taxa, but even within them. In a group such as the mammals there is significant variation in the rate of inferred molecular evolution between different lineages, and this applies to the globins as well as cytochrome c. Rodent proteins appear to have evolved about twice as fast as the mammalian average, whereas the rate in primates is so slow that all the amino acid sequences of cytochrome c, α- and β-haemoglobin are identical in humans and chimpanzee, even though it is assumed they diverged several millions of years ago (there is debate about how long), which should have been enough time for at least some substitution to occur. Results such as these have led some workers to conclude that any molecular clock ticks so erratically that it is of no value.

As more data were acquired it also became apparent that, not only were the branches on the molecular phylogenies of unpredictable length, even the arrangement of the branches was equivocal and no longer convincingly paralleled morphological evolution. In an extensive study of cytochrome c, based on sequences from 87 species, a parsimonious tree based solely on the degree of similarity between the sequences differed from a conventional phylogeny so much that the investigators postulated various gene duplications and 'gene expression' events along some lineages in order to overcome the worst anomalies (Baba *et al.*). Similar difficulties were found with the globins. In a detailed investigation of myoglobin, several different approaches were employed for grouping the sequences in some sort of phylogeny, but all produced at least some associations which seemed anomalous in terms of accepted morphological groupings (Joysey).

Part of the problem is that, once even a moderate number of sequences are available, the number of possible trees is so large that it is not feasible to assess them all, even with computer analysis. It is therefore necessary to give the analysis a reasonable starting point, which of course is based on wider information available to the investigators, including morphological phylogenies: in other words the accepted morphological tree becomes a guide for constructing molecular trees. This is a perfectly reasonable approach to take and I am not criticising it, but it certainly undermines any idea that the molecular phylogenies are *independent* evidence for the evolutionary scenario. Even more telling is that in most cases the credence given to a molecular phylogeny is determined by its compatibility with morphological phylogenies; in effect investigators end up seeking ways to make molecular trees look reasonable in the light of morphological trees, and, where necessary, 'explaining' outstanding anomalies.

A related issue is that different proteins give different phylogenetic trees, which makes any claim that the molecular phylogenies support an evolutionary scenario seem rather hollow. An interesting example of this comes from cytochrome c and the globins. Early studies showed that the cytochrome c of birds is more similar to that of reptiles than to mammals, and this was advertised as clear support for the evolutionary scenario of birds evolving from a reptilian line. But avian haemoglobins are more similar to those of mammals than reptiles; so this is explained in terms of birds and mammals both being warm-blooded. And then there is the rattlesnake cytochrome c which does not seem to be related to anything else at all.

Adaptation is also used to account for some other discrepancies: for example it is considered unsurprising that the myoglobin of seals (pinnipeds) is similar to that of the whales and dolphins (cetaceans), even though these two groups are thought to have diverged approximately 50 million years ago, because of their similar diving habits. But to set against this sort of explanation is the fact that the whale cytochrome c is identical to that of the camel.

What these examples illustrate is how different emphases are employed as necessary to account for the various discrepancies that arise – but the accumulation of *ad hoc* explanations becomes increasingly unconvincing.

Much more perplexing (in terms of an evolutionary account) are the non-consistent relationships of some plant proteins. *Arabidopsis thaliana* (a type of cress) is generally regarded as a model species of the flowering plants and is used for many investigations of their biochemistry. It was, therefore, somewhat disconcerting to find that the amino acid sequence of its cytochrome c was significantly more like that of yeast than of most other higher plants; and the situation is all the more enigmatic because other features of the *A. thaliana* cytochrome c gene such as the control regions, intron structure and codon usage[8] are typical of the higher plants. To add to the confusion, some other proteins such as histone H3 also resemble that of yeast, but others are not; and some other plants have also been found to have atypical protein sequences.

Facts such as these certainly detract from the weight that can be given to the molecular phylogenies as circumstantial support for evolution; but, not surprisingly, they generally do not find their way into evolutionary textbooks.

Validity

Even the validity of the molecular phylogenies is now in doubt. From quite early on in their construction it was recognised that the number of differences between the sequences of two present-day proteins will be less than the number of changes that have taken place between them since their presumed divergence. There are two main reasons for this: first there are

---

[8] Recall that in most cases the same amino acid can be coded by several codons. It is often found that a group of organisms seems to have a preference for using a particular codon for a given amino acid, though the other codons will work.

back-mutations where a sequence mutates at one site but then mutates again at the same position and this happens to return it to the original amino acid (in fact, where any two substitutions have occurred successively at the same site, the intermediate will be unknown); and second are parallel mutations, where both sequences happen to change in the same way. These phenomena are sometimes referred to collectively as convergent evolution.[9] Even construction of the early parsimonious molecular phylogenies, with relatively few sequences, required the inference of some parallel and back mutations. There is debate as how best to estimate the actual number of changes that would have occurred historically, based on present-day differences. The most common approach[10] indicates that where actual differences are less than ten then the additional changes are relatively few, but by the time there are twenty-five differences then the total changes are over fifty, and so on, as the number of actual differences increases the number of presumed changes increases all the more.

At first it was thought to be just a question of having to adjust the estimated number of changes that occurred between two lineages but that there were no implications for the arrangement of the branches. However, a few researchers began to point our that, due to convergent evolution, the number of unknown (and unknowable) changes in the past is so large that it casts doubt on the reliability of any inferred historical sequences. Although many still use the principle of parsimony to estimate ancestral lines, others recognise that this cannot be relied upon to fill in the gaps because, as neither mutation nor natural selection has foresight, there is no reason to believe that evolution would have taken the most direct route from one sequence to another. Because of this and other doubts about the rates of molecular evolution, several workers now consider molecular phylogenetic trees invalid, except for linking closely related species where the number of differences is small.

However, even within closely related species, as the number of known sequences grows, increasingly it is thought that the differences between them primarily reflect the limits of variation within the prevailing functional constraints, rather than their ancestry. This is seen especially clearly with rubisco. The workers who reported the relative conservancy of the amino acid sequence of rubisco also tried to determine a phylogeny on the basis of the differences that do occur, but they found that they could not construct a consistent tree. They concluded that for many positions the 'amino acids simply toggle among a small handful of residues[11], but do not continue to diverge.' This well illustrates that with more data being available, the early simplistic interpretations are just no longer tenable.

---

[9] This is molecular convergent evolution; in Chapter 9 I will discuss convergent evolution in a morphological sense.

[10] Using a Poisson distribution to account for the occurrence of random mutations.

[11] Essentially, the amino acid side chains.

A similar picture emerges with the many different types of bacterial cytochrome c, for which a phylogenetic tree has been described as a starburst – various sequences of equivalent proteins in each cluster, but the clusters widely separated from each other with no discernible phylogeny to link them.

And a further challenge to the credibility of phylogenies based on a superficial resemblance of amino acid sequences comes from the discovery of a mammalian variant of cytochrome c. The molecular phylogenies for cytochrome c were constructed by comparing the amino acid sequences of mitochondrial cytochrome c which occurs in all eukaryotic cells, now referred to as somatic cytochrome c. However, a variant cytochrome c has been found in some mammals, including man, which is expressed only in testicular tissues, probably only in germ line cells. It is not a pseudogene, but a fully functional protein which is properly synthesized along with the usual somatic cytochrome c. The amino acid sequence of testicular cytochrome c differs from somatic cytochrome c at various positions, typically involving about 15 amino acids, and some of these substitutions require multiple nucleotide changes. It is of course generally assumed that testicular cytochrome c must have branched off from the standard (somatic) cytochrome c molecular tree at some point. However – and this is the interesting point for the present discussion – the sequence of testicular cytochrome c bears no relation to the sequences that are supposed to have arisen in the course of evolution of the somatic cytochrome c (Mills). In other words, although it is presumed to have diverged from the somatic cytochrome c at some point, there is no evident place where it could have done so.

What all of this means is that on both the large scale, linking major taxa, and on the small scale, there is serious doubt about the validity of phylogenies constructed on the basis of amino acid sequences. It highlights the fact that the phylogenies are but the product of theoretical exercises – which at first seemed convincing, but as more data emerge they become increasingly less credible. Nevertheless, they are propagated because biologists are taught that they are meaningful and, because they fit in with the evolutionary outlook, there is little motivation to challenge them. In similar vein, although when phylogenies are first published the technical account generally describes the assumptions and approximations etc. employed by the investigators, all too often when they are used in textbooks such details are omitted, and the impression is conveyed that the phylogenies are much more reliable than is the case. Similarly, where molecular phylogenies agree with morphological groupings they are used, but where they do not they are ignored.

Finally, it is all too easy to be drawn into thinking that the rates and patterns of amino acid 'substitutions' are real phenomena. It is worth reminding ourselves that the raw data are simply the present day amino acid sequences, and proposed 'substitutions' are merely the evolutionary way of trying to explain actual differences between those sequences. Where 'anomalies' arise, all that is actually meant is that there is not a consistent relationship between the degree of difference in amino acid sequence and the estimated time since a presumed divergence from a common ancestor. But we need only expect consistent relationships if we assume an evolutionary origin; the fact that inconsistencies arise is a clear reason to question that hypothesis.

Gene families

The above paragraphs have dealt primarily with phylogenies where they relate to just one type of, or 'equivalent', protein i.e. those with a comparable function in different species. I now consider the phylogenetic trees which include gene duplication and divergence to produce two or more equivalent proteins with different functions (widely referred to as 'homologous' proteins), such as myoglobin and haemoglobin. Many proteins, because of similarities in their amino acids sequences, are assumed to have arisen by gene duplication from a common ancestral source; in fact gene duplication is thought to have been a frequent means of generating new genes, and generally increasing genome size and complexity.

When gene family trees are presented, such as in Figure 5, with the positions of the divergences shown and especially if including estimated numbers of amino acid or nucleotide substitutions on each branch (comparable to those shown in Fig. 4), at first sight they can appear convincing. But a little thought about the implications of these trees which incorporate gene duplication and divergence soon reveals some fundamental difficulties. I will illustrate these with the globin tree which, as mentioned above, was the first to be proposed, and is by far the most common example cited in evolutionary texts.

The genetic family trees are only theoretical

In the first place it should be recognised that the gene family trees are merely the outcome of a theoretical exercise and lack substantiation: they are inferences based on a comparison of present-day amino acid sequences, and there is no independent evidence for them. A similar exercise could be carried out for *any* two proteins. That is, it would be possible to take two distinct proteins (whether or not they have some common amino acids, and whether or not they have related functions) and propose a scheme as to how they could have arisen from a common ancestral sequence – an ancestor which would be assumed to have been a hybrid of the two, i.e. to have some amino acids in common from both actual proteins. This could be done easily for any two proteins of similar length, and could readily be extended to include substantially different lengths and/or species' variations of the two proteins.

This can be illustrated with words. For example, the words 'protein' and 'peptide' could be considered homologous because they have a related meaning, they are the same length, and have some common letters – in fact the proportion of common letters (and in the same place) is about same as for the amino acids in myoglobin and haemoglobin. By analogy with the protein phylogenies, it would be easy to propose a common ancestral combination of letters from which they have both been derived – there are several equally likely contenders such as 'prptede' or 'peotidn' – all of which are closer to both 'protein' and 'peptide' than these are to each other. However, 'protein' and 'peptide' do not have a common origin, their etymologies are quite different.[12] In other words, the fact that a tree can be constructed linking 'homologous' proteins is no evidence whatever of actual common ancestry.

Returning to the globins, although they are portrayed as having many common amino acids, the similarity between the same polypeptide from different species is much greater than between the different polypeptides: After presenting the basic amino acid data on the globins, Dickerson and Geis, who carried out substantial work on the globins, stated

> The most striking observation ... is that the hemoglobins and myoglobins do make up quite separate subfamilies, with the differences within one subfamily (fewer than 110 changes) being smaller than those between myoglobin and haemoglobin sequences (110-132 changes). Further, the invertebrate globins differ more from one another and from all the vertebrate chains (113-146) than the vertebrate myoglobins differ from either hemoglobin chain. [*Hemoglobin*, Ch. 3][13]

Even the α and β chains in the same species differ by at least 75 amino acids (e.g. in man) and up to 89 (for shark), which is well over 50% of the sequence (Dickerson and Geis make the overall number of differences, i.e. including gaps, to be 85 and 100 respectively).

What this means is that, whilst phylogenies can be constructed for each type of globin polypeptide with relatively few amino acid changes between most vertebrate species (in the same sort of way as that for cytochrome *c*, though with similar reservations and anomalies), when it comes to linking the phylogenies for different polypeptides, i.e. by deriving them from a common ancestor, inevitably there are very substantial gaps which must be bridged, requiring numerous amino acid substitutions. It is axiomatic that the tree for each polypeptide must include the highly conserved amino acids of that polypeptide, so the bridge from one polypeptide to another must at least involve the change of one set of conserved amino acids for another set – and there are several significant differences between them. It is worth restating that even though molecular phylogenies such as that depicted in Figure 5 show these major divergences and various sub-

---

[12] Protein is derived from *protos* meaning first, and peptide from *peptikos* meaning capable of digestion (Oxford English Dictionary).

[13] The data referred to scores a difference for each gap in the alignment of the sequences as well as for differences in amino acids.

sequent sub-branchings there is absolutely no independent evidence for the existence of these ancestral sequences or proposed intermediates. They are inferred solely on the basis of *assuming* a common ancestry and then deriving a route of polypeptide evolution, typically the most parsimonious one, to fit the known present day amino acid sequences and consistent with the observed pattern of conserved amino acids.

The biological activity of proposed intermediates

Secondly we need to look a little more closely at the nature of the proposed intermediates.

The standard model for gene divergence is as follows: It is accepted that a single copy of a gene cannot make a substantial jump in terms of its sequence because the chance of jumping to something significantly different and yet useful is too small. However, if a gene duplicates, then while one copy retains the original function (with the usual limitations on how much it can mutate, although permitting some modification), the other copy is freed to 'experiment' – it is able to mutate freely and 'try out' alternatives, and may hit upon another useful function. Importantly, this scenario means that along the family tree leading up to a divergence, and for at least one branch immediately afterwards, biological activity must be maintained; even for the other branch it would be unrealistic to have an unduly large jump or run of sequences without activity. And these constraints must be true for each gene divergence, e.g. for the split between myoglobin and haemoglobin and then between the various haemoglobin chains. This fact is usually completely ignored when molecular phylogenies are proposed.

The divergence between the α and β chains of haemoglobin presents particular difficulties because, following duplication of the initial haemoglobin gene, each copy has to diverge, and diverge more or less simultaneously, in *complementary* ways in order to produce a functional tetramer.

First is the introduction of hydrophobic amino acids. A major difference between myoglobin and the haemoglobins is that, whereas myoglobin has mainly hydrophilic amino acids on the outside of its folded structure (because it is a water soluble protein), both haemoglobin chains have some hydrophobic amino acids, which are necessary to permit association of the subunits. It is therefore proposed that the early haemoglobin was a monomer like myoglobin, with exterior hydrophilic amino acids, and that some of these were substituted by hydrophobic ones when the two different chains evolved. Dickerson and Geis suggest that some of these substitutions occurred before the gene duplication, i.e. that the early haemoglobin evolved into a tetramer before diverging into α and β chains (or possibly a dimer with just one of each chain). But this is not really a viable suggestion because the hydrophobic amino acids which occur on the outside of the polypeptides and interact with the other chains are different and in different positions in the α and β chains (whereas they would be expected to be in similar positions if the hydrophobic amino acids arose before divergence of the haemoglobins). This is precisely because they are complementary: the hydrophobicity is not required in the same place for each polypeptide, but at those sites which contact the other chain(s). For example, the α chains are not able to associate with each other: in the pathological condition, β-thalassaemia (where β chains are not produced), unassociated α chains tend to degrade, so there is no haemoglobin available, and the extreme form of this condition is usually fatal. On the other hand, β chains do associate (e.g. in α-thalassaemia) and have a comparable affinity for oxygen as myoglobin; but that is the problem – although some haemoglobin is available, the β chain tetramer will not release its oxygen to the tissues, and this condition too is invariably fatal before birth (life is sustained in the first few months prenatally by production of the ζ chains).

This leads us to the fact that it is not just being able to associate as a tetramer that matters: the *raison d'être* of tetrameric haemoglobin is the cooperativity between the chains which provides the sigmoid oxygen affinity curve described in Box 7.1. Many amino acids in both chains are essential for the correct functioning of this mechanism. This is illustrated by the various pathological disorders that can be attributed to a change in just one amino acid, the effect being to disrupt this cooperativity; and, importantly, many of the amino acids concerned in this interaction are at different positions in the polypeptide sequences, or are different amino acids at an analogous position. This evidence, and that of the thalissaemias, shows that half-baked solutions will not do – many amino acids need to be right in *both* chains for haemoglobin to be viable.

In summary, for family trees to be credible, most if not all of the putative ancestral sequences must be functional; but this presents a major stumbling block in the production by divergence of proteins with different functions. To get from one set of conserved amino acids to another is either an unlikely big jump, or the intermediates must have biological activity; but the latter seems unlikely because it contradicts what we know about conserved amino acids.

Intrinsic difficulties with duplication and divergence

Gene duplication is cited so often in evolutionary literature to explain observed genetic structures that it is easy to be given the impression that it is almost an everyday occurrence, or even happens on demand. But gene duplication requires a non-routine event, such as unequal chromosomal crossover, to occur at just the right place so that duplication of the appropriate gene results. Next, taking on board the comments in Chapter 5 about the fate of most mutations, there is a very high probability that the duplicated gene will not survive. In particular, until the duplication acquires a new useful function, it confers little or no selective advantage (because a simple duplication offers nothing new) so its spread through the population will be substantially by random drift, with the expectation that before long it will be lost. In fact, once it has lost its old function but not yet acquired a new one it is likely to impair fitness, which will hasten its demise. In effect this puts a limit on the time available for the duplicate to mutate to a new useful sequence. Also, during this stage, it is just wishful thinking that it will mutate only in a beneficial manner. As Kimura put it:

> This process that facilitates the production of new genes will, at the same time, cause degeneration of one of the duplicated copies. In fact, the probability of gene duplication leading to degeneration must be very much higher than that leading to production of a new gene having some useful function. [*Neutral Theory*, Ch. 10]

And even if a new useful sequence did arise, its prospects of survival are still bleak, and it would probably need to recur many times before it could spread through the population.

Further, it is not only that the polypeptide-coding part of the gene may mutate such that the protein end-product is useless: while the polypeptide has no useful function there is no selective advantage in maintaining the integrity of the gene regulation (in fact some advantage in not doing so, to save waste) so the gene itself may cease to be transcribed. This is one suggestion for how pseudogenes have arisen. When this occurs, no amount of mutation within the polypeptide-coding region(s) is of any value at all; the only hope is for a fortuitous (and very improbable) reverse mutation to restore the gene's functionality.

Gene regulation

Again, by focusing on the similarity of structure and function of haemoglobin and myoglobin it is all too easy to overlook the fact that they are synthesized in completely different tissues – haemoglobin in bone marrow and myoglobin in muscle. This means that, following the proposed gene duplication which gave rise to these separate lines, the substantial changes in the polypeptide-coding part of the gene had to be accompanied, at more or less the same time, with changes in gene regulation so that each protein is produced in the right tissue. Unfortunately we do not yet know enough about the regulation of gene expression and cell differentiation to spell out just what is involved in this. But it is evident that for successful production of the two types of molecule (even, for the sake of argument, accepting that a single polypeptide could perform the function of haemoglobin, rather than requiring a tetramer) requires an incredible coincidence of changes in gene structure and gene regulation. However, although I have read many accounts of the evolution of the globin genes, in none have I seen any reference whatever to the fact that myoglobin and haemoglobin are synthesized in different tissues and that additional changes are required in the genome to account for this. Neglecting such obvious implications of the proposed gene divergence is to bury one's head in the sand.

Multiple duplications

Despite the considerable difficulties involved in deriving two genes with distinct functions from a common ancestral gene, this phenomenon is advocated freely for the production of the globin genes.

Perhaps the most disconcerting discovery since the globin phylogenetic tree was first constructed was that of a myoglobin in lamprey which is a jawless fish (Agnatha, see Ch. 10). This is because it had previously been concluded from a comparison of amino acid sequences that the earliest globins were monomeric and led to the predominantly monomeric haemoglobins of the various invertebrate lines such as insects and molluscs, and of the jawless fish, and only then did the globin gene duplicate to give the separate myoglobin and haemoglobin of the higher (jawed) vertebrates. As shown in Figure 5, the occurrence of a distinct myoglobin in lamprey requires that the duplication to provide myoglobins and haemoglobins (with all that is involved in achieving this) must have occurred at least twice! – in the jawless fish as well as the jawed vertebrates.

Similarly, it is proposed that each of the α and β chains also duplicated to provide the ζ and ε chains which are expressed in very early embryonic stages. Of additional interest here is that similar embryonic haemoglobin chains occur in birds, designated π (pi) and ρ (rho) respectively. The ζ and π chains are so different from the corresponding α chains (but relatively similar to each other) that it is assumed the divergence to produce these embryonic forms is extremely old – preceding the divergence between birds and mammals. However, modern-day ε and ρ chains are much more similar to modern mammalian and avian β chains respectively, so it is presumed that the split to give these embryonic versions occurred at least twice – in each of the mammalian and avian lines.

In addition, it is presumed that after the initial duplication of haemoglobin to produce the α and β chains, the β chain had to duplicate again to provide the γ foetal version which (in conjunction with the normal α chain) has a higher affinity for oxygen; and of course that this was accompanied by changes in gene control such that γ is expressed during pregnancy but synthesis of the β chain is suppressed until birth. It makes such good sense to have a foetal version of haemoglobin which can extract oxygen from maternal blood, that, with an evolutionary backdrop to the modern way of thinking, it is all too easy to assume that mutation and natural selection could readily find this sort of solution. But such a simplistic outlook totally ignores the biochemical implications which we are beginning to unravel.

What these inferred scenarios highlight is the way in which so much weight is given to a phylogenetic interpretation of comparative amino acid sequences, and without any regard to the implications. It is not just that the proposed genetic changes are highly improbable, but as we learn more of what is involved (and some of this is outlined in the next chapter) it is increasingly apparent that the nucleotide changes necessary to implement the scenarios far outnumber the nucleotide differences between the amino acid sequences that the scenarios are trying to explain. In other words, the proposed solution is worse than the problem!

A further example of this comes from the foetal haemoglobin in the cow. The bovine γ chain is more similar to the bovine β chain than is generally the case between β and γ chains in other mammals. Consequently – blinkered by their focusing exclusively on comparative amino acid sequences – Dickerson and Geis concluded that the bovine γ chain is not a 'true' γ chain (even though it is expressed antenatally), but a later duplication of a β chain, which has nevertheless acquired a foetal function. Given what we already know about gene regulation, it seems highly likely that the changes required to ensure antenatal rather than postnatal expression vastly exceed any similarity between the β and γ sequences.

If nothing else, we can hope that elucidation of the molecular mechanisms of gene regulation will at least provoke a serious rethink of the current widespread simplistic phylogenetic interpretations of comparative amino acid sequences.

The difficulties involved in gene duplication and divergence are such that it seems the only reason why the phenomenon is advocated is because of the commitment to an evolutionary explanation for the origin of genes: despite the difficulties, this seems a less improbable route than starting from scratch. Somewhat paradoxically, biologists are willing to accept the statistical argument that it is too unlikely for two proteins to arise by chance with similar sequences, but unwilling to give weight to the statistical argument that they are too unlikely to arise at all by chance.

Short primordial proteins

Because of the hopelessly small probability of obtaining a biologically active macromolecule of realistic size, a necessary feature of generally accepted theories relating to the evolution of macromolecules is that the earliest forms were very much smaller than present day proteins. However, there are substantial objections to this sort of scenario.

Folding

First, there is a fundamental problem relating to the practicalities of polypeptide folding which, as has already been noted, is essential for protein function. A thorough account of the factors which need to be taken into account in folding a polypeptide to produce the biologically active protein can be found in various texts, such as *Structure in Protein Chemistry* by Jack Kyte. Here I can only summarise the main points.

You may recall from the preceding chapter that a protein will fold only if it has an appropriate mix and distribution along its sequence of suitable hydrophilic and hydrophobic amino acids, so that the hydrophobic ones can be packed together in the interior and leave the hydrophilic ones on the exterior. It is important to emphasize that for this to work the 3D fit, especially of the hydrophobic side chains in the interior of the folded molecule, must be exceptionally close so that many weak short-range forces can be established between them.[14] Recent studies have shown that the packing of atoms in the interior of biologically active proteins is the most compact known in organic chemistry – even more compact than in crystals of the amino acids concerned! This gives a fair indication of the specificity required – that the amino acids need to be the right shape, size and position so that they can interlock closely, and without straining the polypeptide backbone which links all the amino acids together. The energetics of protein folding are surprisingly finely balanced and Kyte concludes that for the hydrophobic effect to be adequate to stabilise a folded structure, a typical polypeptide requires a minimum length of about 70 amino acids. [15]

So, whilst there are short polypeptides which have biological activity in the sense of being hormones, transmitters or regulators, no proteins have anything like catalytic activity or a structural role unless they are large enough to fold into a 3D structure, in fact very few enzymes are shorter than about 100 amino acids. Descriptions of the proposed evolution of proteins from small polypeptides focus on the α-helices and β-sheets; they point to these as showing how some order can be formed in shortish polypeptides. But these arrangements arise only as part of a larger protein structure, i.e. a sequence of ten amino acids which may form an α-helix in a larger protein, will not do so in isolation.

An interesting illustration of this minimum size for folding comes from the hormone insulin. You may recall that insulin comprises two polypeptides of 21 and 30 amino acids (a total of 51) which are covalently joined by two disulphide bridges. If the two polypeptides are obtained separately and then mixed under typical physiological conditions, active insulin does not result – the two polypeptides seem unable to adopt the right configuration. However, as already mentioned, insulin is produced via proinsulin which is a single polypeptide of 84 amino acids. Proinsulin spontaneously folds into the correct configuration, the disulphide bridges are formed to stabilise the structure, and only then is the intervening C chain enzymatically excised to leave biologically active insulin.

It is also instructive to consider the case of the globins: These of course include a haem group which contains iron. Iron (chemical symbol = Fe) can exist in two oxidized forms: ferrous in which it has lost two electrons to become Fe^{2+}, and ferric where it has lost three electrons to become Fe^{3+}. When oxygen is available, iron readily oxidizes to the ferric form, e.g. common rust is ferric oxide. But the iron in myoglobin and haemoglo-

---

[14] Such as Van der Waals bonds which are effective only at a very short range.

[15] For readers familiar with the concept of entropy: the change in energy by associating the hydrophobic amino acids needs to be sufficient to counter the loss of entropy in going from a randomly orientated polypeptide to a tightly packed and ordered folded protein.

bin is in the ferrous oxidation state (even though their function is to carry oxygen, and in the presence of oxygen the iron would normally be expected to oxidize to ferric), and this is critical for its physiological role. According to Dickerson and Geis:

> The purpose of the heme and the polypeptide chain around it is to keep the ferrous iron from being oxidized (metmyoglobin, with a ferric iron, does not bind oxygen), and to provide a pocket into which the oxygen can fit. [*Hemoglobin*, Ch. 2]

In myoglobin and haemoglobin, the ferrous iron is prevented from being oxidized to ferric by the specific chemical groups of the amino acids surrounding it, and the way in which the iron is held in relation to the rest of the haem moiety. However, it is a fragile state of affairs, evident from the comments made by Dickerson and Geis in connection with obtaining crystallized haemoglobin:

> The met [oxidized] forms were experimentally the easiest to obtain, and the deoxy states also could be crystallized and studied with careful experimental techniques. The oxy forms proved more intractable: Unless extreme care is taken, O_2 oxidizes the heme iron from Fe^{2+} to Fe^{3+} rather than simply binding, thus yielding the unwanted met form of the molecule. [*ibid.*]

Not surprisingly, the amino acids around the haem group are generally highly conserved. And, as further confirmation of this delicate situation, many pathological conditions relating to haemoglobin arise because, due to changing just one amino acid, the polypeptide no longer retains the haem group correctly and allows the iron to oxidize. The amino acids concerned need not be adjacent the haem group, but may occur in various parts of the polypeptide chain, but replacement of which changes the positioning of the amino acids next to the haem so that they are no longer able to interact correctly with it.

Given this criticality of function and sensitivity to its disruption, and that the amino acids involved in retaining the correct position of the haem group occur throughout the molecule, it is very difficult indeed to conceive how a primitive functioning myoglobin- or haemoglobin-type molecule could arise with only a few amino acids, certainly not less than that required to permit polypeptide folding around the haem group. It is worth noting that this conclusion is based solely on a consideration of the proteins' primary function of reversibly binding molecular oxygen, not from any of the refinements of this function which the molecules also exhibit, such as cooperativity.

We can carry out this analysis for myoglobin and haemoglobin because we have a good understanding of their mechanism of action. It seems very likely we will come to similar conclusions for other enzymes as we acquire more detailed knowledge of how they function.

Critical amino acids occur throughout the polypeptide

A further major problem with the idea that proteins could evolve from small polypeptides is that the essential amino acids occur throughout the protein molecule. We have already seen that this is true for cytochrome *c* (p 154), and a similar picture emerges with the globins, or, for that matter, with any other protein one cares to consider.

It is not just that amino acids throughout the length of the molecule are necessary for the overall folding, but even the amino acids contributing to key parts of the enzyme such as binding prosthetic groups (e.g. haem) or constituting the active site, are also scattered through the sequence. If proteins had evolved from short polypeptides then one might have thought that at least these critical amino acids would still be grouped together, because to disperse them during the course of subsequent evolution would require restructuring of the protein, which would incur the same sort of improbability that the suggestion of small polypeptides is trying to overcome. In fact, to progress from short polypeptides, where the critical amino acids are necessarily close together, to longer polypeptides where they are dispersed, requiring some measure of activity at all of the intervening stages, seems to me to be overall even less probable than getting the required polypeptide in one go.

So when textbooks propose that proteins evolved gradually, adding amino acids one at a time, starting with very short polypeptides, it totally ignores what we know of protein chemistry. A typical example, taken from a leading textbook on evolution (Strickberger), is to illustrate the process by evolving the word 'EVOLUTION' by starting from just one letter and adding one letter at a time (selecting from a 'bowlful' of just ten possible letters, not even twenty to represent the number of possible amino acids), and assuming that at each stage (even when there are only one or two letters) the molecule has useful activity. This is not just wishful thinking, it is misleading – a gross misrepresentation of the facts.

Protein building blocks

A related idea for acquiring full size proteins is to build them up by piecing together short sequences of polypeptide.

A popular suggestion is the duplication of amino acid (nucleotide) sequences within a gene: wherever there is any indication of similarity in different parts of an amino acid sequence it is presumed they have arisen by internal duplication. However, what we need to recognise in the present context is that internal duplication in this way does not provide a route up the mountain. Whether or not amino acid sequences have been duplicated within genes in the past, this does not mitigate the improbability of biologically useful macromolecules. This is because mutation and natural selection have no prior knowledge of the (potential) utility of a duplicated sequence; that is, whatever the mechanism for duplicating sequences within a gene, it will operate on all sorts of sequences, not only the ones which, once duplicated, will prove useful. In other words, internal duplication is merely a mutation – just another way of producing random sequences. That is, there can be any number of sequences arising by duplication and the chance of having a useful one is no better than that for an amino acid sequence produced by some other means. A short sequence will be selected preferentially for duplication only if it has utility in its own right so that it is preserved by natural selection. But in most cases, all this does is set the problem back a step rather than solve it.

A favourite example to illustrate sequence replication within a gene is ferredoxin, which functions as an electron carrier somewhat like cytochrome c. Its polypeptide chain is folded over rather like a sandwich, with the top and bottom halves being very similar to each other, and four iron atoms in between. Because of the close similarity of the two halves it is confidently assumed that the molecule arose by duplication of an earlier half-molecule; and because there are repeats within a half, it is further assumed that this arose by earlier duplications of even shorter sequences. But such a scenario ignores the obvious question about the viability of the putative earlier molecules – and unless they had utility they would not have obligingly waited around for a fortuitous duplication event. A cursory inspection of the ferredoxin molecule shows that both halves of the molecule are necessary to hold the iron atoms. In view of the importance of sandwiching the iron atoms between the two halves, it seems doubtful that even half the molecule would work, never mind shorter fragments.

Finally, a brief comment is in order about the suggestion that large multifunctional proteins might have been built up by combining functional domains. Whether or not this has occurred, the important point to note is that domains are generally at least as large as small proteins – typically around 150 amino acids, but often larger. We have already seen that proteins of this size are very improbable structures, so there is no need to discuss this option further in the present context.

The blind watchmaker with foresight!

It seems appropriate to make some comment here on the speculative model of protein evolution proposed by Richard Dawkins in *The Blind Watchmaker*. The essence of his evolutionary process is that of cumulative selection: accepting that the complexity of biology cannot possibly be achieved in a single improbable jump, he argues that it can be achieved gradually by a series of small steps each of which has a much better chance of occurring.

When it comes to macromolecules, using haemoglobin as his example, Dawkins accepts it is too improbable to obtain a fully functioning protein in one go but proposes the following: Starting with a random sequence of amino acids of the right length, one assumes there is some way of detecting whether any amino acids in the sequence are the 'right' ones, and further assumes that for any that are 'right' there is some way of fixing them – of protecting them from mutating to something different while further random changes are made at all the other positions. With successive mutations progressively more correct amino acids arise (and, with his model, once a right one is found it is never then lost) so, as he points out, with a fortuitous cumulative selection of this sort it usually does not take very many rounds to arrive at a correct sequence.

If nothing else, this scenario at least avoids the general problems of short polypeptides, and faces up to the fact that critical amino acids can occur at various sites along the sequence. However, despite many comments in his book that natural selection has no foresight, this is exactly what Dawkins is imputing to it here; and even in subsequent discussions he offers no alternative mechanism. If there is any rationale at all behind his model it would be that if you get 1% of the sequence right then you can expect to have 1% activity, or at least some meaningful level of activity, and so on. But this is a totally false notion: the evidence is that you need the vast majority of the sequence to be right to have anything remotely useful; but, of course, getting the majority of the sequence right requires the extremely improbable event that he is trying to circumvent. And he does not even raise the question of how in the first place the length of nucleotides coding for the polypeptide is recognised as a gene and not just a random sequence, or even of why it should be destined to become haemoglobin rather than any one of thousands of other proteins.

—

What we need to recognise is that the various mechanisms outlined above are proposed solely to try to circumvent the daunting improbability of biological macromolecules. Any straw is grasped in the hope it may take us a step up the mountain. However, all of these approaches rely at some stage either on a fortuitous preservation and duplication or combining of subunits which by themselves have no value, and/or in some way imply prior knowledge of potential usefulness.

THE EVOLUTION OF MACROMOLECULES

I started this chapter by emphasizing that the biochemical understanding of biology which has emerged over the last half-century means it is no longer adequate to describe evolution solely in terms of morphology. For the theory of evolution, in its widest sense, to stand, it has to be viable at the biochemical level too – to provide an adequate explanation for biological macromolecules, ultimately for meaningful genes.

No doubt many molecular evolutionists will protest that they can perceive evolution at the biochemical level. But I put it to them, that all they demonstrate are relatively minor changes which occur within and sometimes between genes – but they do not provide a convincing account of how useful genes originate. And I repeat that I am not referring here pri-

marily to prebiotic chemistry: higher organisms require many macromolecules which are absent from lower ones, so evolution of the higher organisms would require the origin of those macromolecules within the context
of mutations to genetic material. Similarly, it is generally assumed that the
prebiotic phase applied only to the formation of some sort of replicating
system, and that most of the enzymes and structural proteins that constitute the fabric and metabolism of even lower organisms were obtained by
mutation of genetic material rather than arising abiotically.

However, there is a *prima facie* case that biological macromolecules are
extremely improbable structures and cannot realistically be attained at
random. Many biologists now recognise this, but argue that the improbability can be overcome by macromolecules developing progressively, i.e.
by evolution at the molecular level. In the preceding pages I have explained why this sort of proposition is not convincing, and I elaborate on
this in the next chapter.

I shall now return to my initial point regarding the improbability of
macromolecules. I am not going to present any more calculations, but
merely draw attention to points which emphasize, firstly, the specificity of
useful macromolecules, and secondly that acquiring useful genes is a
much more formidable process than is generally recognised.

Specificity of amino acid sequences

From what has been said already, it will be apparent to readers that a
functional protein has to fulfil several requirements.

Of prime importance is the ability of a protein to fold upon itself into a
well-defined 3D structure – generally packing hydrophobic amino acids on
the inside with hydrophilic ones on the outside – because without this folding none of the other criteria has any hope of being met. It may sound
relatively straightforward simply to have a mix of hydrophilic and hydrophobic amino acids and arrange them appropriately, but to be effective the
fit has to be exceedingly close. At present, we are unable to design a protein that will fold – quite apart from it having any biological utility. In fact,
even when we know the amino acid sequence of a biological polypeptide
which we know does fold, we are currently unable to predict *how* it will
fold. Because of the freedom to rotate of most of the bonds within a protein, the number of distinct positions that a polypeptide can adopt vastly
exceeds even the possible number of amino acid sequences for the same
length of polypeptide. This means that trying to solve the folding of polypeptides is far beyond the capability of today's computers, and workers in
this area are trying to identify rules for polypeptide folding which will help
them to focus on the most promising options. The difficulty is that an
amino acid sequence that will fold is not readily distinguishable from a
random amino acid sequence, and there are not even short amino acid sequences that are reliably associated with specific structural features such
as secondary structures.

In all enzymes there are at least some amino acids in the sequence that are unconnected with active sites and the like, but appear to be critical solely for the purpose of enabling the protein to adopt an appropriate 3D conformation. Further, the energetics of protein folding are finely balanced and the ability to fold is readily lost; incomplete polypeptides often lack sufficient information to fold properly. For example, a truncated polypeptide of bovine ribonuclease (124 amino acids) that is missing just the last six amino acids is unable to produce a folded protein with enzymatic activity. There is no doubt that, as some reviewers have put it, 'a good fold is a rare fold', and the criterion for folding, by itself, is probably enough to defeat any attempt at finding a protein randomly.

However, folding alone is of course insufficient. It is obvious that for a protein to have any biological utility it must make intimate contact, selectively and constructively, with other molecules – perhaps substrates on which it acts, perhaps other macromolecules with which it interacts. These interactions require the folded protein to have parts of its exposed surface to be of appropriate shape and chemical character (e.g. positively or negatively charged) to match the other molecule(s). An indication of the specificity required for this is that the mated surfaces of two polypeptides are as densely packed as the hydrophobic cores of proteins. Further, it is of no use to bind the other molecule in any haphazard fashion, but these surfaces must bind an appropriate part of the other molecule, and hold it in an appropriate orientation with respect to other reactants and/or the enzyme's own active site.

Which brings us to the active sites. Enzymes are astonishingly good catalysts, and (as outlined towards end of Ch. 6) they achieve this by having appropriate reactive groups positioned in such a way as to interact with the reactants, to facilitate the desired reaction. Hence, active sites require specific amino acids, with just the right (type of) side chain to be present at a fairly precise location to enable appropriate interaction with the substrate(s). So it is not surprising that generally the active site is responsible for several amino acids being absolutely invariant. And, of course, it is the requirement for precise 3D positioning of these amino acids which imposes constraints on the rest of the protein molecule. Related to the idea of the active site is that many enzymes require some non-protein moiety to function, e.g. cytochrome c and haemoglobin which use haem, and hence must have appropriate amino acids in the right places to bind these.

Another of the remarkable properties of many proteins is the way in which their activity involves changes in conformation. We have already seen something of this in the co-operative interaction of the haemoglobin chains and how this is modified by DPG (Box 7.1), and will see several more in due course. Almost all enzymes have their activity modified one way or another by binding other molecules. All of these capabilities require not only surface recognition sites for the modulating molecule, but also the right internal structure for their implementation.

So far in this summary of their specificity I have considered proteins simply in terms of their amino acid sequences. When we turn to the genes which encode them, the problem of their specificity is compounded. First, the intervening processes of transcription and translation mean that the link between genotype and phenotype is indirect, and underlines the fact that utility of a gene can be determined only retrospectively, e.g. by natural selection, and there is no possibility of a gene somehow being encouraged to mutate in an advantageous way. Further, not only must the core base sequence encode the right amino acid sequence, but the wider base sequence (mainly upstream of the codons) must include appropriate regulating sequences, such that the overall sequence is recognised as a gene and transcriptable, and to ensure that genes are transcribed in the right tissues and at the right time. I shall say more on the mechanisms of gene regulation in the next chapter. Whilst the prime focus of these comments is proteins, this is a convenient point to mention that similar considerations apply, of course, to the genes coding for such as the ribosomal and transfer RNAs which are generally even more conservative than those for proteins.

Fisher's multi-dimensional adaptive peak

It is interesting that Fisher's comments on the effect of random mutations on adaptation (mentioned above in connection with the neutral theory) are pertinent to the specificity of macromolecules, even though his comments preceded knowledge of the structures of proteins and nucleic acids.

You may recall that he showed there is a 50% chance that a random mutation will effect an improvement in fitness, provided the change caused by the mutation is small enough. He demonstrated this conclusion by considering a 3-dimensional space of fitness and adaptation, but in Box 7.3 I have illustrated his argument by starting with just one dimension. From the one-dimensional case it can readily be appreciated that, where fitness depends on only one factor, any random change in that factor has an equal chance of moving the organism towards or away from the optimum, provided the change in fitness caused by the mutation is not so large that it overshoots to well beyond the optimum point. With two factors affecting fitness (equivalent to two dimensions), the approximately 50:50 chance applies only for small changes, because there is a possibility that the mutation will take it sideways (figuratively speaking) rather than directly towards or away from the optimum. And as the number of factors affecting fitness increases there is a greater possibility for lateral move ment (in multiple dimensions), and the probability that a random change will effect improvement is approximately inversely proportional to the number of factors.

Box 7.3   Fisher's model of adaptation

Fisher (1930, Ch. 2) envisaged adaptation in three dimensions (degrees of freedom), but to appreciate his point it is probably easier to start with one.

One dimension

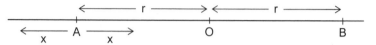

The horizontal line represents an organism's fitness in its environment, with the point of optimum fitness (adaptive peak) being at O, and the organism's current fitness is at A. Assume the organism acquires a mutation, one that is random in the sense that it is equally likely to move it in either direction along the line of fitness, and the magnitude of the change in fitness is represented by the distance x. It is apparent that if the mutation has a relatively small effect, then it is equally likely that the mutation will improve or impair fitness, i.e. the chance of improvement is a half. This is true for mutations which give a change in fitness up to the value of 2r (including when x equals r). Once the change in fitness exceeds 2r, then even if the change is in the direction towards O, it will actually be carried beyond B (the point on the other side of the optimum of equal fitness to A), resulting in it being less fit than at its starting point. This is the basis of Fisher's conclusion that, for a reasonably well adapted organism, the chance of a small mutation being beneficial is about one half.

Two dimensions

Now consider two dimensions, which would correspond to fitness being dependent on two independent factors. Fitness is represented by a plane with the organism's current position being A in relation to the optimum at O. A mutation resulting in a change of fitness of magnitude x will now take the organism to a point on the circle of radius x. It can readily be seen that less than half of the circumference of this circle lies within the circle of radius r centred on O, so the probability of effecting improvement must be less than 0.5. For comparison with the one dimensional case, when x equals r, the chance of being closer to O, i.e. to have effected an improvement in fitness, has fallen to one third (1/3), compared with a half in the case of one dimension.

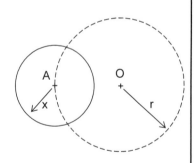

Three dimensions

For three dimensions, consider the circles in the above diagram to be spheres instead, so a mutation of magnitude x will take the organism to a point on the surface of a sphere of radius x centered on A. For comparison with the previous cases, the probability of a mutation of magnitude r effecting improvement has fallen to one quarter (1/4).

continued...

Box 7.3 *continued*

Multiple dimensions

For many dimensions, the probability of a mutation of a given size effecting improvement is approximately inversely proportional to the number of dimensions (factors) concerned. In particular, for a mutation of magnitude r, the probability of effecting improvement is $1/(n+1)$, where n is the number of dimensions.

Fisher pointed out that, once fitness depends on several factors, it becomes increasingly important that the various factors are consistent with each other, and it is much more likely that fitness will deteriorate as a result of a random change in any one of them, even a rather small change, than it will be improved by it. He illustrated this with a microscope, stating that a significant change to any of its components (e.g. position or shape of a lens) is far more likely to impair its performance than improve it, 'while in the case of alterations much less than the smallest intentionally effected by the maker or operator, the chance of improvement should be almost exactly half.' Note that this makes the size of changes with any chance of effecting improvement very small indeed.

We can readily visualise adaptation in a spatial way only up to three factors, corresponding to our familiarity with a 3D world. But Fisher recognised that 'The representation in three dimensions is evidently inadequate; for even a single organ, in cases in which we know enough to appreciate the relation between structure and function, as is, broadly speaking, the case with the eye in vertebrates, often shows conformity in many more than three respects.' And where fitness does depend on many factors, because of the close agreement required of the various factors, the adaptive peak (recall the concept of adaptive topology introduced near the end of Chapter 5, but now we are thinking in terms of getting the right combination of nucleotides in a gene rather than the right combination of genes) becomes very tightly defined and with very steep sides, i.e. this is another way of saying that even small changes are likely to throw it off its adaptive peak.

This concept is especially relevant to the structure of biological macromolecules because they consist of long sequences of amino acids or nucleotides, with each position being a factor that affects fitness, which means they have very many, even hundreds of, degrees of freedom (dimensions). Accordingly, their adaptive peaks are very tight indeed, so it is not surprising that we find most mutations are detrimental. The particular importance of this in the present context is that it emphasizes that the adaptive peak of a functional macromolecule stands out distinctly from the surrounding plateau of totally non-functional sequences, with no indication of which direction to move in order to find the peak. If the topography around an adaptive peak were gently sloping towards it, then it might be argued that a protein could gradually evolve towards the peak – mutations may have a 50:50 chance of improving performance and in time could move towards the peak through the action of natural selection. But Fishers' work provides a theoretical basis for what we have since learnt from experience – that most of the amino acid sequence has to be right for the protein to have any function at all.

The gene steeplechase

It is evident that the calculation with which I started this chapter was exceedingly overoptimistic in terms of how rapidly new sequences might be generated. That was because I wanted to emphasize a point – that no matter how favourable a scenario one envisages, it just is not feasible that useful biological macromolecules can be obtained by chance. It is now time to consider a more realistic picture. Mainly it is to note that it is not sufficient just to produce variations, but to be of any evolutionary value useful ones need to be adopted by the species' population at large, and, as we saw from the outline of population genetics in Chapter 5, there are various hurdles in the way of achieving this. As indicated above, in so far as the following comments need to be given a context, it is primarily in the sense of new useful genes arising to enable evolution of lower organisms to higher ones.

With this sort of scenario, it is apparent that the available resources are limited to the prevailing genetic material. However, not only is the total amount of this strictly limited, but only a relatively small part of it is actually available for trying out new sequences to search for new useful functions. Obviously, this is partly because some of it is tied up in providing the genes required for the organism's present functioning – tampering with material which encodes essential proteins or RNAs would generally be detrimental and therefore excluded by conservative natural selection. Also, when it was first realised that much DNA does not seem to be expressed, this seemed to provide a promising breeding ground for trying out new sequences; but we have now found that most of this is occupied by multiple copies of the same sequences – so clearly it is not used in this way.

So far as the rate at which new variations are produced is concerned, it is no longer relevant to think in terms of the rate at which polypeptides or nucleic acids are synthesized. Instead, we are totally dependent on the occurrence of mutations, which are thought to arise mainly from miscopying when DNA is replicated. Despite the critical role of mutations in evolution, in fact there is much uncertainty as to their rate of occurrence, partly because it is so low. The generally accepted figure is about 10^{-9} per generation for each base position, which, for a gene of 1000 nucleotides (coding for about 300 amino acids) gives a mutation rate for the gene of about 10^{-6} which is in broad agreement with the figure arising from the early genetic studies with e.g. *Drosophila*. It is the slow rate of mutation that makes it so unlikely that you could get even a handful of specific mutations at about the same time and place; and certainly not enough to, for example, convert a monomeric haemoglobin into two co-operating chains.

When it comes to considering the fate of mutations, a point that cannot be overemphasized is the overriding criterion that the base sequence in question be recognised as a gene, having appropriate sequences to enable its transcription. Without this, no matter how potentially useful a sequence may be (e.g. coding for a viable protein or RNA), it is utterly useless. This means that in order to have any hope of utility arising from a random sequence of bases requires, not only that part of the sequence code for a potentially useful macromolecule, but the incredible coincidence of having a potentially useful sequence associated with appropriate control sequences, arising at more or less the same time. Getting either one aspect right by itself is not sufficient. It is clear that a potentially useful sequence will not wait while the control sequences arise: because it is unexpressed there is absolutely no mechanism (such as natural selection) to conserve it, and it will simply mutate to something else. And the other way around is almost as bleak: i.e. if a control sequence arises which leads to transcription of part of the DNA, but the resulting product is useless, then there is no basis for retaining the controlling sequence (and some advantage in losing it, to avoid waste), so it is likely to degenerate by mutation long before there is any realistic hope of the coding region mutating to something useful.

It is worth noting that the situation where part of the DNA is translated but the product is useless is equivalent to a neutral mutation (or slightly detrimental). So, quite apart from the possibility of its loss due to subsequent mutation, it is most likely to be lost by the vagaries of inheritance (recall the fate of a mutation from Ch. 5). Further, even when the product is useful, most will be lost for the same reason – and so requires this incredible coincidence to recur many times before there is a reasonable chance that it will be adopted by the species. And, even if this extraordinary run of favourable coincidences did somehow arise, it would then take probably many thousands of generations for the new gene to spread through a reasonable proportion of the population such that the improvement could then act as a basis for further evolutionary progress.

I think that the general scientific community just has not been prepared to face up to the overwhelming odds against acquiring new useful genetic material. Of course, an attempt to improve the odds is to point to gene duplication as a means of getting some of the sequence right, ready-made; but, as I pointed out in discussing the case of the globins, it is far more likely that a duplicate gene will degenerate than be transformed into a new useful gene. And, it should be noted, even this theoretical route is available for only some genes.

SUMMARY

In this chapter I have demonstrated the high, and in some cases extreme, specificity of biological macromolecules and that, as a consequence of this specificity, the probability of obtaining a macromolecule with useful biological activity is so remote that the idea must be rejected. Although comparative amino acid sequences are frequently cited as evidence of protein evolution because of the molecular phylogenies which can be constructed from them, in fact they give no indication of how biologically active proteins arose. On the contrary, the picture emerging from protein families, the neutral theory, crystallographic data and detailed examination of enzymes which are considered to have evolved early on – all point to proteins being essentially complete and fully functional when they first appear. It is of course possible to suggest that proteins were evolving in the very early or even prebiotic period (so far back that there is no discernible trace in present day proteins); but that is just speculation – there is no evidence for it, and such a route cannot account for proteins which have arisen only in the higher organisms which are considered to have evolved relatively recently. Furthermore, the popularly presented views (even in serious textbooks) of proteins evolving from short polypeptides just is not tenable because of the minimum length of around 70 amino acids required for a polypeptide to be able to fold, and even the probability of obtaining something useful as short as this is much too low.

However, in spite of this weight of evidence, no doubt some readers will point to the operation of natural selection and the fossil record as proof of evolution – as proof that the difficulties raised in this chapter have in fact been overcome, even though we do not know how. I shall address these issues in Chapters 9 and 10 respectively. Before doing so, the next chapter develops the biochemical argument by looking at some further issues which are especially difficult to accommodate within an evolutionary framework.

8

Chance and Complexity

In the preceding chapter I presented the basic argument that biological macromolecules are extremely improbable polymers due to their high degree of specificity in the midst of unlimited potential diversity. In particular, biologically active proteins are such tiny and isolated islands of utility in a boundless ocean of possible but useless amino acid sequences, that it is not credible they could be happened upon by fortuitous drifting around. I introduced this point of view primarily with cytochrome c and the globins because they are well known and, being the proteins usually used to illustrate molecular evolution, it was important to point out the weaknesses in the evolutionary accounts that are built around them.

Some readers may already have been persuaded that there is a case to answer; but evolution is so entrenched in the modern mind that I am sure many more will remain totally unconvinced. Perhaps they will argue there is always the possibility that a 1 in 10^n improbability has nevertheless happened (and subjectively give undue weight to the possible billions of years and inhabitable planets). But the previous chapter was only scratching the surface, presenting the basic argument with relatively simple cases. In this chapter I will extend it to illustrate more of the complexity of biological systems – of individual macromolecules and of interactions between them – a combined complexity which completely defies any possibility of an evolutionary origin for biological systems.

One of the aspects I want to convey in this chapter is the impressive capability of some biological proteins, not just in terms of catalysing chemical reactions, but also as molecular machines. It is all too easy to think of enzymes only as catalysts, but the function of some is to carry out mechanical movement. This is, of course, effected by means of chemical reactions, but the end result is solely mechanical change – there are no new compounds, except perhaps products from the hydrolysis of ATP or similar used to power the mechanical change (see Box 8.1). An example is helicase which unwinds and separates the strands of DNA in order to expose the bases, and we will meet others which perform much more elaborate tasks than that. We are still in the early days of deciphering the action of these enzymes, but as we do so, and identify the structure and workings of their several active sites, I am sure it will challenge even further any notion that such molecules could arise blindly.

However, I am conscious that no matter how impressive a single enzyme may be, no matter how intricate or elaborate its mechanism of action, some will still argue that such complexity could develop progres-

Box 8.1   Adenosine triphosphate (ATP)

Whilst the subunits of DNA are nucleotides – comprising a nitrogenous base, deoxyribose, and a phosphate group (see Box 6.1) – the compounds used for DNA synthesis are trinucleotides, i.e. they have two extra phosphate groups. These are attached in turn to the preceding phosphate by high energy bonds, and the energy of removing them is used to form the bonds involved in adding the nucleotide to DNA. Adenosine triphosphate is one such trinucleotide.

ATP also has a wider role in cell metabolism. The energy of its additional phosphates is used to enable all manner of biochemical reactions which would otherwise be energetically unfavourable. Usually, only one of the extra phosphates is removed, to leave adenosine diphosphate (ADP). A few examples are mentioned in this chapter.

An important part of cell metabolism is the regeneration of ATP from ADP, using the energy from food, such as carbohydrate. In this, glucose is oxidized to carbon dioxide (CO_2) and at the same time oxygen is converted (reduced) to water. Whereas the same net reaction takes place immediately if glucose is burned in air, biological systems effect it through a series of intermediates, linking the reactions to the production of ATP; i.e., instead of losing all of the energy as heat, some is stored as ATP. It takes place in three stages:

First is the conversion of glucose to pyruvate, in a sequence of reactions called glycolysis which results in the net production of two ATP molecules (see Figure 12 for more details). The enzymes which carry out glycolysis occur throughout the cytosol, i.e. not connected with an organelle.

Mitochondria are the most abundant organelle in most eukaryotic cells, and it is here that the second stage takes place. As well as their outer membrane, mitochondria have an inner one, dividing the organelle into an inner matrix and an inter-membrane space. The inner membrane has many deep invaginations, so that its area is greatly increased and such that the mitochondrion looks packed with layers of membranes.

Pyruvate (from glycolysis) is transported into the matrix where it is converted to acetate. The latter does not exist as such, but is attached to a carrier molecule called Coenzyme A, so is present as acetyl-CoA which is a key intermediate in many biochemical syntheses as well having a role in energy metabolism. In the mitochondrial matrix acetyl-CoA is combined with oxaloacetic acid to form citric acid which, by a series of eight reactions known as the citric acid (or Krebs) cycle, is oxidized back to oxaloacetic acid, with the loss of two molecules of CO_2. Of particular interest in the context of this chapter is that all of the enzymes involved in the cycle are closely associated with each other, such that the substrates are passed directly from one to the other without being free in solution. Importantly, in the course of the cycle four carrier molecules (mainly NAD) are converted to their reduced form (NADH). Glycolysis and the conversion of pyruvate to acetate also generate these reduced compounds.

continued...

Box 8.1 *continued*

Which brings us to the final stage in which these reduced carriers are re-oxidized. This involves a series of proteins which act as electron carriers: each is reduced (gains electrons) by oxidizing its predecessor, and in turn is oxidized by (loses electrons to) its successor; and the final member of the series reverts to its oxidized state by converting oxygen to water. Most of these carriers are located in the inner mitochondrial membrane and, as they pass electrons along the series, hydrogen ions (H^+) are removed from the mitochondrial matrix, resulting in a concentration gradient of H^+ across the inner membrane. Also situated in the inner membrane is a protein complex, comprising many subunits, which allows the hydrogen ions to pass through it back into the mitochondrial matrix and, as they do so, it converts ADP to ATP.

An approximate analogy to this system for the production ATP would be a chain of people on a flight of stairs, passing buckets of water to each other, with the one at the top pouring the contents into a tank. A pipe drains water from the tank, returning it to ground level, but passing it through a water-wheel which can be used to do useful work, such as a turbine to generate electricity.

Cytochrome *c* is one of this series of enzymes, by far the smallest. Unlike the others, it is not membrane-bound but free in the intra-cellular space where it shuttles between two different sub-groups of the other enzymes.

sively. So I will also discuss here situations which we often find in biological systems that present a major obstacle to this generally accepted evolutionary explanation. One of the notable features of biological systems is the interaction and even mutual dependence of many components, as is well known in ecological terms at the level of the whole organism. We will see that this also occurs at the biochemical level, and this is important when it comes to accounting for the structure and function of macromolecules.

There are many instances where several proteins are required to implement some function within a cell, whether acting concurrently or consecutively, and the proteins concerned have no other role. It is very difficult to envisage how systems such as these evolved, because it would seem that many if not all of the components need to be present for any one of them to have any use. That is, if a particular function requires even just two components, then there is no selective advantage in having one with its activity (whether partial or complete) without the other being present with at least some of its activity. This is because natural selection does not have foresight: one part-formed protein is not going to wait for its companion to develop – if there is no utility, no fitness advantage, then its amino acid sequence will simply continue mutating blindly, and the likelihood is that any activity that it might have had will be lost. Hence, some significant activity is required of both components before there is any selective advantage for either.

It is worth pointing out that if the chance of finding a useful macro-molecule is say 1 in 10^{40}, then the probability of acquiring two is 1 in $10^{40} \times 10^{40} = 1$ in 10^{80} (not 1 in 2×10^{40}); and to obtain them together so they may interact is even less likely, considerably so. What are the chances of two distinct macromolecules arising, with complementary activity, at more or less the same time and place? This is the sort of question that proponents of evolution must face up to, and must provide a reasonably viable explanation if their case is to stand. To my mind, though I find the argument for just one macromolecule carries much weight, it is the examples of concerted activity of biological macromolecules, in many cases involving several proteins, not just two, that make the case conclusive. And there are many such examples; here I can outline only a few.

This concept of proteins acting in concert is taken even further when we consider multimeric proteins, i.e. where a functioning protein complex comprises multiple polypeptide chains (which may be the same or different); a situation which is in fact much more common than functioning proteins comprising just one polypeptide chain. For example, consider the single polypeptide of myoglobin: its role is interactive (e.g. its affinity for oxygen needs to be greater than that of haemoglobin), but it has no necessary direct or physical interaction with other proteins. But when it comes to vertebrate haemoglobin, there are four polypeptide chains which interact with each other in at least two distinct ways. First, the contacting surfaces need to be complementary, not just in terms of overall shape, but also in terms of hydrophobicity and surface charge, so that the chains can aggregate (and recall from the preceding chapter that this is usually an extremely close fit). Second, the interaction must enable the co-operative binding of oxygen – the binding of oxygen to one polypeptide causes a conformational change in that chain, which induces a conformational change in the other chains, which in turn promotes their binding of oxygen. Because of such interactions of the subunits, the sequence of one haemoglobin chain limits the viable or working options for the other – in fact they limit each other in this way. So, where proteins require multiple subunits in order to function (which is most of them), all of the essential components need to arise at more or less the same time and place, with compatible structures and capabilities to enable the interaction(s) between them. In this chapter we will see many examples of multimeric proteins where these sorts of limitations apply.

Hence, whilst some will argue that evolution does not have to find unique solutions (and with this I agree), in view of constraints such as those I have mentioned, it is not valid to presume the opposite extreme that the infinite number of possible amino acid sequences means evolution is bound to find a solution somewhere.

I will now describe some examples of the kinds of interaction I have been referring to. It should be noted that they are not from obscure little-used biochemical pathways; on the contrary, most are crucial to the functioning of cells, or are typical of how important biological systems work. It is only in the last few decades that we have begun to appreciate just how complex the processes of life are; it is this complexity that challenges the theory of evolution which was developed and generally adopted before we were aware of what biology entails at the molecular level. Whilst some of the molecular mechanisms are impressive, my reason for describing them is not to evoke a kind of subjective sense of wonder, but rather to emphasize that life at the molecular level is remarkably intricate, and necessarily so, and to provoke objective questioning of the evolutionary explanation for this complexity. As readers progress through this chapter I invite them to keep asking: Given the complexity and consequent improbability of each of the macromolecules involved, is it really realistic that these mechanisms could have arisen by chance, even in a progressive manner by fortuitous (opportunistic) trial and error? What routes might have been possible? Is any realistic? And how likely is it that it could actually have happened?

This chapter necessarily includes descriptions of various biochemical systems, and I am conscious that this will be new and rather technical ground for some readers. But I make no apology for this detail, because the complexity of biochemistry is the reef on which the theory of evolution founders.

DNA REPLICATION

Let's start by taking a closer look at the replication of DNA. As we have seen, the basic idea for duplicating DNA, with each strand being a template for the other, is very elegant, and is one of the reasons why the proposed double helix structure with its complementary strands was readily accepted. It is conceptually so straightforward that one might be lulled into thinking that DNA replicates so easily it could almost duplicate itself. However, a closer look soon shows that necessarily there is much more to it than first meets the eye; and as we start to find out how it is actually achieved in biological systems, we begin to appreciate just how complex a process it is.

First, because DNA in all living systems is either of medium length and circular (in which case there is no free end from which it can proceed) as in bacteria, or linear but very long (so it would take much too long to duplicate in a single stretch from one end to the other) as in eukaryotes, there needs to be some way of starting duplication within the body of a chain. This leads immediately to the next hurdle that, before duplication can begin, it is necessary to untwist and separate the two strands over a short region, and progressively thereafter as duplication proceeds. Then, because of the structure of nucleotides, a DNA strand can be synthesized in only one direction. And because the two strands of a double helix are oriented in opposite directions (see Box 6.1) only one strand can be duplicated continuously; the other strand must be duplicated discontinuously i.e. by a series of short lengths of DNA which are synthesized backwards (relative to the overall direction of duplication) and then linked up. These are termed the 'leading' and 'lagging' strands respectively, and the point at which synthesis takes place is called the replication fork. That sets the scene; the following paragraphs (and see Fig. 6) outline how DNA is replicated in practice, mainly based on what has been found in bacteria (*E. coli*), which has been more fully worked out, at present, than for eukaryotes.

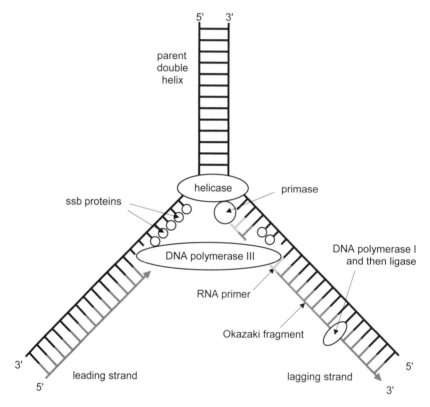

FIGURE 6. Replication of DNA: the major components at a replication fork.

To begin with, there are stretches of DNA which specifically function as origins for replication. There may be as few as one in a bacterial DNA, but hundreds or even thousands in a eukaryotic chromosome; and duplication of both strands proceeds in both directions from these origins. In bacteria the replication origin comprises seven distinct nucleotide sequences, each of 9 or 13 base pairs, spread over a total length of about 250 base pairs. These sites are recognised by a protein (designated dnaA), several of which bind to a single origin area and cause local distortion of the DNA helix, preparing it for uncoiling and strand separation. The DNA:dnaA complex is then recognised by a further protein, helicase (dnaB), which attaches to the DNA and progressively uncoils and separates it, using energy from ATP to do this (and its initial binding to DNA requires the assistance of a further protein, dnaC). As soon as the strands separate, small (ssb) proteins bind to the exposed bases to prevent rejoining of the strands while they await duplication.

DNA polymerases cannot initiate a new nucleotide strand, they can only extend one; so to get started a further enzyme, a primase, is required which attaches a short RNA primer to a strand of DNA. This enzyme is then released and DNA polymerase III (which is the principal enzyme for synthesizing DNA) takes over. DNA polymerase III has two catalytic sites and it extends both DNA strands more or less concurrently. Discontinuous synthesis of the lagging strand requires of course the repeated use of primase to reinitiate DNA synthesis on that strand. As the DNA segments (called Okazaki fragments) are synthesized, the RNA primers are removed by a separate enzyme, DNA polymerase I, which also fills in the gaps left by removal of the RNA (this enzyme also has an important role in DNA repair), and a further enzyme, a ligase, joins together all the bits. This overall process, especially with the discontinuous synthesis of the lagging strand, may sound cumbersome; but co-operation between the proteins – of which there are at least 20 different types – is such that, in bacteria, replication can proceed at the astonishing rate of up to 1000 base pairs per second!

There is one more feature of the polymerase to mention. Part of the overall polymerase III complex is two identical subunits, each shaped like a half-circle, which join together (with the help of smaller proteins) to form a ring which encircles a double helix of DNA. Without these subunits the polymerase will work, but keeps detaching from the DNA after relatively few base pairs have been added; with the ring, the polymerase stays in place and can put together a strand of up to half a million base pairs at a stretch.

So far as eukaryotes are concerned, the basic approach to DNA replication seems to be more or less the same. Perhaps the main difference is that the leading and lagging strands are synthesized by two different DNA polymerases. The polymerase synthesizing the lagging stand is closely associated with a primase to facilitate repeated initiation of the many fragments which need to be made, and possibly this polymerase with its primase initiates synthesis of the leading strand. But there is a further problem for eukaryotes with their linear chromosomes: the unidirectional nature of DNA synthesis (i.e. only in the 5' ⇨ 3' direction), and that all DNA polymerases need a primer, means that the last few bases on the lagging strand cannot be synthesized (because they need to be approached from upstream, but there are no upstream bases), and there would be a progressive shortening of the chromosomes at every round of duplication without some means to compensate for this. The solution is a telomerase ('end enzyme') which incorporates a length of RNA. Part of its RNA matches and pairs up with the last few bases of the DNA strand which will act as template for the lagging strand, but the RNA extends beyond the end of the DNA, and the enzyme adds additional bases to the DNA using its own RNA as a template. After this addition, the lagging strand is completed; and, although the last few bases of the template cannot be duplicated, because these originated from the telomerase, no information is lost from the chromosome.

There is one further practicality which requires consideration regarding DNA replication, though not actually part of the synthesis, and that relates to coiling and looping of the DNA helices. As duplication proceeds, the continuous untwisting of the chains at the sites of synthesis means that the parent DNA in front of the duplication fork becomes progressively more twisted. This becomes critical when two duplication forks converge from different directions – the DNA in between would become so coiled as to prevent its continued duplication. The problem of excess coiling is overcome by a set of enzymes called topoisomerases (Type I). These are U-shaped proteins which clamp on to a stretch of DNA, cut through one strand and, while holding the loose ends with the arms of the 'U', pass the uncut strand through the break in the other, and then rejoin the cut strand. A related problem is that double helices can become entangled, e.g. it is unavoidable that duplication of circular DNA results in the two daughter DNAs being interlinked. Type II topoisomerases are able to make a break right through a double helix, pass another double helix through the break, and then rejoin the first. Although these topoisomerases are not directly involved in synthesizing DNA, it should be emphasized that they are absolutely essential for the viability of both prokaryotes and eukaryotes, i.e. loss or corruption of these enzymes is lethal.

The above description has conveyed something of the complexity of DNA replication, and I trust dispelled any simplistic notion that replication could happen more or less automatically. This complexity is evident in the overall scheme with its interplay of many proteins, and in the function of some of its individual members, especially the polymerases and topoisomerases.

To avoid misunderstanding, I emphasize that I am not suggesting that the manner of DNA replication by biological systems is the only way it could be done, that evolution must find some unique mechanism. But on the other hand there seems to be no doubt that, whatever the solution, there must be a substantial minimum degree of complexity required. And in the context of the theory of evolution, the question has got to be asked explicitly: how reasonable is it to believe that a mechanism of this sort of complexity could arise by a trial and error process based on randomly generated mutations? It seems inconceivable that even some of the individual components could arise in this way, still less the suite of co-operating components.

It is worth saying a little more about the topoisomerases. Their activity includes binding a chain of DNA, chemically severing one or both strands, securing the loose ends while the other strand or helix is passed through the gap, and then remaking one or more nucleotide bonds. And, of course, the enzymes need to 'know' when to act. In fact, the topoisomerases have a wider role in controlling the degree of supercoiling of DNA – not only do they relax overcoiling caused by DNA duplication, but when appropriate they are able to increase the DNA coiling, with this latter process powered by ATP. In each of these applications the right piece of DNA must be passed through the break, and in the right direction; at present we have no idea how these 'minor' but essential details are determined. It is certainly appropriate to regard the topoisomerases as molecular machines, and it will be fascinating to see how they carry out their dextrous feats, once this has been elucidated. Can such intricate molecular mechanisms have arisen by chance permutations of amino acids driven by random nucleotide mutations? The stock answer is 'Yes – fine-tuned by hundreds of millions of years of evolution'; but increasingly this looks more like an expression of commitment to an evolutionary explanation rather than an objective assessment of the facts. In the previous chapter I demonstrated that the available resources are woefully inadequate to achieve a random solution to this sort of problem, even for a simple protein. How much more unattainable are the large complex units we have seen here.

Then consider the eukaryotic telomerases which include a length of RNA as well as the catalytic protein components. This exemplifies that many enzymes are not just polypeptides but incorporate other chemical groups which are essential for their activity, and illustrates that any random searching for meaningful proteins has to include exploring possibilities for combining with other types of molecules – such as haem or RNA. (The RNA incorporated in telomerase is not random but coded by a specific gene.) The latter point immediately raises the further problem that those additional compounds are not usually readily available but need to be synthesized, which almost always requires enzymes, which have to evolve in the hope that a use can be found for their end-product – and so the so-called reasoning goes on. At what point do we realise that this sort of 'explanation' is just trying to patch up a worn-out theory which really ought to be thrown out instead?

The replication of DNA also introduces the class of proteins which interact directly with DNA, especially those with the key role of recognising specific base pair sequences. These include the proteins (dnaA) which bind the origination sites for DNA replication and, as we shall see shortly, those that are central to the regulation of transcription. It is important to note that this is not base-pairing – the proteins recognise the base pair sequence from the outside of the double helix which remains intact. Obviously, in order to recognise a specific base pair sequence in this way requires a close fit between the 3D shape of this particular part of the helix, and that of the protein molecule, and interaction of complementary chemical groups. As we have seen already in connection with the shapes of proteins in general, the relationship between the primary sequence of amino acids and folded 3D structure is very indirect and relies crucially on the particular sequence – even small changes can totally alter the configuration of the molecule. Thus there is no way that the base pair sequence of the DNA binding site can direct the amino acid sequence of the binding protein; and, from an evolutionary perspective, we have to rely on random variations in nucleotides of the gene for the DNA-binding protein to produce an amino acid sequence which will bind the totally different sequence of nucleotides at the binding site. Then, of course, the role of these proteins is not to bind DNA and do nothing, they must fulfil a useful role too: sometimes this is to alter the conformation of the DNA e.g. to permit a subsequent protein to act on it, sometimes it is to interact directly with other proteins e.g. helicase with dnaA. Either way, at the same time as the DNA-binding protein evolves its unlikely task, it is necessary for at least one other protein to arise with a complementary function.

TRANSCRIPTION

As mentioned in Chapter 6, the term transcription refers to the process of making a strand of RNA to match part of the cell's DNA, irrespective of the type or subsequent role of the RNA that is produced. The regulation of transcription is particularly important because in both prokaryotes and eukaryotes the primary means for regulating the levels of proteins in the cell is to control production of the relevant mRNA. This is not surprising as it would generally make sense in terms of the cell's economy to control the first stage of the process. Other methods are used as well, but we need not consider them here.

Before embarking on further discussion it is worth noting the distinction that is generally made between so-called 'structural' proteins and 'control' or 'regulatory' proteins. The term 'structural' refers to what might be called the desired end-product, even though often this will be an enzyme rather than part of the 'structure' of the cell; and 'control' proteins are those whose principal role is in regulating production of the structural proteins. In general it is a useful distinction to make in discussions such as this, although the demarcation is not always clear-cut.

Transcription in prokaryotes

The enzyme that performs transcription is called RNA polymerase, and in bacteria there is just one form of this, which necessarily is used to produce all types of RNA. It is worth pointing out that this protein is among the largest in the cell, comprising four subunits (three different, but two copies of the smallest), having a total of about 3500 amino acids. A further subunit (σ, sigma) is required for locating the starting position on the DNA, but once RNA synthesis is under way, this subunit is released. As transcription proceeds, the polymerase separates the two DNA strands to expose the bases, makes an RNA copy of one strand for an appropriate length (allowing the DNA strands to rejoin behind it), and then releases the DNA and its newly synthesized RNA. So this enzyme alone, by virtue of its size, interaction between its subunits, and capability, poses a major challenge to any proposed evolutionary origin; and bear in mind that this is in the simplest forms of organism that we know. But what makes transcription really interesting are the mechanisms used for its control, even in the lowly bacteria.

The lac operon

The first proteins for which the control mechanism was worked out are three connected with the metabolism of the sugar lactose. One of these proteins (lac Y) facilitates entry of lactose into the cell; the largest (lac Z) is the enzyme β-galactosidase which carries out the initial split of lactose into glucose and galactose, which compounds can then be readily utilised as part of the cell's routine metabolism; and the third (lac A) is another enzyme, although currently we are not sure as to its role. In bacteria, the genes for these three proteins are next to each other, in series, with the only base pairs between them being those necessary to mark the end of one gene and the beginning of the next. Immediately upstream of the genes is a section of DNA known as a control region, and immediately upstream of that the DNA codes for another protein (lac I) – a 'control' protein which is involved in regulating transcription of the three 'structural' proteins. The complete stretch of DNA including the control and structural genes is known as an operon, and is illustrated in Figure 7.

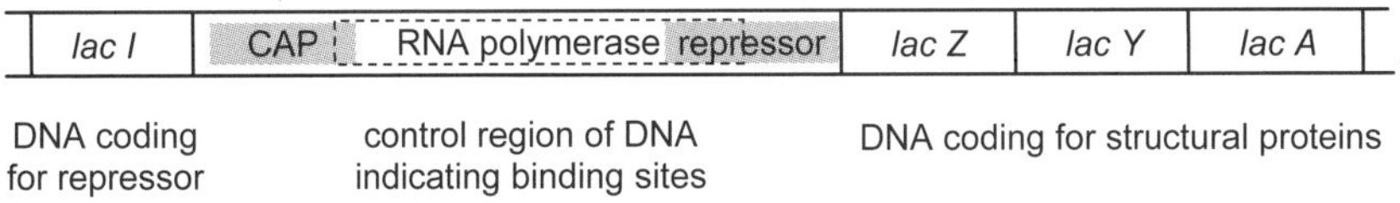

lac I	CAP ┊ RNA polymerase repressor	lac Z	lac Y	lac A

DNA coding for repressor	control region of DNA indicating binding sites	DNA coding for structural proteins

FIGURE 7. The *lac* operon. The lengths of DNA are not to scale: the control region is actually much shorter than any of the protein-coding regions.

The control region includes base pair sequences which comprise sites for the binding of three different proteins (none of the structural proteins). Most importantly, it includes the base sequences that are recognised as an initiation site by RNA polymerase, i.e. RNA polymerase can bind here via its σ factor and commence transcription from it of the three structural genes. However, at the downstream end of this initiation site is a base pair sequence that is recognised by the control protein. In its native state, this protein binds to the DNA here and prevents the RNA polymerase from doing so – hence it is called a repressor protein. But this protein will also bind the compound allolactose (a minor product of the action of β-galactosidase on lactose), and when it does so, it changes shape so that it can no longer bind the DNA. So this arrangement works as follows: when lactose is absent, the repressor prevents production of the mRNA, and hence of the proteins; but when lactose is available, some is converted to allolactose which inactivates the repressor, and allows transcription of the structural genes, permitting synthesis of the relevant proteins, and enables metabolism of the lactose. This, of course, is exactly as it should be – a very sensible arrangement to produce enzymes only when there is a use for them. But it is obvious that the efficacy of this control mechanism depends critically on the structure of the control protein, and it is worth pausing for a moment to consider its requirements.

First, of course, the control protein must be able to recognise and bind to a specific base sequence in DNA – and I have made comments already about how specialised a capability this is, and how unlikely to arise fortuitously. The repressor protein must also bind allolactose, and in such a way that it is no longer able to bind DNA. (In isolation, it may seem relatively easy to disrupt the DNA-binding capability, but we will see below that in other cases, where it is appropriate, the additional compound is required to *enable* DNA binding.) What makes the control protein so remarkable, of course, is the significance of the stretch of DNA it binds. It is the coincidence that the piece of DNA that the *lac* repressor binds happens to control expression of the *lac* genes which are relevant to the metabolism of the compound that affects the DNA binding of the repressor. And note that the location of the length of DNA the control protein binds is important: it must be upstream of the relevant structural genes but not upstream of their RNA polymerase binding site – which limits it to a very short stretch. In contrast, the gene encoding the control protein must be upstream of the control region so that the repressor is always available to play its part in the control process.

But that is not all. Recall that the control region has binding sites for three proteins. This is to enable a further tier of control over expression of the lactose genes, such that the genes for metabolising lactose are expressed fully only when the preferred food source, glucose, is unavailable. The third protein (designated CAP) assists the binding of RNA polymerase, but in order to function in this way it requires an additional compound called cyclic AMP (cAMP, which has many roles in higher organisms; one role in bacteria is to act as an 'alert' signal); and cAMP is abundant only when food reserves are low, but we need not go into how this occurs. The effect of this arrangement is that, when glucose is readily available, then cAMP levels are low, CAP does not bind the control site, RNA polymerase binds only weakly, and relatively little expression of the lactose genes occurs even if lactose is present. But if glucose is unavailable then cAMP levels are high, CAP binds the control site, and promotes the binding of RNA polymerase with subsequent transcription of the *lac* genes, subject to lactose being available so that the repressor is inactivated.

Other control systems

I have outlined the *lac* operon because it was the first to be worked out and possibly is still the best understood. There are now many other systems known, over 100 in bacteria, where enzymes are switched on only when the appropriate substrate is available. In each case there is a repressor protein which recognises the appropriate control sequence, but is inactivated by the relevant substrate.

Transcription is also controlled in the opposite way. For example, the compound tryptophan is an essential part of the cell's metabolism and bacteria have the means to synthesize it, but the relevant genes are expressed only if tryptophan is absent. So in this case the tryptophan control protein (now called an activator) will bind DNA only when it is bound by tryptophan, i.e. the presence of tryptophan switches the genes off rather than on.

Finally, you will recall from above that it is the σ subunit of the RNA polymerase that is responsible for identifying the appropriate point from which to start transcription. In bacteria almost all genes use the same sigma subunit, designated σ^{70}. However, a few other types of bacterial sigma factors are known, their use apparently related to the transcription of particular classes of genes, and we will come across an example later in this chapter. So clearly there are multiple control sequences, each with their matching σ factor, which enable a further level for controlling transcription.

Transcription in eukaryotes

When operons were discovered in bacteria, it seemed to make such good sense to group genes having related functions and to have a common control system, that it was expected similar arrangements would be the norm in eukaryotes. However, this turned out not to be the case. As we have begun to unravel the mechanisms controlling transcription in eukaryotes, we have found that it is much more complex than in prokaryotes.

An immediate indication of the greater complexity of transcription in eukaryotes is that all of them, from yeast to higher plants and animals, have three different RNA polymerases, used for the production of different types of RNA. One of them (RNA polymerase II) produces all of the cell's mRNA (i.e. used for protein synthesis), and the other two produce predominantly rRNA or tRNA. All three polymerases have four subunits equivalent to those of the prokaryotic RNA polymerase, plus six additional subunits which are common to all of them, and up to seven others which are unique to the polymerase concerned. Almost all of these subunits are absolutely essential for the correct functioning of the RNA polymerases, to the extent that their loss leads to cell death.

I will focus on RNA polymerase II because most work has been done on the functioning of this enzyme, and it compares most readily with the prokaryotic enzyme, although we will see that its mechanism for the control of transcription is much more elaborate. Whereas prokaryotes require a single sigma factor (usually σ^{70}), the eukaryotic RNA polymerase requires several ancillary proteins (initiation factors), which must be added in the right sequence, in order to recognise an initiation site and put together a viable transcription-initiation complex. Because most of these factors are multimeric proteins, the total complex has at least 40 polypeptides, comprising approximately 20,000 amino acids. As with the prokaryotic sigma factor, the initiation factors are discarded once transcription is under way.

One of these factors is of especial note. It is called the TATA-box promoter because it binds specifically to a TATA base sequence in DNA (see below). Its overall length varies substantially between different eukaryotic species, but the 180 amino acids at its C-terminal end (see Box 6.4) are highly consistent, with 80% of the sequence being identical between yeast and humans, and most of any differences being obvious neutral alternatives. This protein binds the minor groove of DNA (which is unusual, most bind the major groove) and causes a substantial bend of the double helix which is thought to be important to enable binding of RNA polymerase.

When it comes to the regulation of transcription there is still much to be worked out. In so far as we have elucidated the mechanisms to date, there are frequently used patterns and similarities to the mechanisms used in prokaryotes, but also many variations.

As with prokaryotes, in most cases there is a specific base pair sequence a short distance upstream of the gene to be transcribed, and at least for structural genes this usually includes the sequence TATA, which is bound by the initiation factor mentioned above. But in addition to this, for many genes there are several other base pair sequences in the 100 to 200 base pairs further upstream which also have a role in controlling transcription. As well as these, for each gene there are multiple other control regions, each typically 100 to 200 base pairs in length, and often remote from the gene they regulate, being as much as 50,000 base pairs distant, and may be up- or downstream of the transcribed gene, or even in an intron within it!

As might be expected, these control regions are binding sites for regulatory proteins – known as transcription factors. To begin with, by analogy with prokaryotic systems, it was thought that they would be either activators or repressors; but as more details of their interactions have emerged it has become evident that, depending on circumstances, the same factor or regulatory site may act to promote or inhibit transcription. It is thought that the way they act is by interacting with assembly of the transcription-initiation complex (and interacting with each other in their effects on this), with distant sites being be brought into play by the DNA looping back on itself.

So, especially considering the many units which constitute the RNA polymerases, together with the transcription factors which regulate its activity, and not forgetting the central role of the DNA itself, here we have an excellent example of the sort of situation I mentioned at the beginning of the chapter. It is not just that there are many macromolecules which have specific sequences that need to be 'found' by evolution, but most function only in collaboration with others – in isolation they have no utility whatever.

Incidentally, we should not be misled by the term 'factor' into thinking that the eukaryotic transcription factors are small. On the contrary, rather like the proteins contributing to the initiation complex, the eukaryotic transcription factors are large proteins. Typically they have a DNA-binding domain, which locates the relevant control sequence, and one or more activation domains which are thought to be the parts which interact with other factors and/or the initiation complex; and each of these domains is usually hundreds of amino acids long.

Already we have identified hundreds of these eukaryotic transcription factors, and there is little doubt we will find many more. Again, we are only just beginning to learn how they interact with each other and how they affect transcription, but the number of factors and their variety of structures points to the importance of specificity. An interesting observation is that the actions of some regulatory proteins appear to be limited to specific cell types – so we are beginning to see how cell differentiation is effected at the molecular level. Also, it is apparent that one of the ways in which transcription is controlled is by regulating production of the regulatory proteins, and obviously there could be many tiers of this sort of control.

We should not be surprised at the complexity of genetic control in multicellular organisms, to account for how one cell develops into many types with such diverse functions as those which produce hair or bone or muscles, conduct nerve impulses, or become sex cells with only half the genetic material, or red blood cells with no nucleus at all. (And the processes which effect this proceed reliably, an indication of which is the existence of identical twins – where the same genotype produces very similar phenotypes.) Our knowledge of biochemistry has accentuated the evident differences between these types of cells. Those differences alone should have made us question whether such specialisations could have arisen by fortuitous trial and error. Now that we are beginning to unravel the mechanisms of how development and differentiation take place, we see that the picture is even more complex. We may feel that complexity can accrue, progressively working up from lower to higher organisms – but adding just one extra control sequence with its binding protein which has the desired effect requires an incredible coincidence of improbable occurrences. And the evolutionary account requires that such coincidences have occurred many thousands, perhaps millions, of times.

Steroid hormone receptors

Many readers will be well aware that there are various steroid hormones, of which there are three main types. Glucocorticoids (e.g. cortisol) are concerned with regulating the body's metabolism, such as the synthesis and breakdown of proteins, carbohydrates and fats. They are produced in the adrenal glands (next to the kidneys), secreted into the blood (where their concentration is by far the highest of the steroid hormones), and have an effect on a wide range of tissues such as muscle and fat. Mineralocorticods (e.g. aldosterone) are involved in controlling the electrolyte balance (e.g. sodium) of bodily fluids. Like the glucocorticoids they are synthesized in the adrenal gland, but their main targets are the kidneys. Sex hormones tend to be predominantly male androgens or female oestrogens. They are produced mainly in the gonads, and affect many organs to implement secondary sex characteristics, and obviously have an important role in the female menstrual cycle, pregnancy and lactation (milk production).

Why the steroid hormones are of interest here is that they act by affecting transcription of appropriate genes in the relevant target tissue, this effect being mediated by proteins which act as both hormone receptors and transcription factors. All of the receptors share a common general layout that includes a hormone binding domain which is about 250 amino acids long at the carboxyl end of the protein, followed by a DNA binding domain which is about 70 amino acids long, and the amino end varies in length from about 100 to 500 amino acids depending on the hormone concerned. So their overall size ranges from 400 to 1000 amino acids. Clearly, despite these similarities of overall structure, what matters is the specificity of each region – that the hormone binding region is specific to a particular hormone, and the DNA binding region is specific to the DNA control sequence relevant to the action of that particular hormone. Then, although we do not yet know how the variable region functions, there is little doubt that that too will be specific to how the receptor interacts with other control proteins. Finally, it is also apparent that the relevant receptors are produced only in cells of the appropriate tissues, so this in itself requires some higher level of transcription control. That is, a tissue produces receptors which enable it to respond to particular hormones; then when a relevant hormone is present, it proceeds to transcribe the appropriate genes.

Much as I should like to have described some of the post-transcriptional changes to RNA (RNA editing) and some of the details of protein synthesis carried out on the ribosomes – which would have provided many more examples of cooperative activity of proteins and nucleic acids – limitations of space prevent me from doing so, and we must move on.

BIOSYNTHETIC PATHWAYS

Another sort of situation where proteins require several others to be available for any one of them to have any utility is in biosynthetic pathways. Suppose, for example, that we have a short pathway in which an end product, compound E, is made in a series of four steps from a starting substance A via intermediates B, C and D, which we could represent as A⇨B⇨C⇨D⇨E, and requires a separate enzyme for each of the four steps in the overall process. At first sight, it might appear that in a sequence of reactions such as this all four enzymes would need to be present with at least some activity in order for the pathway to arise. That this would be excessively improbable is generally accepted even by proponents of evolution, and it is usually proposed that the pathway did not appear all at once, but that it developed progressively, one step at a time, either forwards or backwards (or a combination of these).

In the back-to-front scenario it is assumed that the compound E is initially available in the environment and is useful to the organism e.g. it might have been an amino acid in the primeval soup (see Ch. 13). However, in time, the stocks of this compound decline. But if compound D is available, then an organism which acquires an enzyme that can convert D to E will be at an advantage: it will survive preferentially, along with its new enzyme which will gradually be adopted by the species as a whole. In time, compound D also declines, but if compound C is available then an organism which has an enzyme which can convert C to D will survive; and so on. In this way the whole pathway is built up until compound A is reached which is either freely available in the environment or already producible biologically. The obvious weakness in this argument is that the time for acquiring a new enzyme is severely limited: while E is readily available there is little if any selective advantage in possessing an enzyme which can produce it from something else (and some disadvantage in producing a redundant enzyme), but when E is used up it is too late. So there is a restricted window of opportunity, when the availability of E is limiting but it has not actually run out, in which the right enzyme can arise. So it is apparent that this sort of scenario completely ignores the improbability of biologically significant macromolecules and relies on the implicit tenet of evolution that if there is a need for something then the something is sure to turn up. Or, even worse, that the need somehow prompts the solution – a kind of Lamarckian way of thinking.

Then, there are many biosynthetic pathways where the end product has never been available non-biologically, so far as we aware; for these we need to propose a forwards scenario. Again, assuming that A is readily available, we speculate that an organism acquires an enzyme to convert it to B which happens to be useful, so possession of the enzyme confers a selective advantage. In time, an enzyme arises which can convert B to another useful compound, C; and so the pathway gradually builds up. At least this scenario avoids the time constraint of the first, because the new enzyme for each stage could be acquired at any time, provided the product for that step had a use at that time. Which brings us to the main difficulty with this scheme (apart from the improbability of macromolecules anyway), that it requires most if not all of the intermediate compounds in a pathway to have had utility in their own right (i.e. not merely as an intermediate to some future product), so that having each additional compound conferred some selective advantage – otherwise there is no basis for retaining the enzyme that makes it. But there are many biosynthetic routes where certainly there is no vestige of such utility, and it seems unlikely that there ever was. A prime example of this is the biosynthesis of steroids.

Biosynthesis of steroids

In animals cholesterol is a key precursor of steroid hormones and as such shares their polycyclic structure; it is also an important component of their cell membranes, especially in higher taxa such as mammals where it may constitute 25%. Other eukaryotes such as plants and fungi use the related stigmasterol and ergosterol in their membranes instead. Prokaryotes do not synthesize steroids, but a few (some mycoplasmas) make use of small quantities which have been synthesized by eukaryotes. It is evident (see Fig. 8c) that the steroids have a sophisticated chemical structure, and not surprising that they do not arise spontaneously but that the only known sources are biological. This means that, of the two types of scenario just outlined, the back-to-front rationale cannot apply, so something along the lines of a progressive build-up of the synthesis would have to occur.

What makes the biosynthesis of cholesterol especially interesting is the way in which the polycyclic structure is made: it is not built up one ring at a time, but all in one go from a linear molecule called squalene which is 'polyunsaturated' which simply means it has several double carbon-carbon bonds (C=C), and it is these which make the ring-formation step possible. Squalene has 30 carbon atoms, as does the initial polycyclic product, lanosterol, which must subsequently be modified to produce cholesterol which has only 27 carbon atoms.

Acetyl-CoA (see Box 8.1) is a source of carbon atoms for many biosynthetic pathways – we could perhaps think of it as like compound A in the above schematic pathway. It is therefore unexceptional that it should be the basic building block for making steroids; in fact, the product of joining together three acetyl groups is the compound HMG-CoA which is also used in many biochemical pathways, so I will start my account of cholesterol synthesis from here. This is also the key point at which biological control of steroid biosynthesis is exercised.

The biosynthetic route from HMG-CoA to squalene and then lanosterol is shown in Figure 8 (a to c). in which each step requires a specific enzyme to carry it out. The first step is removal of the CoA to produce mevalonate,

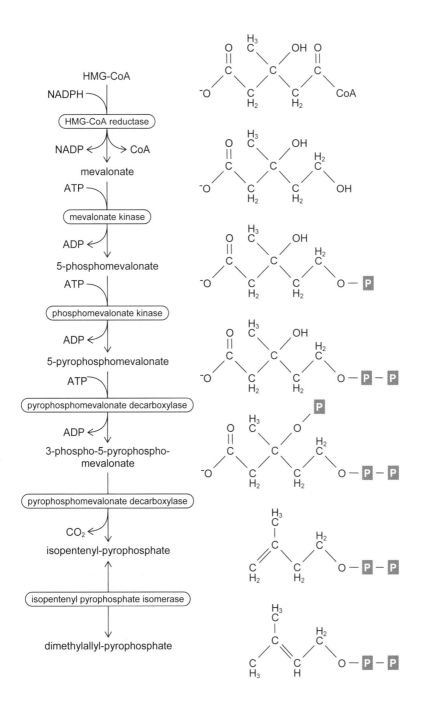

FIGURE 8a. Biosynthesis of cholesterol (1) : HMG-CoA to dimethylallyl-pyrophosphate. P represents a phosphate group.

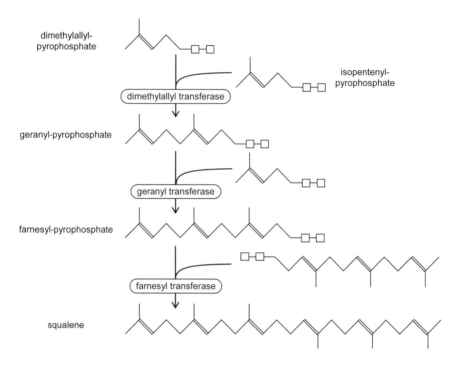

F<small>IGURE</small> 8b. Biosynthesis of cholesterol (2) : dimethylallylpyrophosphate to squalene.

and this is followed by the successive addition of two phosphate groups to one end of the molecule, providing a high energy pyrophosphate (PP) group (comparable with the extra phosphates of ATP, see Box 8.1) which will be used to drive some later steps in the synthesis. Before that, a third phosphate group is added to the middle of the mevalonate and then removed, taking with it one of the carbon atoms (as CO_2), both steps being carried out by the same enzyme, thereby introducing the important double bond.

The ensuing product is isopentenyl PP, and some of this is isomerized (rearranged without adding or removing any atoms) to dimethylallyl PP. Using the energy of its pyrophosphate group, this compound is then combined (head to tail) with a molecule of (unisomerized) isopentenyl PP, to produce geranyl PP which is added to a further molecule of isopentenyl PP to give farnesyl PP. In effect, three isopentenyl groups have been joined in series.

With its important double bond and reactive pyrophosphate group, isopentenyl PP is used by many organisms (plants and animals) to produce a wide range of compounds, ranging from vitamins to rubber. Similarly, farnesyl PP is used in a few different ways apart from the synthesis of steroids, though this is by far the most important. After farnesyl PP, the

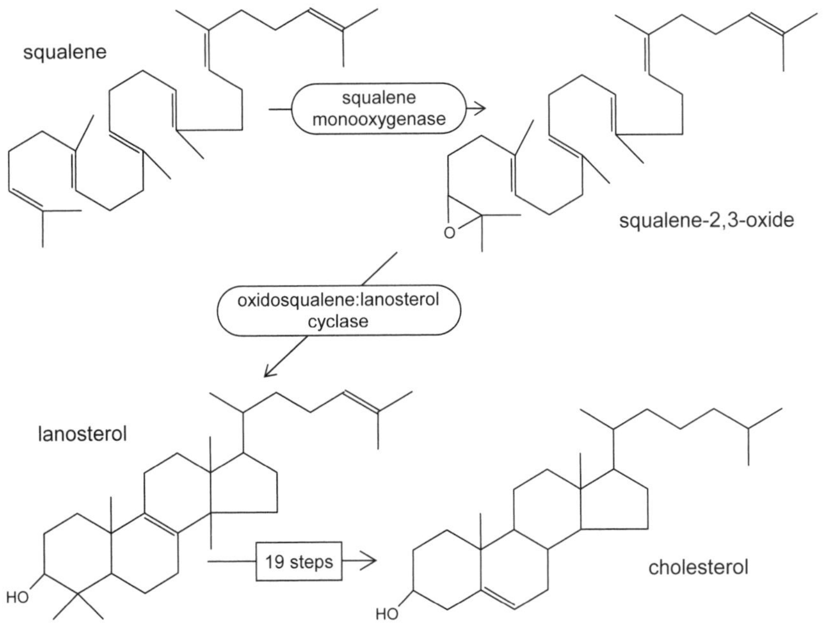

FIGURE 8c. Biosynthesis of cholesterol (3) : squalene to cholesterol.

biosynthetic route is used exclusively for the synthesis of steroid-like compounds.

In the route to steroids, two farnesyl PPs are joined (head to head) with
loss of both pyrophosphate groups to produce squalene, using the enzyme
farnesyl transferase. Although squalene has its double bonds in the right
places to enable the steroid cyclic structure to arise, cyclization does not
happen automatically but the molecule needs to be primed. This is done by
addition of an oxygen atom to one of its double bonds to form an epoxide,
a step which requires molecular oxygen in conjunction with the reducing
action of a compound called NADPH, and is implemented by squalene
monooxygenase. Despite the reactivity of the epoxide group, cyclization
still does not proceed spontaneously, but is mediated by the enzyme oxi-
dosqualene:lanosterol cyclase. This opens up the epoxide group, thereby
initiating a sequence of reactions within the molecule in which, rather like
a domino effect, one double bond after another opens up and rearranges
to form a ring of carbon atoms. The end result is the steroid lanosterol.

Recognising that all of the compounds up to farnesyl PP are involved in
various biochemical pathways, a case could perhaps be made that all of
these intermediates had utility in their own right, and so it might be argued that the synthetic pathway could be built up one step at a time, in the

proposed evolutionary fashion, up to that point. But from farnesyl PP to lanosterol requires three separate enzymes, operating sequentially, with no known utility for the intermediate compounds except as intermediates in steroid synthesis. So production of the polycyclic steroid structure requires at least these three enzymes to arise, each with some significant level of activity and all at more or less the same time. This alone would require one of those extraordinary coincidences which we have met elsewhere, and I think that the formation of lanosterol (even without regard to the elegance of how it is formed from squalene) is more than enough to challenge the accepted evolutionary wisdom about how biosynthetic pathways might have originated.

But the astounding fact which takes the argument beyond any reasonable doubt is that to remove the surplus three carbon atoms and make a few other changes in order to convert lanosterol to cholesterol requires a further *nineteen* biosynthetic steps, each having a separate enzyme! Currently there is no known use for any of the intermediate compounds except as intermediates to cholesterol. Steroid chemistry is complex and there is always the possibility that another role for some of them may be found, but certainly it will be minor compared with that of producing cholesterol, and cannot be used to 'explain' how the pathway might have arisen progressively. Similarly, even if some of the enzymes are found to have additional roles, that would not significantly alter the overall conclusion that an evolutionary origin of this pathway would be an incredible coincidence.

In terms of a possible evolutionary explanation for this biochemical pathway, another significant factor to take into account is that prokaryotes cannot make steroids. (An intriguing twist to this is that mycoplasmas are the simplest organisms and hence thought to be among the earliest to have evolved – yet these are the ones that use the eukaryote-derived cholesterol!) Even more confounding is that, as indicated above, animals, plants and fungi use different steroids, and the point where the pathways diverge is *before* the cyclization step. In all eukaryotic groups the biosynthetic route is common up to squalene epoxide, but then the different groups of organisms use different enzymes to effect the cyclization step, resulting in different initial steroid compounds, and this is followed by different multistep routes to the different final products. So this means that steroid synthesis would need to have arisen, not just since the origin of the eukaryotes but since divergence between the major eukaryotic groups; so its origin cannot be shunted too far into the dim and distant past along with the answers to so many other evolutionary problems. Equally important of course, it shows that the critical cyclization step must have arisen independently at least three times!

I now want to look at biosynthetic pathways from a slightly different angle. So far I have considered their origin primarily in terms of the proposed evolutionary scenario that they arose essentially one step at a time as each advantageous enzyme happened to arise – although I have shown that, especially for the critical cyclization step, it would appear that three enzymes would have had to emerge more or less together. But certain aspects of the biosynthesis of steroids raise serious doubts that the pathway could possibly have arisen in such a haphazard or opportunistic manner – some steps just do not seem to make sense except in the light of subsequent ones.

What comes to mind first, of course, is the synthesis and cyclization of squalene. For example, it is common practice for cyclic or polycyclic compounds to be made in a laboratory from linear unsaturated compounds: based on knowledge of such compounds, chemists can synthesize appropriate precursors – with double bonds in the right places – and arrange appropriate reaction conditions for the desired cyclization reaction to occur. But what is remarkable is that nature seems to have found this way too – supposedly by trail and error, without any 'knowledge' of chemistry and without any prior knowledge of the potential utility of the end product. Even if nature had had some way of 'knowing' that the steroid product was useful (e.g. if the back-to-front approach could have applied), then it would still be extraordinary if it had 'happened' upon squalene as a potential precursor to these compounds. What makes it even more exceptional is, first, that the useful end product (e.g. cholesterol) is far removed from the immediate cyclization product, second that squalene does not cyclize directly but must be primed in some way, and third that different cyclization products can result, typified by the different eukaryotic groups. In the absence of any mechanism for conveying to a prospective evolving organism why and how squalene might be converted to a steroid structure, and the utility of steroids, there appears to be no evolutionary justification for synthesizing squalene. Opportunism just will not do.

A similar though perhaps less impressive example is the way in which a double C=C bond is made by first adding a phosphate group and then using removal of that to carry away a carbon atom. Along the same sort of lines is the addition of a pyrophosphate group (which, you will recall, requires two successive additions of a phosphate, by separate enzymes) which enables reactions up to five steps later to proceed. Or why should the head-to-tail combining of isopentenyl groups stop at farnesyl, two of which are then joined head-to-head, instead of continuing the build-up of isopentenyl groups? The straightforward answer is that six isopentenyl groups all with the same orientation does not give the right chemical arrangement to form a steroid structure, whereas two head-to-head farnesyls does. But that is hardly a satisfactory evolutionary answer!

Biosynthetic pathways such as for producing steroids are a serious challenge to any sort of evolutionary explanation. Most evolutionary texts propose that they arose progressively as outlined above; but at least some admit they are an enigma.

BIOLOGICAL MACHINES

In introducing this chapter I mentioned that one of the notable accomplishments of some biological systems is to carry out mechanical movement – where the movement is not incidental to the catalytic activity of the proteins concerned, but their primary function. We have already seen some examples of proteins acting as molecular machines, such as the topoisomerases, where mechanical operations are carried out on individual macromolecules; we will now look at instances where the concerted action of proteins results in movement on a larger scale. These can range from transporting objects around inside the cell, to locomotion of the cell, to substantial actions of whole multi-cellular organisms. An important feature which will emerge is that coordinated movement not only requires specific sophisticated proteins (which alone are a major hurdle to any evolutionary explanation), but their effective activity requires that they be organized in relation to each other, and their operation regulated. The organization of assembly and control of operation add significant levels of complexity to the overall unit.

Actin and myosin

Early ideas about the composition of the cell were that it was little more than a bag of 'vital jelly' or 'protoplasm', but as we have learned more of its components and how they function, increasingly we have realised that, especially for eukaryotes, the cell is a complex assemblage of organelles and with an elaborate infrastructure. A major component of that infrastructure is a network of fibres made from the protein actin, which was mentioned in the preceding chapter as having a highly consistent amino acid sequence. With only around 375 amino acids, actin is not an especially large protein, and it folds up into a fairly compact globular unit; but these units are assembled together into long and fairly stiff filaments. These filaments are then inter-linked by various connecting proteins into a network or cytoskeleton which extends in three dimensions throughout the cell, and other proteins connect it to various cell structures, especially to membranes. An important role for this cytoskeleton is to maintain the shape of the cell – for example it is the actin network, connected to the exterior cell membrane, which gives red blood cells their biconcave shape. In some cells the synchronised assembly and dismantling of different parts of the cytoskeleton enable the cell as a whole to move, such as an amoeba which travels by extending protrusions of the cytoplasm (called 'pseudopodia' = false feet) forwards, and retracting the cytoplasm behind it.

Much more dramatic movements are effected when actin interacts with a further protein, myosin. This is a larger protein than actin and its folded structure has three distinct regions – a globular head and long tail, the two connected by a flexible neck region. Also, there are usually one or two short polypeptides around the neck which have a regulatory function. The active part of the molecule is the head which has two binding sites, one for actin and one for ATP. When ATP is available, the head of the myosin oscillates, hinged at the neck, moving one way as it hydrolyses ATP and then returning to its resting position as the ADP is released, i.e. carrying out one complete oscillation for each molecule of ATP hydrolysed. Importantly, the hydrolysis of ATP and the associated power stroke occur only if the myosin head is also bound to a filament of actin – if this were not the case, then myosin would consume ATP but achieve nothing. This means that the power stroke moves the myosin (and anything connected to it) a small distance (11–15 nm) in relation to the actin; at the end of the power stroke, the myosin disengages from the actin and returns to its resting conformation where it rebinds the actin filament, but two or three actin molecules further along. It is quite realistic to think of the myosin as stepping along an actin filament.

How this mechanism is used to generate useful movement depends on the type of myosin, of which there are at least three, and its arrangement in relation to the actin. One type (myosin I) has a tail end which associates with membranes, such as those of organelles within the cell: by attaching its tail to an organelle, and with its head associated with an actin filament, organelles are transported around the cell along the actin network.

Muscle

Undoubtedly the most elaborate application of the actin-myosin combination is in muscles. Muscles are built up from bundles of muscle fibres. These myofibres are unusual because they are the result of many cells (possibly hundreds) combining to produce what is in effect a single very large cell but with many nuclei. Within each myofibre are multiple myofibrils which run the length of the fibre, each of which comprises a series of structures, joined end to end, called sarcomeres, which contain the actin and myosin, and are the basic units of contraction.

The arrangement of a sarcomere is illustrated in Figure 9, which shows that it consists of arrays of relatively thin actin filaments intermeshed with relatively thick myosin filaments. One end of the actin is firmly attached to what is called the Z disc which forms the boundary between adjacent sarcomeres, and the other has a capping protein (tropomodulin) which prevents the actin filament from disassembling. In addition, the actin filament and its attachment to the Z disc are supported by a long protein, nebulin, which is firmly bound to the Z disc and extends the length of the actin filament.

The type of myosin (II) used in sarcomeres has tails which associate with each other (rather than with cell membranes for type I): the tails aggregate to form a central filament, with the myosin heads protruding outwards from it. The filaments contain hundreds of myosin molecules and are symmetrical lengthwise (like a canoe paddle) in the sense that all of the myosin tails are orientated towards the mid point of the filament, which gives an array of protruding heads along each end of the filament. The myosin filaments locate between the actin filaments, with the myosin heads in close proximity to the actin.

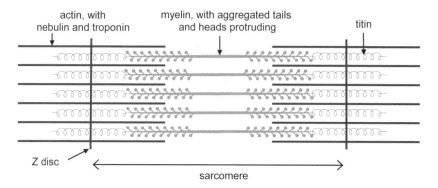

FIGURE 9. Major components of a sarcomere. Note that this depicts a longitudinal section; the array of actin and myelin filaments extends in three dimensions, with each myelin filament surrounded by six actin filaments.

The way the muscle operates is as follows. When the myosin is activated, the action of the myosin heads is to pull the myosin filaments further in between the actin filaments; but because this happens at both ends of the myosin filaments, it is the actin filaments that move; and because the actin filaments are firmly secured to the Z discs, the ends of the sarcomere are drawn together, so the sarcomere itself shortens. In general, not only do all of the sarcomeres within a myofibril activate at the same time, but so do all the myofibrils within a myofibre, so the whole myofibre shortens – and that is the essence of muscle contraction.

In a stretched muscle it is possible for the myosin filaments to be pulled out completely from between the actin filaments, which would clearly disable the contractile mechanism. However, the ends of the myosin are attached to the Z disc by the protein titin which is long and elastic and functions rather like a rubber band: it allows the myosin to move in and out between the actin filaments for normal muscle operation, but also protects against overstretching, and pulls the myosin back in between the actin.

The arrangement and interaction of the actin and myosin are impressive enough, but that is only half the story, the other half is how the contraction is controlled.

You will recall that the myofibres are formed by many cells coming together. One of the specialisations in the formation of the myofibres is that the endoplasmic reticulum (ER) which is part of most cells is modified to provide a sarcoplasmic reticulum (SR). This has ducts which surround the myofibre and pass between all the myofibrils so that every myofibril is in intimate contact with the SR.

Like most cells, resting muscle fibres have a low internal calcium ion (Ca^{2+}) concentration, but unlike most cells a very high concentration of Ca^{2+} is stored in the SR. The motor nerve that controls the muscle fibre contacts the SR, and the effect of receiving a nervous impulse is to open up Ca^{2+} channels in the SR membrane, so that Ca^{2+} diffuses into the myofibrils. It is this increase in Ca^{2+} concentration that causes the myofibril to contract. Calcium ions are actively returned to the SR by a Ca^{2+} pump, driven by hydrolysis of ATP. This pump is unique to muscle tissues – most cells keep their internal Ca^{2+} concentration low by pumping it completely out of the cell, rather than into the ER.

Even that is not the end of the story because the obvious next question is to ask how the Ca^{2+} causes the muscle to contract. This involves another protein, troponin, which wraps itself around the actin filaments. In the absence of Ca^{2+}, the resting state, the troponin covers the myosin binding sites on the actin so that myosin cannot bind. Hence, although ATP is available within the myofibril and binds to the myosin, because the myosin cannot bind actin it is not activated. However, when Ca^{2+} is available, this is bound by the troponin, and the effect of this is to change its conformation so that it no longer occludes the myosin binding sites, myosin binds, is activated, and contraction takes place. When nervous stimulation of the myofibre ceases, the level of Ca^{2+} in the myofibres is reduced by the Ca^{2+} pump, troponin returns to its resting conformation which occludes the myosin binding sites on the actin, and contraction ceases. By having the control of contraction dependent on the reversible conformation of a protein enables very rapid switching between the on and off states.

Having learned something of the various proteins involved in muscle contraction, how credible is it that such a mechanism could have arisen in an opportunistic trail and error manner? Just getting the right proteins – even only those that are unique to muscles – would be an unbelievable task. (Incidentally, the forms of actin and myosin that occur in muscle are unique to muscle.) Unfortunately, we do not yet know how the many nascent muscle cells unite to form a myofibre, or how on the molecular scale the various proteins of a sarcomere are arranged into their ordered array (which does not happen spontaneously but requires some sort of organizing machinery). But when we do, it will surely make the whole system even more complex and totally beyond the reach of any blind evolutionary explanation.

Bacterial flagellum

One of the most surprising discoveries of molecular biology is the structure and operation of the bacterial flagellum which is a long thin filament used for locomotion (swimming). By analogy with that used by unicellular eukaryotes (e.g. protozoa) for movement, it was assumed that the prokaryotic flagellum would also operate in some sort of oscillating, perhaps whip-like, way. However, it became apparent that in fact it works like a propeller, driven by a molecular motor – and I do not mean that it twists first one way and then the other rather like a weight suspended from a string – it really does rotate like a motor. It is such a remarkable device that it would be worth describing just for interest, but it is especially appropriate in the present context as an example of intricate biological structures and mechanisms. The bacterial flagellum is illustrated in Figure 10.

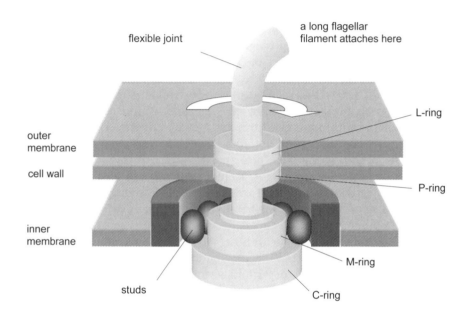

FIGURE 10. Bacterial flagellum. The rotor is pale coloured; the stator is dark and shown in cross-section. It is thought that motive force is generated primarily between the M-ring and the studs. The C-ring is also involved in generating motive force, and it controls the direction of rotation.

The unit that provides rotary motion is embedded in the (inner) bacterial cell membrane. Just like a typical man-made motor, it comprises a rotor which revolves within a stationary component (stator), and interaction between them provides the rotary movement. The rotor is a tube which extends from within the cell, through the cell membrane(s) to the outside, where the flagellar filament (propeller) is attached via a flexible joint. Where it passes through the inner cell membrane, the rotor has at least two rings (M and C) of proteins, which are part of the driving mechanism; and surrounding the rotor at this point is a circle of about ten protein 'studs' which, in association with other proteins, span the inner membrane and comprise the stator – the other part of the driving mechanism. Most bacteria have, in addition to the inner membrane (where the motor is positioned), an outer membrane and a cell wall. Where the rotor passes through these, there are bushes or bearings comprising two further rings of proteins (L and P). The flagellar filament (propeller) is built up from subunits of just one polypeptide, flagellin, which comprises 490 amino acids. The subunits are arranged in a tight helical structure to produce a stiff hollow rod, with each subunit contributing a small propeller blade to the outside of the structure.

As the hydrolysis of ATP is the chief source of energy within cells, and especially because it is used to power eukaryotic motor proteins such as myosin, it was assumed that the power for the bacterial motor would also come from ATP, but it was found not to be the case. Instead, the driving force is usually the same gradient of hydrogen ions (H$^+$)that is used to produce ATP (comparable to that used in eukaryotic mitochondria, see Box 8.1), although in some prokaryotes it is a gradient of sodium ions instead. It is not yet known how the passage of H$^+$ through the relevant proteins causes rotation, but about 1000 are needed for each complete turn, and a typical speed is 300 rotations per second.

Other proteins on the cell side of the motor effect a switching arrangement which alternates the direction of rotation; but I repeat that this is not like an object on a string twisting first one way and then the other, it is more like reversing the angle of the blades in a turbine; and I will say a little more about operational control of the motor shortly.

There are about 20 different proteins (with most present in multiple copies) contributing to the overall structure of the complete flagellum, and a similar number involved in its assembly and regulation of production. Not surprisingly given its intricate construction, the flagellum does not simply self-assemble from a mixture of its constituent proteins, but assembly depends on specialised apparatus for exporting components to their appropriate locations, though the structure and mechanism of this are not yet known. The assembly process is known only in outline, but it begins with the components on the inside of the cell membrane and the hollow rotor; subsequent proteins for constructing the flexible joint and filament are fed through this tube and progressively assembled on the outer end.

Due to its relative simplicity, it has been possible to elucidate some of the genetic control of flagellum synthesis too. There are about 40 genes involved, including those for the structural proteins, grouped into several operons, and organized into three tiers of genetic control. The first tier seems to be regulated primarily by cell growth – it comprises just one operon and it appears likely that the resulting proteins are involved only in regulating the second tier of genes, none actually being incorporated into the flagellum. Conversely, most of the second tier genes encode proteins which are structural and/or involved in transport and assembly; but, interestingly, many of them also act as regulators for the third tier genes. For example, the second tier genes include a sigma factor required for transcription of (some of) the third tier genes, and a further second tier product is part of the flagella assembly machinery, but the latter also binds the sigma factor and deactivates it. Hence, in the early stages of using the second tier genes, although the sigma factor is available, it is disabled from activating the third tier genes; but as flagellar construction progresses, the assembly line protein is lost from the cell through the nascent hollow flagellum, so its inhibiting effect is lost and transcription of third tier genes proceeds.

Chemotaxis

The structure and basic functioning of the bacterial flagellum is impressive, but the complexity is compounded when we see something of how the flagellum is controlled, especially in chemotaxis – the movement of bacteria towards a favourable stimulus or away from an unfavourable one.

The usual rotation of an individual flagellum is counterclockwise, interspersed by brief bursts of clockwise rotation. Usually there are about five to ten flagella on a bacterium, and when they all rotate counterclockwise, they group together and impart a steady directional motion to the bacterium. But when any one of them rotates clockwise, they disperse around the cell (bearing in mind the very small size of bacteria this dispersal and grouping of the flagella occurs very quickly) and the bacterium somersaults in more or less the same place. In the absence of an external stimulus, a bacterium moves in a given direction for about one second, tumbles briefly, and then moves off again in a different direction; taken together, these movements result in a more or less random walk.

The basic idea of chemotaxis is that a bacterium swims towards a favourable chemical stimulus or away from an undesirable one. It is of course a little more involved than this, because it is not simply the presence of a stimulus that matters, but its changing concentration. In effect, what the bacterium does is compare the present concentration with that of a few seconds ago: increasing concentration of an attractant prolongs the directional movement, and increasing concentrations of repellent leads to tumbling, and vice versa.

The presence of a stimulus is detected by receptors which project into the surrounding medium, but also extend into the cell, such that the binding of a stimulant on the outside results in a detectable change on the inside. Via a series of reactions which I am not going to describe, this change affects the switching mechanism at the base of the flagella, promoting or inhibiting clockwise rotation as appropriate. There are different types of receptor, responding to different chemical stimuli. They have an extraordinary range of responsiveness, extending over about 100,000-fold change in concentration, and the comparison of present with recent concentrations is carried out by interaction of cellular signals which proceed at different rates.

So, in this small device of ordinary bacteria we see a remarkable level of sophistication – in the device itself, its manner of assembly, its operation, and in the control of that operation. All told there are possibly less than fifty different proteins involved – not many compared with some other biochemical systems. At the level of the complete structure it is a complex device and it seems incredible that it could have arisen by chance; when we consider the improbability of each of the individual components involved in its construction and operation, there should be no doubt left at all.

COMPLEXITY

In the preceding chapter I showed that biological macromolecules are much too improbable to have arisen by any sort of random process. The generally accepted evolutionary explanation to overcome this improbability is to propose that, although today's proteins appear very specific (and therefore improbable), it is unreasonable to require that this situation arose all at once, but proteins have arisen progressively – efficient modern proteins gradually evolved from primitive ones by trial and error. However, quite apart from the fact that there is no evidence for constructive protein evolution, there remains the fundamental problem that any putative early protein must at least be large enough to fold – which automatically means it is highly improbable.

Further – the point I have addressed more fully in this chapter – any proposed evolution of proteins is absolutely dependent on the putative early proteins having some measure of activity which can confer at least some advantage and be selected for, in order to provide a stepping-stone to further improvements. But most proteins do not have any utility in isolation, so any hypothesis of proteins evolving gradually requires several part-formed proteins to arise more or less simultaneously, which, of course, requires incredible coincidences. In this chapter I have demonstrated that the cooperative action of proteins – and hence their mutual dependence in terms of having utility – arises in all sorts of ways. For example, there is the concerted activity of many proteins and nucleic acids involved in DNA duplication and translation. Similarly, the proteins that implement mechanical movement are like parts of a machine, with most if not all of the parts being required for the machine to operate at all. This also applies to the proteins which act sequentially, as in biochemical pathways, because in so many cases the intermediates do not have utility in their own right. An extension of this, although I have not elaborated on it, are the hormones and neurotransmitters (used by nerve cells) where macromolecules with complementary function need to be made in different cells, sometimes quite remote from one another. And underlying all of these is the complexity of how gene expression is regulated which, even in prokaryotes and especially in eukaryotes, requires an array of complementary genes and specific interactions between nucleic acids and proteins.

The point of all this is to emphasize that biological systems are much too complex to be accounted for by the standard evolutionary explanation – the trial and error incremental accumulation of favourable variations – because they cannot be built up piecemeal. This is a theme developed by Michael Behe in his book *Darwin's Black Box*. Behe uses the concept of an irreducible minimum requirement for a functional system, which he illustrates first with a simple mechanical device such as a mousetrap, and then moves on to biological systems that require several components in order to work, comparable with those I have mentioned here. This is a useful point to make, because Darwinism relies on progressive evolution, and Behe stresses that if you need several components for a system to work, then there is no benefit in having one without the others, or even all but one of them. (Remembering that natural selection is blind, so evolution will not develop the components separately and save them for potential future use.) Behe's argument is useful so far as it goes, but loses impact because he does not also take into account the improbability of macromolecules. For example, if you have a large box of mixed items of clothing and you have to select from it blindfold then it may take many selections before you have a matching set, but you will probably get there before too long. Similarly, if you had an assortment of ready-made proteins and all you needed to do was to select a dozen to construct a biochemical pathway, then this could probably be done in a reasonable number of attempts, or within a limited time-scale. The improbability lies primarily in acquiring the individual proteins; but having recognised that this is the primary difficulty, it is compounded by the cooperative actions of proteins which completely scuppers any idea that individual proteins could evolve gradually.

We have also begun to see in this chapter that it is not sufficient just to have the right components, but they also need to be arranged appropriately in relation to each other. Just as it is with a mechanical device such as a mousetrap – a mere collection of base, spring, catch etc. will not trap a mouse – the parts have to be set up properly; or an assortment of watch components must be assembled correctly in order to provide a functional timepiece: this is also true of molecular biology. For example, the principal proteins which comprise muscle, such as actin, myosin, nebulin, troponin and not forgetting the huge titin, are remarkable in themselves; but a disorganized mass of these proteins would be quite useless: to provide an effective contractile mechanism they need to be connected in a suitable way, such as in a sarcomere.

One of the outstanding feats of biological systems is self-assembly – a capability which currently we can achieve to only a very limited extent with our own machines. For example, the parts of a watch are first manufactured and then must be assembled by a watchmaker; but biological systems have no such artisan and have the remarkable ability not only to synthesize the necessary macromolecular components, but also to assemble them internally into the right configuration – using molecular mechanisms which themselves have to be controlled appropriately. At present we know very little of how proteins are assembled to produce larger structures, though we are beginning to understand something of this in relatively simple cases such as construction of the bacterial flagellum. When it comes to muscle tissue where hundreds of nascent cells unite to form the myfibres with their shared myofibrils and sarcoplasmic reticulum, we have no idea of how this is achieved on the molecular scale. Perhaps in another decade or two we will have found out how this is done; and from what we do know so far about the control of transcription and cell differentiation, we can confidently anticipate that many control proteins and DNA sequences will be involved.

One of the points I made in Chapter 7 and have sought to elaborate here is that evolutionary biologists have focused too much on superficial resemblances at the molecular level, in terms of the similarity of many common proteins, but have neglected the crucial control mechanisms. In particular, I pointed out the nonsense of basing supposed evolutionary relationships on differences of a few base pairs in a structural gene whilst ignoring the hundreds or even thousands of base pair differences in the control sequences (including the genes coding for the control proteins) for those proteins. This simplistic view was perhaps excusable to begin with when we knew so little, but now that we are gaining some idea of what is involved it can no longer be sustained. The control mechanisms cannot be ignored from an evolutionary perspective; they will be no less improbable than the structural proteins and perhaps much more so.

A new look at an old chestnut

In this chapter I have drawn attention to some of the remarkable complexity of the chemistry of life. Complexity in terms of the structure of individual macromolecules, of their direct interactions with each other, and of how many macromolecules function cooperatively, whether simultaneously or sequentially or even at a distance. Further, although at present we can see only a glimpse of what is involved at the molecular level in assembling larger functional biological units, there is little doubt that the systems will be very complex, perhaps outstripping anything we have seen so far. This complexity, which ranges from the level of the molecule to that of the whole organism, is a serious challenge to the generally accepted evolutionary explanation for biology – to my mind, totally refutes it.

So where does this lead us? If the complexity of biology cannot be accounted for by natural processes (and evolution is the only serious contender) then inevitably it points to the idea of design – the concept which most biologists had thought and hoped had died a century or more ago. Indeed, perhaps many readers will be dismayed at such a suggestion, and see this as a retrograde step – back to the anachronistic argument of Paley and his natural theology. However, I would not want you to think that all I am doing is advocating the old-fashioned argument for 'design' but merely couched in modern (i.e. biochemical) terms – a sheep in wolf's clothing (!): there is a fundamental difference between the design argument as advocated by such as Ray or Paley and what I am presenting here.

Paley pointed to the exquisite form and function of biology, notably the organs of the human body, especially the eye. He said that such sophistication and evident functionality should be compared with a watch. A watch is an intricate construction of many specifically engineered components arranged in a specific way, and performs a useful function. Further, it is evident that the crafting of the components and their placement in relation to each other is with the express purpose of performing that function, and he maintained that similar evidence of purpose can be seen in biology. So, just as it is inconceivable that a watch could arise spontaneously, but has to be designed and made purposefully, so biological structures must have been designed and made. Paley's argument, based on morphological features, is reasonable, and many at the time were convinced by it, and some still are. But in essence it is a subjective argument – as Richard Dawkins (1986, Ch. 2) puts it, the Argument from Personal Incredulity. What this type of argument relies on is the subjective perception that biological form and function could not just happen.

The argument that nature manifests design was seriously weakened by Darwin's proposal that natural selection working on spontaneous variations could progressively mould and adapt organisms to their way of life so effectively that they appear designed, though in fact they are merely the product of natural processes. This view was strengthened by the substantial amount of work done in the first half of the 20th century which showed that natural selection is a real phenomenon in nature, and has become the accepted explanation for the elaborate structures of biology. With Neo-Darwinism it was recognised that phenotype is determined by genotype, and hence that the source of useful variations is meaningful genes. By then, evidence for the adaptive action of natural selection at the morphological level was so convincing that biologists assumed it could be equally effective at the molecular level. In other words, that selection operating on chance variations (mutations) in genes could account for the production of useful genes in the same sort of a way that natural selection of large-scale variations can produce useful morphological adaptations. This remains the prevailing view.

However, what I have demonstrated in this and the preceding chapter is that such a view can no longer be sustained in the light of what we have learned about molecular biochemistry over the last few decades. Even individual biological macromolecules are much too improbable to have arisen opportunistically. Further, the integrated action of many macromolecules first excludes the possibility that individual macromolecules could acquire their activity progressively, and second reveals a degree of complexity that is well beyond the scope of any sort of evolutionary explanation. The more we find out, the more untenable it is that biological systems could have arisen by some sort of trial and error process: selection on random variations does not offer a satisfactory explanation or mechanism at this level. The salient point is that this is essentially an objective assessment, not subjective; and that is the important difference between this conclusion based on our modern understanding of molecular biology and Paley's based on 18th-century morphological biology.

The 18th-century naturalists knew biology only on the morphological scale, but it was evident to them that many biological structures are sophisticated and exquisite. Indeed, to many, life itself seemed wonderful and mysterious. So their limited knowledge meant they felt that biology could not arise naturally – there must be a designer. In the context of their superficial knowledge of biology, Paley's argument was entirely reasonable; but he could have been right or wrong – in his time there was no way of knowing either way. It was not until we acquired in-depth knowledge of biology – notably of the chemical structure of biological structures, and something of how biology works at the chemical level – that we have become able to judge the strength of his case. It might have turned out that biological structures could arise spontaneously and permit adaptive and/or progressive evolution in the way Darwin envisaged. In other words, as we came to understand how biology works at the chemical or molecular level, this knowledge might have explained how complexity on the large scale can arise. But the inescapable conclusion from modern molecular biochemistry is that it hasn't; in fact exactly the opposite has happened. As I have outlined above, the more we have found out about biology at the molecular level, far from explaining complexity, what we see is increasing complexity; and because we now know a great deal about chemistry we can conclude that there is no way the complexity of biology could have arisen simply by chemistry. Whereas our previous lack of knowledge had meant that a natural explanation seemed possible, now we know it is out of the question.

For the sake of clarity, I emphasize that, although much of molecular biology is impressive, I am not simply transferring the subjective appreciation of biology at the morphological level to the molecular level. I am not saying 'Look at the complexity of molecular biology: doesn't it look wonderful – it must have been designed.' Rather, I am saying: 'Look at the complexity of molecular biology, and, because we know much of the underlying chemistry, we can objectively conclude that it could not possibly have arisen by chance.'

This conclusion is underlined by the fact that biology works and, so far as we can see, complexity at the molecular level is necessary in order for it to work. That is, complexity is not incidental, but is an essential aspect of molecular biology. For example, one need only think of what is required to replicate DNA, or to produce a myofibre in the right tissue, to realise that to achieve these reliably through molecular-scale mechanisms (without some vitalistic overseeing watchmaker to piece together the various com-

ponents) must entail substantial complexity. The more we elucidate biological mechanisms, the more we see that this is so.

Importantly, this requirement to function (not just appear impressive),
and our assessment that it functions is substantially objective rather than
subjective. Whilst there can be much debate about the functionality or
otherwise of various biological structures at the morphological level, it is
generally straightforward to assess functionality at the molecular level,
e.g. whether or not an enzyme catalyses a reaction. For example, a classical argument for teleology was that of adaptation – perhaps citing the
various species of the Galapagos finches with their different beaks which
seem to be adapted to different feeding habits, so the beaks look designed
for their function. But in response to this it can be argued that a bird feeds
according to the shape of its beak – if its beak were a different shape then
it would feed differently. So the argument for design, or purpose, or teleology in biology is weak at the morphological level. But at the molecular
level we have an unequivocal case. To cite just one example: there is an
absolute requirement for bacteria to have some means of separating the
interlinked DNAs which result from its replication, and it is unquestion-
able that a type II topoisomerase does the job (and nothing else does, and
it does nothing else).

In summary, as originally formulated, even going back to Aristotle, but
especially exemplified by Ray and Paley, the argument for design was essentially subjective; but what I am suggesting here is an objective assessment based on the functionality and improbability of biological structures
at the molecular level. I shall return to the question of design in biology in
due course, but it is now time to take a closer look at the morphological
evidence for evolution – natural selection and the fossils.

9

Variation and Variability

Evolution and Natural Selection

Although Darwin was not the first to propose that life had evolved through a progression of increasing complexity, his major contribution was that he presented evolution as a coherent and credible theory by proposing a directing mechanism – natural selection. He recognised that species are not uniform but vary in all manner of attributes, and argued that those individuals within a species which, for whatever characteristic or combination of characteristics they possess, are fitter – these survive better than those without those characteristics. Consequently, in general, and assuming of course that they are heritable, the preferred characteristics are more prevalent in the next generation, and are preferentially perpetuated. Fitness can be interpreted quite broadly, but ultimately, so far as evolution is concerned, fitness is primarily in terms of producing offspring which in turn are capable of reproducing.

Darwin did not observe selection in nature. What he did observe was the breeding of various plants and animals by man, and he argued that a similar mechanism (in principle) could operate in nature, driven by the competition between organisms in their struggle for survival. Only, instead of selection for characteristics of the breeder's choosing, nature could 'choose' or 'select' solely on the basis of fitness. Successive selection for preferred variations, even though each may be minor in itself, could lead to substantial changes in the morphology of the species.

One of the strengths of Darwin's theory is that it provides a convincing explanation for adaptation: progressive minor changes can increasingly modify a species to its particular way of life. In this way it can account for the many instances where organisms are aptly suited to their environment and/or interaction with other organisms. More than any other aspect of biology, adaptation is probably the one that had prompted suggestion of 'design' or teleology from Aristotle onwards, and in natural selection Darwin provided an alternative natural explanation. This is seen as one of the major triumphs of Darwinism.

Darwin extended this idea by proposing that where different groups of individuals within a species acquired different preferred features, this could lead to divergence – the appearance of distinct races – and ultimately the formation of multiple species. He further extrapolated that, by the gradual assimilation of small changes, organisms could progressively increase in complexity, such that over a prolonged period of time the whole array of living organisms could arise from primitive organisms.

Some have criticised the very concept of natural selection, saying that it is a circular argument: 'survival of the fittest' simply means that the fittest survive; and what determines which survive? — only that they are in some way fitter. But this is not my view. On the contrary, I think Darwin was right to identify the significant and important role of natural selection in nature. That members of a species are not uniform but vary is evident, and that some do survive better than others of the same species at least in part due to variation has also been demonstrated (melanism in some moth species, mentioned in Chapter 5 and discussed further below, is a clear example of this). It is also known that at least some of the variation is due to heritable factors, so it is reasonable to assume that in selecting between phenotypes, some selection of genotype is also taking place – otherwise artificial selective breeding could not be effective. So there seems no good reason to doubt that natural selection, in essentially the sense that Darwin introduced the term and modern biologists generally use it, does operate in nature. I think that identifying natural selection was a valuable contribution that Darwin (and Wallace) made to biology and especially, in due course, to our understanding of ecology.

However – and this is where I part company with Darwin and the generally accepted scientific view – I strongly challenge the virtually unlimited extrapolation made in support of evolution as a whole, based on observed instances of natural selection. As mentioned at the beginning of Chapter 7, Darwin had no difficulty in supposing that small variations could accumulate to produce dramatic changes because he had no knowledge of the implications of doing so. The same sort of extrapolations have been made since Darwin's time, right up to the present day, but they should no longer go unchallenged in the light of what we now know of subcellular biology. Clear cases of changes to species arising from natural selection are seen as convincing evidence for the whole evolutionary process – of how complex organisms have gradually developed from simpler ones. It is this extrapolation which I maintain is completely unfounded, and which I contest here.

'Evolution by natural selection' — the phrase is so well known that the two issues of evolution and natural selection are generally considered inseparable, part and parcel of the same thing. But they are not the same. Right at the beginning of his book *The Genetical Theory of Natural Selection* Ronald Fisher, who did so much pioneering work on population genetics and in so doing helped establish the synthesis of Darwinism and Mendelian genetics, stated that:

> Natural selection is not Evolution. Yet, ever since the two words have been in common use, the theory of Natural Selection has been employed as a convenient abbreviation for the theory of Evolution by Natural Selection, put forward by Darwin and Wallace. This has had the unfortunate consequence that the theory of Natural Selection itself has scarcely if ever received separate consideration.

The distinction between natural selection and evolution is something many present-day biologists seem to have lost sight of, an oversight that needs to be redressed. Fisher's concern was that the close tie between evolution and natural selection meant that the focus on evolution resulted in the phenomenon of natural selection itself being neglected. My concern is that the association of evolution with natural selection means that evidence of natural selection is taken as proof of evolution in the widest possible sense.

So in this chapter I review some frequently cited examples that demonstrate natural selection at work, and related phenomena such as diversification and speciation. I discuss what can reliably be concluded from these observations, and what cannot. In particular – and this will be a recurring theme – the occurrence of natural selection at the morphological level is no evidence whatever that the problem discussed in the preceding two chapters of the extreme improbability of biological macromolecules has been overcome, for example by natural selection operating at that level.

ADAPTATION AND SPECIATION

Natural selection in action

A closer look at industrial melanism

As mentioned in Chapter 5, before the middle of the 19th century the peppered moth, *Biston betularia*, was known only as a light-coloured (pale) form, but thereafter a dark (melanic) variety was found and over the succeeding 50 years or so the species became predominantly dark-coloured, at least in industrial areas. Haldane calculated that differential selection between the light and dark forms of about 50% would be needed to account for such rapid change. At the time this was seen as a major triumph for the new science of population genetics and substantiation of evolution through natural selection; and these observations probably remain the most popularly-known case of natural selection and support for evolution.

The difference in fitness was attributed to selective feeding by birds, and this was demonstrated by Bernard Kettlewell (1917–97) in the 1950s. He released pale and melanic moths in industrial and non-industrial areas and observed their fate. In industrial areas, where tree trunks were dark, birds preferentially captured the normal variety; whereas in non-industrial areas they tended to select the melanic ones. The observed differences in predation by birds accounted for the relative survival of the two forms, and in turn accounted for the shift in population from light to dark.

Yet that is only half the story. November 1952 saw the infamous smog that resulted in several thousand premature deaths in London's population. This event more than any other led to the introduction of the Clean Air Acts and similar legislation which brought about a progressive reduction of air pollution throughout the UK over the succeeding twenty to thirty years. Interestingly, alongside this environmental improvement there has been an increase in the pale coloured moths: the trend of the preceding century has been reversed. Not only is this a useful indication of the effect of pollution control (a subject in which I have a professional interest), it also serves to demonstrate even more clearly the process of selection in natural populations. As the blackening of tree trunks has diminished, it has become more protective for the moths to be pale rather than dark, and there has been a consequent return to the pale colour.

Further, it should be noted that industrial melanism did not occur only in the peppered moth: in Britain alone there are over 100 species of Lepidoptera (butterflies and moths) in which melanic forms increased during the industrial period, and many more in other industrial regions of the northern hemisphere. It has become evident that long before industrialisation melanic forms were a normal part of many species of Lepidoptera, but often occurring as only a very small proportion of the population. Indeed, it is now clear that melanism was not a new phenomenon in the 19th century, possibly prompted by industrialisation, but that genes for melanism occur in many species (and not only Lepidoptera), and have been present for a very long time. Kettlewell called them paleogenes in recognition of their antiquity.

So what can we conclude from these observations? Undoubtedly the change in relative abundance of the pale and melanic forms of B. betularia and various other moths over the last 150 years is a clear demonstration of natural selection (whether exclusively due to differential bird predation or for other reasons[1]). And the outcome – the change of predominant coloration depending on the environment – can certainly be seen as an adaptation. Also, in so far as the theory of evolution includes changes in gene frequencies as a result of natural selection, and includes consequent changes to phenotype, it is appropriate to use the term 'evolution' to describe those changes.

However, we need to be quite clear about what we mean by 'evolution' in cases such as these. So far as the moths are concerned, all that occurred was a change of the frequencies in the relevant populations of the gene(s) responsible for coloration. Extrapolation beyond that is not justified. For example, in a recent textbook on evolution it is stated that 'If the process that operated in the nineteenth century in a single species of moth had been continued for the thousands of millions of years since life originated, much larger evolutionary changes could be accomplished' (Ridley). This is a wholly unjustified statement. The 'process that operated in the nineteenth century' was natural selection operating on an existing pool of genes. The development of complex organisms from simple life forms re-

---

[1] Recent studies have shown that differential predation by birds is not the only selective factor; in at least some species the melanic forms seem to be associated with a generally hardier phenotype. But the precise basis for selection is not important to the present discussion.

quires the production of genes, i.e. the generation of biologically significant macromolecules – an entirely different process from that demonstrated in the moths.

So all of the evidence regarding the moths provides no insight at all
concerning the origin of genes, no evidence whatever of a way around the
problem of the prohibitive improbability of useful macromolecules. As
Kettlewell so clearly points out, the genes responsible for melanism in the
moths have been around for a very long time. Production of useful genes
is crucial for evolution. Evolution of complex life is not a mere extrapolation of changes in gene frequencies such as occurred in the moths. Change
in gene frequencies is one part of the evolutionary process, but only one
part; it requires other completely different processes too.

Kettlewell recognised that for many species of Lepidoptera all that had
occurred was an increase in frequency of the genes that give rise to the
melanic forms. But for others, including *B. betularia*, because there was
no record of the melanic form being seen beforehand, he assumed the
necessary genes must have arisen by mutation at the time of the industrial
revolution. However, we should note that Kettlewell preceded knowledge
of the biochemical basis of melanism – when it was thought that new useful genes could arise by a simple mutation (I comment further on this later
in the chapter). We now know something of the biochemistry of melanism
– the chemical structure of the pigment melanin and how it is biosynthesized. This is a clear example of how modern biochemical knowledge has
upset previously held evolutionary views. Most biologists who are aware
of the biochemistry recognise that a system such as this which involves
multiple enzymes and control proteins could not possibly arise in so short
a period of time, and certainly not simultaneously in many different species scattered around the globe, and now accept that the genes were already present. Even the above-cited author accepts this, which is why it is
so inappropriate that he should have made the statement quoted above.

However, this is the popular perception: that the well known example
of the peppered moths demonstrates the whole process of evolution, even
proves the theory of evolution to be true. Unfortunately, it is not surprising
that such is the popular view when professional biologists perpetuate this
exaggeration.

Divergence within a species

North American house sparrow

Coincidentally, at the same time that melanic forms of the peppered moth
were proliferating within the UK, the house sparrow was spreading across
North America. The house sparrow is not native to America, but was introduced in 1852 on the east coast at Brooklyn, New York; and over the
next 100 years it colonised most of the continent.

What is especially interesting about this, and what has caught the attention of evolutionary biologists, is that, as the sparrows spread out, quite distinct subspecies or races appeared – varying especially in colour and size – some so different from the originating birds that they may not be immediately recognised as house sparrows by European ornithologists. What appears to have happened is as follows. Because the species was expanding into a large virgin territory, in many instances there were just a few pioneering individuals dispersing to new areas and colonising them. Their small number meant that they carried only a subset, or unrepresentative sample, of the gene pool available within the initial population; and this led to their offspring differing from the parent population. Also, of course, as the birds spread out in various directions, many of the colonising flocks carried different subsets of the immigrant gene pool, the effect of these separate colonisations being the production of a range of morphologically distinct races.

Many biologists see the change in form of the sparrows as an example of evolution, and Ernst Mayr has cited this case of the sparrows as an example of rapid evolution – to counter the common view that evolution is generally so slow as to be imperceptible in a human lifetime. Further, Mayr has emphasized the importance of geographical separation in the formation of distinct species, and he sees the divergence of the sparrows as a first step towards full speciation.

As with the moths, I do not object to such use of the term evolution, provided we are clear about what it means in this context, and what it does not mean. The sparrows are a good example of divergence as a result of segregating genes from a common gene pool, and this divergence could lead to speciation. And no doubt many other subspecies could have arisen than those that actually have appeared if the genes had segregated differently. But, as with the moths, although this is evidence of evolution in terms of morphological change due to different combinations of genes, this is no evidence whatever of evolution in terms of the origination of useful macromolecules, in particular of useful genetic material.

Adaptive divergence within a species

The sparrows seem to be an example of divergence of a species into different races or subspecies due primarily to chance or random drift. But clearly an important aspect of the theory of evolution is adaptation, in particular the selection of preferred morphologies to suit a habitat or mode of life. In support of this, there are cases where a clear gradation of characteristics occurs across a species' range, the gradation corresponding with changes in one or more features of the environment. In fact the situation with the moths is a case in point[2] – melanism did not spread uniformly

---

[2] It is not quite this straightforward: one of the more recent observations which has challenged the traditional explanation of the cause of selection is that melanism is prevalent in the non-industrial area of East Anglia.

across the whole of the moth's range, but generally concentrated in and around industrial areas.

Yarrow plant

A frequently cited example where there is a well-defined gradation of characteristics across a species' range is the yarrow plant, *Achillea*. The morphology of the plants varies from the coastal region in California to the Sierra Nevada at an altitude of 3000 metres. In general, the plants found at low level grow larger and more luxuriantly than those living on the higher ground, and there are also differences in the plants' growing seasons. But it is apparent that these differences are not merely a direct consequence of the different growing conditions on a genetically uniform species – there are genetic differences between the plants occurring in different locations. This is evident because when plants are taken from their usual location and grown in another they generally do not thrive so well as in their native habitat – to the extent that some of the plants from low-lying areas die when planted at a significantly higher altitude. The morphological differences are sufficient to warrant the plants being designated as different species – *A. lanulosa* near the coast and San Joaquin Valley, and *A. borealis* in the Sierra Nevada. But even within these designated species there are differences of growth response. Probably a more realistic picture is to see the plants as constituting a fairly continuous gradation from those suited to low lying habitat to those suited for altitude.

This is clear evidence that, although derived from a common stock, the genes have segregated, and not in just a random fashion, but genes have been selected to suit the local environment. That is, although all the plants are considered to have arisen from a common ancestor, at each generation variations in the mix of genes will have arisen in all the populations; those mixes which suited a particular environment will have been selected; and, of course, different mixes will have suited different environments. Progressively, the different groups of plants will have diverged, each becoming more suited to its particular environment until there are substantial differences between them. In other words, the plants have developed a gradation in their genetic makeup to match the range of habitat.

Adaptive selection and segregation of gene combinations is an important aspect of the evolutionary process – it illustrates both the adaptive role of natural selection and the potential for divergence leading to the formation of new species. Species such as the yarrow plant, in the process of diverging in this way, are therefore seen as demonstrating evolution. But in a similar way to my previous comments in relation to the moths and sparrows, whilst species such as the yarrow plant demonstrate gene segregation, that is all they demonstrate. There is nothing to suggest that anything other than gene segregation has occurred; in particular, there is no evidence regarding the origin of the genes that have been segregated.

Ring species

A feature of the theory of evolution is that divergence within a species, whether arising by chance (genetic drift) or by adaptive divergence, ultimately can lead to the formation of distinct species. An important point is that the formation of new species is merely an extension of divergence within a species, it is not a wholly new mechanism. This continuum from divergence within a species to the formation of new species is commonly illustrated by means of what are called 'ring species'.

Ring species occur where a particular species diverges, i.e. there is a gradation of characteristics, usually morphological, across its range, and the two ends of the range come together. The result is a gradation of characteristics across the range, but then an abrupt change across the boundary where the two ends of the range meet. Taking any small area within the range, any change of characteristics is so small that there is no question that it is just one species, but the accumulation of small changes across the whole range adds up to a substantial change, such that the two terminal forms placed side by side would often be classed as quite different species. Sometimes there is a fairly clear demarcation between the two ends, sometimes the two forms intermingle or exist side by side, and sometimes they hybridize to some extent.

Gull

In Western Europe two gull species, the Herring Gull and the Lesser Black-backed Gull, live side by side: they are distinct in appearance and behaviour and only rarely interbreed. Across the Atlantic, the American Herring Gull is considered to be the same species as its European counterpart, but the European Herring Gull does not extend eastwards beyond the Urals. However, the Lesser Black-backed gull does extend eastwards right across northern Asia, and as it does so, it becomes progressively more like the Herring Gull until by the Bering Strait it is considered by some to be the same species as the American Herring Gull. So we see with these gulls that there appears to be a gradation of characteristics all around the Arctic, with change most rapid in the Siberia region, and an extensive area of overlap of the distinct forms in western Europe.

Here we have a snapshot of speciation in progress – the divergence of characteristics within a species to the extent that different subpopulations appear and behave as distinct species. This is a necessary step in the evolution of separate lineages, and the occurrence of ring species is seen as supporting the evolutionary scenario. But it should be recognised that even divergence to this extent can be wholly accounted for in terms of the segregation of genetic material that was present in the original parent population; there is still no evidence for the formation of new genetic material.

Speciation

Ultimately, of course, diversification and segregation within a species may lead to complete speciation – the formation of wholly distinct species. To be effective, some measure of genetic isolation is necessary. Mayr emphasized the importance of geographical separation to achieve this, and other biologists, especially Dobzhansky, identified alternative mechanisms.

Galapagos finches

A particularly famous example is, of course, the finches which Darwin observed on the Galapagos Islands. In view of their close similarities – they are all sparrow-like and dark brown or black – yet evident differences especially in terms of their bills and mode of eating, Darwin wrote of them 'one might really fancy that from an original paucity of birds on this archipelago, one species had been taken and modified for different ends.'

Usually 13 different species of Galapagos finches are recognised, distinguished mainly by their bill and eating habits: There are six ground finches which feed mainly on seeds though two feed off cacti, three tree finches which are insectivorous and one which is vegetarian, another two with habits like woodpeckers, and a warbler. One of the 'woodpecker' finches is particularly interesting because of its use of a 'tool' – a twig or cactus spine to poke insects from crevices. Despite these significant differences, their close relationship has been confirmed by genetic studies, on the basis of which some biologists have suggested that they should all belong to the same genus, though usually they are categorized into two.

It is recognised that migration between the Galapagos Islands can occur but will be infrequent for these small birds in view of the distances involved. This means there is only limited breeding between the various island populations, enabling distinct morphological groups to arise, and it is not hard to see how speciation could have occurred readily between the islands' populations. A number of studies have been carried out to try to explain how the present species might have arisen from a common ancestral population, based on diversification between dispersed groups and with some subsequent hybridization. Hybridization occurs regularly in the wild between some 'species' of the ground finches.

The finches are the dominant birds on the islands, and it seems that in the absence of competition from other birds, they have diversified to occupy many ecological niches. In particular, their different feeding patterns mean that there is less competition between them than there would be otherwise, and various species may live side by side.

So here we have a good example of an ancestral species diversifying into several present day species which are distinct in terms of morphology and behaviour. However, there is still no reason to believe that anything other than gene segregation has occurred, and this view is supported by the close genetic similarities.

EVOLUTION IS A MULTI-STEP PROCESS

In the preceding section I have outlined the sorts of observations that demonstrate natural selection, especially in terms of producing an adaptive response, and divergence such as leads to the formation of new species. All of the examples are taken from standard evolutionary texts in which they are cited as clear evidence of evolution. However, as I have already indicated in commenting on them individually, the sense in which it is true that they demonstrate evolution is limited. This is because they demonstrate only part of the evolutionary process – that of segregating genes or changing the frequency of genes – but not the whole process of evolution which necessarily involves the origination of those genes.

It is evident that evolution comprises three substantially distinct processes. First is the production of genetic material, which is the fundamental basis of *variability*, i.e. the genetic means for enabling phenotypic variation. It is, of course, generally assumed that this arises through mutation; but, as we have already seen, this is in fact a very unlikely explanation. Second is the production of *variation* by the shuffling of that genetic material; this would be primarily through the random distribution of genes between gametes at meiosis, and the random combining of gametes at fertilization. And finally is the *selection* of those genotypes which confer greater fitness to the individuals concerned, so that advantageous genes and gene combinations are propagated in preference to others.

All of the examples in the preceding section are wholly explicable in terms of generating variation from, and selecting subsets of, the available genetic material, but provide no evidence of the production of the fundamental variability – of the genes themselves. Let me expand on this.

Evolution by segregating genes

Whereas the classical view of species was that they are essentially fixed, the existence of variability in natural populations is now beyond question, indeed the demonstrable breadth of variation within some species has surprised even the biologists.

The diversification of the house sparrow as it colonised North America is a clear example of how variability can become evident in what initially appeared to be a substantially uniform species, given the right conditions. It shows how markedly different morphological individuals can arise simply by taking different combinations or subsets from the initial gene pool. In the case of the sparrows, the divergence resulted in several distinct races arising from a common stock. The different races are equivalent to the varieties that are known in other birds, such as the carrion and hooded crows, many domesticated animals and plants,[3] and the various human

---

[3] Plant varieties can also be obtained by multiplication of the genetic material (polyploidy).

racial groups.

Segregation to produce distinct forms can arise by chance (genetic drift) and/or by the effect of selection. For the sparrows it seems that segregation arose mainly by random sampling of the available genes rather than by any definite selection of advantageous gene combinations. On the other hand, with the yarrow plants there is an obvious association of morphology and growth patterns with geography, which clearly points to the selection of specific gene combinations that suit the local environment, i.e. a form of adaptation. (And, of course, there can be other factors than geography that may determine preferred gene combinations.)

Clearly, continued segregation, whether due to genetic drift and/or selection, can ultimately lead to full speciation, as has been the case for the Galapagos finches. I have commented previously that speciation is not simply a question of having sufficiently different morphologies but that various genetic isolating mechanisms can be involved too. It is apparent that different species can arise from a common stock, whether the definition of species is based predominantly on morphological differences or loss of interfertility.

As an alternative to speciation, a species can remain as one but change its morphology (or other less conspicuous characteristics) due to shifts in gene frequency under selection pressure. This was illustrated by the moths, although in that case the changes in gene frequencies were largely confined to industrialised areas rather than across the whole of the species' range. Also with the moths, the changes in gene frequencies were reversible, the melanic form did not completely displace the pale form which was able to return when the environment changed. But it is quite conceivable that a particular allele or group of alleles could be lost altogether from a population – whether by chance and/or selection, although for statistical reasons it is quite hard to lose alleles completely unless the population is rather small.

The overall theory of evolution includes all of these mechanisms – adaptation (whether of all or part of a species), divergence (whether random or adaptive), and speciation (whether in terms of distinct morphology or loss of interfertility). All of these were worked out in principle and well understood in the first half of the 20th century. The theoretical population geneticists could describe how gene frequencies may change due to random variations and/or selection pressure; and the field biologists provided ample examples of such processes operating in nature. And, as I have said before, the successful and complementary union of the theoretical and experimental studies gave great support to the theory of evolution as a whole; by the middle of the 20th century Neo-Darwinism reigned supreme in biology, and with good reason, for it had substantial supporting evidence.

However, what I must emphasize is that all of the above mechanisms (and standard evolutionary texts may fill several chapters discussing elaborations of these main themes) fall within the production and selection of variations; none relates to the production of variability. Put another way, all of these processes are to do with changing the distribution of genes, but nothing to do with the source of those genes. In other words, the extensive innate genetic variability of most natural populations enables substantial adaptation and/or divergence solely by selecting appropriate subsets from the available gene pool (such as divergence of the yarrow plants) and/or changing the relative frequencies of genes within the pool (such as melanism in the moths). But the successful demonstration of these mechanisms sheds no light whatever on the source of that genetic variability.

When evolutionary scenarios are presented, many do of course include the production of variability in their accounts. For instance, a typical description of speciation may start with separation of a uniform population by a geographical barrier into subpopulations, and then the two separated populations develop independently, eventually becoming distinct species. Divergence of the separate species includes differing segregation of genes that were present in the original uniform population, and the assumption is usually made that the separate subspecies diverge even further by acquiring different new genes through mutation. But this is only an assumption – there is no evidence for it actually occurring – the observations are entirely consistent with gene segregation alone.

Indeed, as already mentioned at the end of Chapter 5, as the extent of available variability became apparent during the 20th century, many biologists expressed the view that evolution does not need any more variability, there is already more than enough in natural populations. And, of course – the point I have been emphasizing all along – scenarios which presume the acquisition of new useful genes completely ignore the prohibitive improbability of it happening. This important point – that what we observe in nature is the segregation of genes rather than their production – is further emphasized by what we know from domestic breeding, which Darwin saw as an analogy for natural selection.

Domestic breeding

Most wild populations exhibit such extensive variability that it seems almost any feature or characteristic can be selected and the plant or animal bred to accentuate or diminish that feature. This is the principle behind domestic breeding, and its success is seen in the enormous range of phenotypes that can be produced from a common stock. The various breeds of dog illustrate this particularly well. They are so diverse that Darwin thought they must have descended from several original species rather than only one, although we now know that they are all derived from the wolf. Also, all dog breeds are interfertile in principle, although physical incompatibilities such as a wide disparity in size may present insurmountable difficulties in practice.

It is recognised (at least now that we have a 20th-century understanding of genetics) that domestic breeding is selecting genes that give the desired response and/or discarding those that give the opposite – in effect taking subsets of the original gene pool. But there is a limit: when all of the alleles promoting the selected characteristics are present in maximum frequency, and all those detracting from them are excluded, there is no more potential for further selection.

The historian Michael Ruse is a keen advocate of evolution and has written many articles on the subject. In his essay *Darwin's debt to Philosophy* he argues that the general scientific view in mid 19th century was that domestic breeding actually pointed away from providing or explaining an evolutionary mechanism precisely because it was so well known that the scope for breeding was limited. It is evident that Darwin recognised this limitation as he wrote 'It certainly appears in domesticated animals that the amount of variation is soon reached', but he still felt that the selection carried out by breeders was a useful illustration in principle of the process of natural selection.

This limitation is referred to more recently by Richard Dawkins (in the context of arguing against there being an inherent resistance to evolution):

If anything, selective breeders experience difficulty *after* a number of generations of successful selective breeding. This is because after some generations of selective breeding the available genetic variation runs out, and we have to wait for new mutations. [*The Blind Watchmaker*, Ch. 9, emphasis in original]

Here Dawkins refers to the limits which selective breeding encounters – limits imposed by the currently available genetic material. What is noteworthy is that he then states explicitly that further selection can be achieved only if and when there is constructive modification to the genetic material. This is the tacit assumption of evolutionary biologists, but rarely stated so plainly. But there is no evidence that it occurs. It is sometimes argued that for present populations, being the product of a long history of evolution and already possessing much variability, it is not surprising that further constructive mutations are extremely rare, especially within the relatively short period in which we would have recognised them if they had occurred. But this is no evidence whatever that they will arise in time, and the improbability of useful biological macromolecules is a *prima facie* case that they never will.

Breeders of course recognise that what they are doing is selecting subsets of the wild population, and often in developing a strain with a particular favourable characteristic they lose something they would rather retain. To restore this they go back to the wild populations and seek to reintroduce the desired lost feature by judicious interbreeding (or, these days, perhaps by appropriate genetic engineering). Important examples of this are attempts to regain natural resistance against various pests for a wide range of domesticated crops. And, of course, this is why there is such concern at the loss of natural habitats and their species – these are invaluable repositories of useful genetic material, and if lost will be lost for ever.

So domestic breeding re-emphasizes the point I introduced earlier: most natural populations have enormous variability, the potential to produce an extensive range of phenotypes, whether by natural or artificial selection. But we are well aware that there are limits, and there is no evidence for the production of the new variability which is the essential first step for evolution. Yet the belief persists that useful (or potentially useful) mutations have arisen and will continue to do so, and I think it is instructive to review why this is so.

Misleading variations and mutations

To begin with, when the likes of Buffon and Lamarck began to challenge the accepted fixity of species, they attributed morphological changes either to the impact of the environment or to the effect of use and disuse. In other words, in some way mutations were directed – either by the environment, or by the organism itself. In the absence of a proper understanding of genetics, this sort of explanation seemed to make sense and persisted right through the 19th century. For instance, when it became apparent that the melanic forms of the peppered moth were increasing so dramatically a link with industrialisation was soon made. At the time it seemed perfectly reasonable that smoky pollution could have caused the darkening of the moths' wings, and several theories were advanced as to how this might have occurred.

So far as Darwin was concerned, whilst admitting that he did not know the origin of variations, he felt that an organism's mode of life was an important driving force behind generating them. But he also saw variations as arising undirected, and relied on natural selection to retain advantageous ones. It was mainly with Weismann, who clearly distinguished between somatic and germ-line cells, that it became generally accepted that mutations occur randomly.

Almost immediately with this insight came various observations which suggested that mutations had an important role in evolution. First were de Vries' experiments with the evening primrose which quite spontaneously and repeatedly produced offspring which were very different from the parent plants. De Vries assumed the different forms were due to mutations, and it was observations such as these that led him to believe that macromutations – mutations having a substantial phenotypic effect – were a major means of evolutionary advance. As we saw in Chapter 5, along with Bateson, he was one of the main contestants in the debate in the early part of the 20th century between those who supported a key role for macromutations rather than Darwin's gradual approach.

Then, in the first couple of decades of the 20th century, through his work with *Drosophila*, Morgan came across mutations which affected almost every part of the fly's anatomy – such as eye colour, number of bristles, shape of wings etc. Whilst some of these mutants were monstrous – such as legs developing in place of antennae – and no unequivocal advantageous mutations were actually identified, the range and scale of morphological changes resulting from mutation was enough to convince most biologists that this was how evolution occurred: mutations arose which caused morphological variations that affected fitness, and the beneficial ones were propagated. The mutations also occurred at a frequency which, in conjunction with natural selection, could support long-term evolution.

However, all of these observations preceded an adequate biochemical understanding of genetics, when genes were merely discrete entities which *somehow* caused the phenotypic effect. With the knowledge of what genes are and how they work, especially that most functional genes encode a specific protein, we can now explain these early observations. We now know that the odd behaviour of the evening primrose was not due to recurrent mutations but to unusual gene linkage groups which were recombining in the normal way, but were lethal when homozygous, and produced a variety of heterozygotes. (For a summary see Sturtevant.) Similarly, we now realise that the phenotypic effects of the mutations in *Drosophila* are due entirely to the corruption of functional genes, not to the production of new ones. This is readily illustrated for the different *Drosophila* eye colours, explained in Box 9.1.

So, in retrospect, we can see that biologists were misled by the observations of de Vries and Morgan into thinking that constructive mutations occur fairly readily. Unfortunately, although we now have a correct understanding of these, and realise the erroneous conclusions that were based on them, the impact of the early misunderstanding of these mutations is still with us: The general perception remains that useful new genes can be generated spontaneously, even though we have no examples to support this.

Of course, the existence of genetic variability is seen as evidence that constructive genes can and do arise by mutation. A comment often found in evolutionary texts is that, although we cannot actually point to any constructive new genes, there must be some in order to account for existing variability and for evolution to have occurred. But clearly this is circular reasoning, and should no longer go unchallenged.

Unfortunately, a clear recognition that useful new genes do not arise has also not been helped by the prominent role of population geneticists in the formulation of Neo-Darwinism. For example, Ernst Mayr was fond of emphasizing that (in contrast to the three steps mentioned above) evolution proceeds in two stages:

Evolution through natural selection is (I repeat!) a two-step process. The first step is the production (through recombination, mutation, and chance events) of genetic variability; the second is the ordering of that variability by selection. [1978]

Box 9.1 *Drosophila* eye pigments

The normal 'wild type' eye colour of *Drosophila* is a reddish-brown (brick red). This is due to two principal pigments, ommochrome which is brown and drosopterin which is red, and various minor components related to drosopterin which are predominantly blue or yellow. The pigments are synthesized by the following (somewhat simplified) pathways:

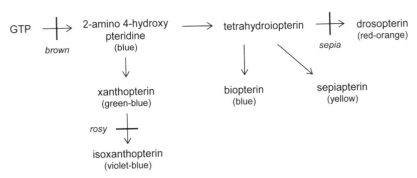

(This is the synthesis that Beadle helped to work out, see Ch. 6.)

These pathways indicate the colours of some of the pigments and intermediates, and the reactions that are disrupted in various mutant strains. Note that when a reaction is blocked, not only are the pigments downstream not synthesized, but there can be a build-up of the intermediates immediately before the disruption which may also affect eye colour.

So, for example, if both synthetic pathways are disrupted early on, so that no pigments are produced, then the flies have white eyes. If only the ommochrome route is affected then the eyes are vermilion (due to drosopterin and related compounds such as isoxanthopterin); conversely if ommochrome is synthesized but none of the pteridine pigments, then the eyes are brown. If only drosopterin synthesis is prevented then the eyes are sepia (due primarily to ommochrome and isoxanthopterin) and become progressively darker as excess biopterin and sepiapterin are produced; and if only isoxanthopterin is affected then the eyes are rosy due mainly to ommochrome and drosopterin. Many other eye colours arise depending on exactly which step(s) in the syntheses are affected.

The important point is, of course, that whereas early geneticists thought the mutants with different eye colours were examples of constructive mutations – involving the production of new pigments – we now know they are due solely to the disruption of the normal production of the usual pigments. That is, they are deleterious mutations, arising from the corruption of viable genes, not the production of new ones.

Mayr's prime concern was to distinguish between the random events that produce different genotypes and the non-random selection of advantageous ones. He therefore lumped together the origin of genes by random mutations and the random shuffling of genes which occurs in the course of reproduction.[4]

Similarly, it is quite common to see the effect of mutations treated along with that of new genes entering a population due to the influx of migrants – because, to the population geneticist, either is an equally valid source of new genetic material for the population in question. But it is quite evident that the influx of ready-made genes is completely different from the production of new useful genes by mutation.

It is, therefore, highly misleading to lump mutations with either recombination or migration in this sort of way. In particular, because there is no doubt that recombination produces new useful variations, and no doubt that migration can introduce new genes into a population, linking recombination or migration with mutation gives the mistaken impression that there is no doubt mutations produce new useful genes.

Hence, so far as Neo-Darwinists were concerned, there was little if any doubt that mutations are the source of genetic variability and, as the chemical nature of genes became known, it was taken for granted that mutations could account for the origin of genes – of meaningful nucleotide sequences. Although it was soon recognised that *prima facie* genes are very improbable, the strength of the prior opinion was such that insufficient weight was given to such 'theoretical' objections, and it was assumed there must be some satisfactory evolutionary explanation.

Resistance to antibiotics and insecticides

In considering the preceding paragraphs, some readers may have thought of a prominent exception to the general point I am making – that of the resistance which has arisen over the past 50 years or so to various chemicals used to control diseases or pests, notably antibiotics and insecticides. There seems little doubt that in such cases organisms have a new and useful capability, and the acquisition and spread of resistance is commonly regarded as exemplifying evolution. So clearly we need to take a look at this.

The principle on which modern antibiotics work is to identify some biochemical feature of e.g. a pathogenic bacterium which is clearly distin-

---

[4] This oversimplification is perhaps excusable because there is occasional overlap between them, such as when chromosomal crossover occurs between two different versions of the same gene, which can produce a third version. Where this happens then the effect of the crossover is, in this respect, equivalent to a mutation. This is, however, a rare occurrence compared with the more usual source of new sequences arising from mutations due to miscopying, and the role of recombination in shuffling genes is far more important than in generating new sequences. In the vast majority of cases mutation and recombination are wholly distinct processes.

guishable from that of its host, and to find or devise a chemical compound which will selectively disrupt that part of the pathogen's biochemical machinery while leaving the host cells unscathed. Penicillin is an excellent example of this because it interferes with synthesis of the bacterial cell wall which is a feature completely absent from animals. On the other hand, many other antibiotics (e.g. tetracycline) make use of the fact that, although both bacteria and humans have ribosomes, there are significant biochemical differences between those of prokaryotes and eukaryotes; so it is possible to find compounds which will prevent the synthesis of bacterial proteins, but not those of higher organisms. Similarly, although insects have a nervous system which uses essentially the same components such as nerves, synapses and neurotransmitters as higher animals, there are sufficient differences between the macromolecules concerned to be able to target the insects' relatively selectively. Thus, for example, DDT and pyrethroids prevent proper functioning of the voltage-gated sodium channel[5] which is essential for the conduction of nerve impulses.

In mid 20th century, the discovery of selective antibiotics and pesticides to treat infections and infestations was seen as the dawn of a new era, with disease and pests firmly under control; and we have gained substantial benefits from the use of these chemicals. However, many readers will be well aware that resistance arose remarkably quickly and spread rapidly. The early response to this was to use alternative chemicals; but invariably resistance has arisen to whatever has been used instead, with resistance to one chemical sometimes automatically conferring resistance to others, and we are beginning to run out of options. So how do bacteria and insects etc. become resistant to antibiotics and insecticides?

Modification of target

Given the rationale of how antibiotics and insecticides work, it is not surprising that resistance can arise through modification of the target macromolecule so that either the chemical does not bind at all, or at least that its adverse effect is much diminished. For example, the most common way in which insects become resistant to DDT is by a change in the amino acid sequence of the protein that forms the relevant sodium channel.

One of the notable observations with this mechanism of resistance is that exactly the same few modifications seem to appear in many different species and geographical locations. In particular, so far as the resistance to DDT is concerned, in every known case the modification is to replace the amino acid leucine at position 1014 in the sodium channel (note the size of these proteins) with phenylalanine. Some insects have an additional mutation in the same protein (though it has not been precisely characterized at the time of writing) which enhances the resistance, but it does not seem to occur independently of the first. Although some of this uni-

---

[5] I should like to have explained what this is and how it works, but I had to be selective!

formity can be accounted for by the spread of resistance between different insect populations, it also seems certain that the same modification has arisen several distinct times, indeed may recur almost wherever DDT is widely used.

However, this is not surprising. Recall from the end of Chapter 7 that the mutation rate for a *particular* nucleotide (base) is in the order of 10^{-9}, (such that the chance of *any* mutation in a gene of average length (~1000 base pairs) is about 10^{-6}). Many insect populations are comparable to 10^9 in number, so there is a good chance that most possible single-base mutations will be present in a large insect population. The vast majority of these will be seriously detrimental, if not fatal, and will be lost rapidly whenever they occur; and a few may be more or less neutral and can result in a small number of different forms of the gene persisting in the population (i.e. protein polymorphism). Normally it would require even an advantageous mutation to recur many times to be adopted by the population. But if an insecticide is present and a mutation confers resistance to it, the mutation's chance of survival are exceptionally good because most of the non-resistant insects will be killed, leaving only resistant ones to propagate. Hence, subsequent populations will be dominated by the resistant forms and resistance will spread rapidly. Consequently, it is to be expected that, if there is a suitable single-base mutation, then it will arise before long and be adopted by the population wherever the insecticide is used extensively for a moderate length of time.

However, the mutation that confers resistance by changing the target macromolecule is invariably inferior to the normal 'wild type' macromolecule. That is, in the absence of the relevant insecticide, the resistant strain fairs less well, is less fit, than the normal phenotype. In the particular case where insects are resistant to DDT due to a change in their sodium channel, they are prone to paralysis at slightly elevated temperatures. (Actually, it could hardly be otherwise: if any single-base mutation could produce a macromolecule that were better than the usual form then, by the reasoning just outlined, it must have arisen already (many times) and would have been adopted by the population; in other words, it is almost inevitable that in large populations with short generation times, such as for bacteria or insects, no single mutation could improve upon the prevailing macromolecule, but is likely to be detrimental.) This is, of course, the basis for strategies to overcome resistance which rely on not using the insecticide (or antibiotic) for a while because the normal insecticide-susceptible form, being fitter, should return in due course.

Hence, so far as our wider discussion is concerned, there are two relevant points regarding resistance by target modification:

First – something I shall discuss more fully in due course – is that in most cases resistance of this sort usually requires only one mutation. It just happens that a particular replacement of an amino acid interferes with the action of the insecticide without having an unduly detrimental effect on the protein's function. In fact – and it is not at all surprising when one considers the specificity of biological proteins – there are very few instances of insecticide resistance arising through target modification: resistance to DDT and pyrethroids involves exactly the same mutation; and resistance through modification of other insecticide targets generally involves identical, or a very limited selection of, mutations (ffrench-Constant). Similarly, we know of only a few examples where antibiotic resistance is due to target modification.

Second, and equally important, is that the modified macromolecules that confer resistance are inferior to the normal phenotype. This underlines the fact that what we are seeing here cannot really be considered as demonstrating the production of new useful genes. Rather, it is only under the artificial circumstances of highly toxic compounds being present that it happens to be in the organism's interest to carry a gene which is defective, which actually diminishes fitness, but gives it a relative advantage over its peers which do not have the mutation.

Detoxification of insecticides

Another obvious way of resisting the effects of a pesticide is to have the means to destroy it. Almost all aerobic organisms, prokaryotes as well as eukaryotes, have sets of enzymes known as cytochromes P450 which are used to degrade various organic compounds by oxidizing them in some sort of way. They have a bewilderingly complex chemistry: in any one organism there may be a hundred or more different types, and many of these are able to act on a wide range of substrates and oxidize them in a variety of ways. One of their roles is to degrade endogenous compounds (those produced by the organism itself) such as steroids and other hormones; but, of particular interest to us here, is that the cytochromes P450 are biology's principal means for dealing with foreign compounds (apart from the immune system in higher animals which combats infections and the like).

It would be inefficient to produce the whole range of available P450s all the time and, although we do not yet know the details, it seems the mechanisms are in place to allow appropriate P450s to be synthesized only when there is a reasonable chance of there being a suitable substrate. The principal means of control seems to be by regulating transcription. This is similar to the induction of the enzymes for metabolising lactose (see Ch. 8), but an important difference is that, whereas control of the *lac* operon is very specific to lactose, many of the P450 enzymes are inducible by a range of compounds, and various transcription factors are involved. Neither is there a one-for-one relationship between substrate and enzyme, but any given compound may evoke production of several P450s. In effect, in response to a new foreign compound, an organism scans its repertoire of P450s, and produces a few which are likely to be able to degrade it.

A particularly important role for the P450s in insects is to degrade plant toxins, enabling them to eat plants which would otherwise poison them. A good example of this is the tobacco hornworm. Nicotine is normally toxic to insects (and to higher animals) because of its action on the nervous system; but in the tobacco hornworm nicotine induces production of one or more P450s which degrade it, permitting this insect to live on the tobacco plant.

So it is not surprising that many insects use the induction of P450s as a means of gaining resistance to insecticides. It is particularly common in resistance against pyrethroids and organophosphates. At present we do not know enough about how these enzymes are induced in response to new challenges. It may well be through mutations to transcription factors which, if they produce a desirable result (i.e. inactivation of the toxic compound), in some way are reinforced. Similarly, we know that the substrate specificity of P450s can be modified by changing a single amino acid, and, although we do not yet have any definite examples, it may be that some resistance involves such changes. It seems that the P450 system is organized to generate this sort of variability. But, as with target modification, this enzyme modification involves just one or two mutations which have a reasonable chance of happening, and cannot be extrapolated to the initial production of a useful gene.

Detoxification of antibiotics

Some resistance to antibiotics is gained through target modification, and possibly some through P450 detoxification. However, we have come to recognise that bacteria have an immense wealth of biochemical diversity and flexibility – illustrated in the way they can degrade, even utilise as a foodstuff, just about any organic compound – and, not surprisingly, they employ these resources against antibiotics. This diversity is not so much in individual bacteria, but is held collectively within various interacting bacterial populations. In addition to their chromosomal DNA, most bacteria have one or more shorter lengths of DNA, known as plasmids, in which they carry a range of extraneous genes, and which replicate independently of the main chromosome. The plasmids frequently include what are in effect mutator genes, which actively promote production of variations in their plasmid genes; and the plasmids are transmitted between bacteria, enabling genes to be shuffled between them, comparable to the way sexual reproduction shuffles the genes of eukaryotes. One of the real causes for concern is that some bacteria have been able to accumulate on a single plasmid a battery of genes which confer resistance to a range of antibiotics, so that by transmitting this plasmid they convey multiple resistance.

We can see the application of this biochemical versatility in the mechanism of resistance to penicillin. This is usually effected by means of enzymes called β-lactamases which break a key bond (called a β-lactam bond) in penicillin, such that it is no longer capable of binding and hence deactivating the cell-wall-synthesizing enzyme. At first it was thought that the resistant bacteria had evolved new enzymes – though we should bear in mind that this was the 1940s, when biochemical knowledge was still in its infancy. As investigations have proceeded since then, it has emerged that there are hundreds of different types of β-lactamases, which are scattered throughout bacterial populations – almost every group of bacteria has at least some form of this enzyme. Further, it is evident that they have been in existence since long before antibiotics were used by man, but probably in only some bacteria and normally produced at a very low level.

Hence, when penicillin was used clinically (bear in mind it is a natural compound so some was present in the environment anyway, though only in trace amounts), situations will have arisen where some bacteria which were subject to antibiotic treatment possessed a β-lactamase able to degrade it. These would have survived preferentially, i.e. selection took place, such that subsequent generations would have had an increased proportion of resistant bacteria with their appropriate enzyme.

If that were the full story, then it would be easy to conclude that this form of antibiotic resistance relies exclusively on long-standing enzymes. However, for various reasons which include trying to combat resistance, a range of synthetic antibiotics have been produced which are based on the chemical structure of penicillin but with significant modifications (called penicillin analogues), and a related series of compounds called cephalosporins. In time, resistance has arisen to all of these and, what is interesting from an evolutionary point of view, is that resistance is sometimes mediated by enzymes which appear to be variants of some of the earlier identified β-lactamases. Whilst it is possible that all of the new-found β-lactamases have also existed for a long time, it seems more likely that these variants have arisen fairly recently by mutation from previously existing forms. They differ from the earlier known enzymes by up to four amino acids, with each amino acid substitution being possible as a result of a single nucleotide change. Also – a point I will take up further below – the variants appear to have arisen incrementally; that is, where there are enzymes with three or four amino acid substitutions then at least some of these substitutions are found singly or in pairs in other enzymes. It is also worth noting that the enzymes with multiple amino acid substitutions are not multi-functional super enzymes; rather, different amino acids substitutions tend to have different effects – some giving a broad range of activity, some enhancing activity against a particular type of antibiotic; and resistant bacteria tend to have a mix of enzymes which gives them a range of protection.

Resistance acquisition as evolution

It will be evident from the preceding paragraphs that some resistance to antibiotics, insecticides and the like does not involve mutations at all. This appears to be true for resistance to penicillin itself and some of the penicillin-related compounds introduced subsequently, and may well be the case for much P450-mediated resistance: the necessary enzymes were already present – albeit at a low level and sparsely distributed within the relevant populations. When challenged with the antibiotic or insecticide, those organisms with an appropriate protein survived preferentially and rapidly expanded in subsequent populations. Certainly this was natural selection in action, and may be regarded as evolution, but it illustrates nothing more than the examples considered early on in this chapter.

On the other hand, target modification does involve mutation. However, the result of such mutation is a defective gene which just happens to interfere with the action of the insecticide (or antibiotic), and hence confers resistance. So this can hardly be regarded as constructive evolution.

An interesting parallel may be seen in the way in which haemoglobin-S, which causes sickle cell anaemia, confers some protection against malaria. Haemoglobin-S differs from normal haemoglobin only in that the amino acid at position 6 of the β chain is valine instead of the usual glutamine (arising from a single nucleotide change). You will recall from Chapter 7 that two β chains aggregate with two α chains to form the functional haemoglobin tetramer. The changed amino acid is on the outside of the folded tetramer and, as valine is more hydrophobic (oily) than glutamine, the effect of this change is that the haemoglobin tetramers tend to aggregate, or crystallize, in the very high concentrations of haemoglobin that are present in red blood cells. It is these filaments or crystals of haemoglobin-S which distort the red blood cells, making them appear sickle-shaped, and may rupture the cell membrane. The malarial parasite spends part of its life cycle within red blood cells, where it relies on its high concentration of potassium ions. When the cells are ruptured, the potassium escapes, and this kills the parasite. This is why sickle cell anaemia is prevalent in areas where malaria is widespread: the protection it affords against malaria helps to maintain in the population what is essentially a detrimental mutation (generally leading to an early death when the mutation is homozygous).

However, even if most resistance to antibiotics and pesticides is through the use of old enzymes or target modification, it is clear that there remain at least some examples where resistance appears to be due to constructive changes to macromolecules. A clear example of this is the resistance to penicillin analogues and cephalosporins through modification of the β-lactamases, and at least some P450s may be modified too. So here at last we have the three stages of evolution being demonstrated: mutations produce a modified (new and useful) gene which is then shuffled and selected within the relevant populations to produce a phenotype which is adapted to the presence of a toxic compound in its environment (though not necessarily fitter in the wider sense). As might be expected, many biologists take examples such as these as conclusive proof of the whole process of evolution. But before rushing to accept this, we need to examine just what is being achieved. The important point is, of course, that only a few mutations are occurring – modifying a pre-existing useful protein, not producing a complete working protein from scratch.

From the outset I have not been suggesting that advantageous mutations cannot possibly, in principle or *a priori*, occur. Rather, the point that I have been making is simply that, even though mutations do arise, and could be potentially useful, the specificity of biological macromolecules is such (compared with the virtually infinite number of possible sequences) that a useful sequence is much too improbable to arise in this sort of way. This is, of course, essentially a theoretical argument – applying statistics to our knowledge of chemistry and biology – and no doubt many will find it hard to accept. However, what is especially interesting about the way some resistance arises through advantageous mutations is that it actually provides experimental support that the statistical argument is valid.

Odds of one in a billion do not sound very hopeful, but if you can have a billion attempts then the overall chance of success becomes quite good (about 63%, not the 100% that you might have expected; just as, if you roll a die six times, you have only about a 67% chance of getting a six). Hence, as I mentioned when discussing resistance through target modification, it is quite feasible to find a specific 1 in 10^9 mutation if the population in question is of the same order. Also, once this is found and spread through a reasonable proportion of the population, if a further single-base mutation will enhance the resistance (as it does for DDT) then it becomes feasible to find that mutation too within a moderate time-scale.

Similarly, we know from laboratory studies that the specificity of some enzymes can be modified by changing one or two of their amino acids; and, just as with target modification, the chance of a specific single-base mutation occurring, such as might cause an appropriate amino acid change, is about 1 in 10^9. Bacteria are very small, and the number of pathogenic bacteria within an infected person will generally exceed 10^9, perhaps be as high as 10^{12}; so, if a single-base mutation can enable or enhance resistance, it is to be expected that it will arise and, in the presence of the antibiotic, to propagate before long. This appears to have happened to give rise to the β-lactamase variants which have enabled resistance to penicillin analogues and cephalosporins.

Taking this line of reasoning further, if two mutations are required to implement resistance, but neither alone is effective, then it is necessary to find both mutations more or less simultaneously. (Bear in mind that there is no basis for the first to wait for the second; in fact, as with target modification, it is likely that any single mutation will be detrimental, which is all the more reason why it would be lost.) The odds for obtaining two specific mutations simultaneously are about 1 in $10^9 \times 10^9$ which is 1 in 10^{18} (not 1 in 2×10^9). Although 10^{18} is quite a large number, many bacterial populations are also large, for example it is estimated that there are approximately 10^{14} per person, mainly in the intestine where they contribute to digestion. This means that, despite the poor odds, it is realistic for bacteria to acquire resistance even if it entails finding two mutations simultaneously. But it will not be often that even bacteria find three mutations simultaneously, as the odds of doing so are about 1 in 10^{27}, which number is perhaps comparable with the total world-wide bacterial population (and for this advantageous combination to be selected, it would need to arise where a relevant antibiotic was in use).

However, although it would be exceptional to acquire three or four specific mutations simultaneously, it is of course possible to acquire these incrementally if one or two of these confer some advantage by themselves. In other words, it would be possible to accrue what would otherwise be an improbable resistance provided that it could be built up progressively, with each stage conferring some selective advantage. This is what seems to have happened with the β-lactamase variants which confer resistance to penicillin analogues and cephalosporins.

Perhaps at this point some evolutionary biologists will respond that what I am accepting here is exactly how they propose that biological macromolecules have been built up. That is, if bacteria or insects can find one or two amino acids in a year or two, then does this not show they could find a whole macromolecule before too long, and certainly in the eons of geological time? But such extrapolation overlooks the important point that each stage needs to confer some selective advantage. Once you have a working enzyme, then it is quite feasible that changing an amino acid or two will alter its specificity and/or activity, tailoring it to the organism's immediate needs, and, as we have seen, this can realistically be accomplished by insects or bacteria. However, to obtain a functioning protein from scratch – from a jumble of nucleotides or amino acids – it is necessary to get many, at least a few tens, of amino acids 'right' all at once; but, as we have already seen in Chapter 7, to explore even a tiny proportion of the possible options is well beyond the capacity of the available resources. I accept the principle that biological macromolecules could evolve progressively; but the scenario cannot get off the ground because in practice the first step is much too big – it is necessary for many amino acids to be right for a protein to have any utility at all on which selection could be based. This is compounded by the cooperative action of most proteins which requires that at least partial activity arise simultaneously on the part of multiple proteins – but that is much too improbable to accept.

Also, as I have indicated previously, intuitively we tend to have an exaggerated view of the significance of geological time. It is interesting to note that, coincidentally, the frequency of individual mutations (about one in a billion) is comparable numerically with the years of geological time (a few billion). What this means is that, in terms of finding multiple mutations simultaneously, geological time would enable us to find approximately one more mutation than can be found in one year with currently available resources. That is, if the large size of bacterial populations and their short generation time means they can find three simultaneous mutations in one year (which is probably about their upper limit), then comparable populations would have a good chance of finding four simultaneous mutations in the whole of geological time, but probably not five. That helps to keep time in perspective.

The acquisition and spread of resistance has been hailed by many biologists as a clear case of evolution. A favourite point is to emphasize that many modern pesticides are synthetic compounds which the organisms in question cannot possibly have faced in their previous evolutionary history, yet in a matter of a few years they have evolved resistance to them. What further proof do we need of the power of evolution?

However, it has been illuminating to take a closer look and see what is actually happening. First we have seen that many synthetic agents act in exactly the same way as naturally occurring compounds – such as the natural and synthetic penicillins, or synthetic DDT and natural pyrethroids; and, not surprisingly in view of this, identical mechanisms of resistance are employed. Even where this is not the case, what we see is that organisms have in place a battery of biochemical machinery which they employ for detoxifying a whole range of foreign compounds. Again, it is not surprising that such mechanisms are used against synthetic compounds as well naturally occurring ones. Importantly, their ability to detoxify compounds – even ones they have not met before – is much more a reflection of the capability and flexibility of the detoxification mechanisms that are in place than illustrating the evolution of new functions. So the crucial point is, of course, that such mechanisms rely on an array of macromolecules which are already available rather than the production of new ones. Some resistance may involve the minor modification of one or two of these, but, though it labours the point, I repeat that extrapolating from this to the evolution of whole enzymes from scratch cannot be justified.

It suits evolutionists to have a 'black box' approach to phenomena such as resistance. By not looking too closely at the biochemistry, they can assume that here are examples of new genes arising out of the blue – just as they claim they have arisen in the past. A similar point might be made of melanism: when we did not know what was involved it was easy to assume that a new characteristic had emerged in response to environmental pressures. Once we realised the biochemistry that is necessary to achieve this, it was out of the question.

Macroevolution requires constructive mutations

In the first part of this chapter we looked at the adaptation and diversification of species, and divergence to two or more separate species. As mentioned when discussing these, I do not quibble with using the term 'evolution' to describe such examples, but we should recognise that, whatever evolution is occurring, it is only in a limited sense – essentially shuffling and segregating genes. We must now take the matter further, to examine evolution on the larger scale – of more complex 'higher' organisms from simpler forms.

This sort of transition is now generally referred to as macroevolution, but some confusion can arise because the term was coined by Richard Goldschmidt (1878–1958) to refer to something rather different. He was satisfied that some gradual evolution of a species – such as the development of melanism in moths – could take place, and referred to this as microevolution. However he believed that new species could not arise gradually, but required a substantial jump, and he termed such a transition as macroevolution. He suggested that the jumps occurred through macromutations, resulting in his infamous 'hopeful monsters', most of which would perish, but a few would survive, and this was how substantial evolutionary progress occurred. It is clear that this sort of idea came from observations such as those of de Vries and Morgan on the evening primrose and *Drosophila*, as mentioned above. However, it is now evident that speciation can occur gradually. This is used by many biologists to argue that there is no material distinction – that macroevolution is but the accumulated effects of microevolution. But the important distinction that I want to emphasize is that microevolution involves only shuffling and segregating genes, whereas macroevolution would require the origination of new ones. And I will expand on this here.

Readers will recall that Darwin's theory of evolution by natural selection centres on the fact that species exhibit variations, those individuals that happen to possess advantageous ones reproduce in preference to those that lack them, and in this way beneficial variations are propagated. Darwin assumed that, along with selection taking place, surviving individuals in some way continued to acquire new variations; basing this assumption simply on the fact that natural populations do exhibit variations, and therefore presuming that variations do arise readily. In this way – by the continued generation of new variations, and their assimilation through natural selection – he argued that species change in three main ways: First, a given species can progressively adapt, to become increasingly suited to its mode of life – which can account for the exquisite adaptations found in biology; second, individuals adopting different variations can lead to divergence and thence new species; and finally, over the millions of years of life on earth, variations have arisen which, in at least some lineages, have enabled a progressive increase in their complexity and this is how the higher organisms evolved from lower ones, i.e. how macroevolution took place. It was easy for Darwin to make this sort of extrapolation because he knew nothing of the underlying biochemistry and genetic mechanisms, nothing of how variations arise.

With Neo-Darwinism, it was recognised that the source of variations is genetic variability, and that selection of fitter phenotypes implies selection of the relevant genes. Also, it became evident that a corollary of selecting advantageous genes is that less advantageous ones tend to be lost; so an automatic consequence of continued selection taking place is that there is a progressive loss of genetic variability. This is readily apparent with domestic breeding. On the one hand, with the discovery of extensive genetic variability within most natural populations, it was concluded that substantial evolution in the sense of adaptation and speciation could be sustained at least in the short to medium term without new constructive mutations. But on other hand it was realised that in the long term it would be necessary to compensate for the loss of variability due to selection; so a major assumption of the modern theory of evolution is that variability is topped up by continued mutation. In fact it is assumed that current variability represents a balance between new variability arising by mutation and some being lost by selection (and some can be lost by genetic drift). Importantly, as with Darwin, it is also assumed that, at least in some lines, new genetic variability, arising through mutation, has been such as to enable increased complexity and evolution of the higher organisms, i.e. macroevolution.

As I have pointed out already, even this Neo-Darwinian view was formed before there was adequate knowledge of what genes are or of how they work. Because of their limitations, Darwin and the Neo-Darwinists did not appreciate that macroevolution requires the production of new macromolecules, each of which is highly improbable. We can now see that the Neo-Darwinian position, although it has persisted through the last 50 years, has not faced up to the advances that have been made in molecular biology in that period. Now that we are gaining some idea of the complexity of biology at the molecular level, and especially of what would be involved in lower organisms evolving into higher ones – e.g. more cell types and more macromolecules – the Neo-Darwinian presumption that mutations can provide what is required looks increasingly unrealistic.

In this chapter I have sought to emphasize that, not only do I see selection between organisms having differing fitnesses as a reasonable assumption, but also that there is clear evidence that this process of selection occurs in practice, both naturally and artificially. It is also clear that the genetic variability of most natural species enables substantial changes in their morphology to occur, and hence that species are not fixed. Not only does this permit significant adaptation to occur, but it is also possible that divergence within a species, for whatever reason, can lead to its division into two or more species, whether defined in terms of morphological differences or loss of interfertility. However, whereas the theory of evolution assumes that the reduction of variability due to adaptation and speciation is compensated by new mutations, there is no evidence for this. Of particular importance so far as macroevolution is concerned, there is no evidence for the mutations that would be required to enable an increase in complexity – for higher organisms to emerge from lower forms. The most hopeful examples on offer are the occasional advantageous mutations that confer resistance to antibiotics and the like; but these are clearly of limited significance and, if anything, add weight to the case against macromolecules arising rather than explaining how they might.

Unfortunately, there has been insufficiently clear thinking on the part of evolutionary biologists in relation to this. Because there is clear evidence of evolution in the sense of adaptation and divergence, they have assumed that evolution in the sense of increased complexity must also occur, not recognising sufficiently that the former processes can be achieved solely by the segregation of existing genes, whereas the latter requires new ones.

The crux of the issue is that, whilst there is no doubt new *variations* occur, I contend there is no evidence that new constructive *variability* arises. (I say 'constructive' to distinguish from the variability that can arise through deleterious mutations, such as those found in *Drosophila* by Morgan.) Despite the similarity of these terms, I trust my earlier comments will have satisfied readers that the distinction between variation and variability is substantial and not merely splitting hairs. Indeed, I think that one of the reasons for the current acceptance of the theory of evolution is that too often this distinction has not been adequately made, and the ease with which new variations arise (either naturally or through breeding) has given the mistaken impression that variability – new genetic material – is also readily available. We now know that new variations arise solely through shuffling the available genetic material, not from producing new material, but this fact is all too readily overlooked.

Given that mutations and the evolution of the higher organisms is generally such a slow, long-term process, it might be thought that we cannot assess whether or not the proposed ongoing constructive mutations actually occur. But there are some clear pointers, and I will outline a couple of illuminating examples here.

Anomalous rates of evolution

Ernst Mayr (1967) drew attention to Lake Nagubago which is near to Lake Victoria and has been separated from it for less than 5000 years. Despite this short time for evolution to take place it has many endemic fish species[6] which are quite distinct from those of Lake Victoria. In contrast he comments that the Panama Isthmus has separated the Pacific Ocean and Caribbean for about 5 million years, yet the marine fauna on either side are 'virtually identical' despite their 'colossal' gene pools.

He discusses these observations in the context of illustrating the effect of population size on diversification. In the expansive Pacific and Atlantic Oceans, the gene pools are enormous, on the whole the species either side of the isthmus are freely mating, and the Hardy–Weinberg equilibrium is essentially in effect, so there is little change in gene frequencies, little opportunity for segregation and diversification, so the species remain essen-

---

[6] That is, species which are exclusive to Lake Nagubago.

tially unchanged. Conversely, as mentioned in Chapter 5, population geneticists have known for a long time that random variations are more
likely to be significant in small populations, so we should not be surprised
that random diversification can occur more readily in the small Lake Nagubago. Mayr refers to the rapid evolution of the sparrow in North America in the same discussion.

I entirely agree with Mayr's analysis of these contrasting situations.
What we have in the small lake is evolution in terms of the second and
third steps that we have already discussed in this chapter. Segregation of
genes in this way is the sort of process we can expect to take place over
the course of 5000 years, or even less. But it is important to note that these
observations are in the context of genetic variability already being available – so this is only one side of the coin.

Surely these same observations also tell us something important about
the first step – the production of variability – or, rather, the lack of it. That
is, whilst *segregation* of variability will generally occur more readily in a
small population, variability should *arise* more readily in a large population. The obvious reason for this is that, with more individuals, there is a
larger source for mutations as a whole to arise, in particular of constructive or advantageous ones. Hence, although most mutations are lost
(whatever benefit they may confer), with a larger population there will be
an increased chance of the same (or similar) advantageous mutation recurring so that it will eventually be adopted by the population. In general,
we should, therefore, expect larger populations to acquire more variability,
and generally evolve more rapidly, in the long run.

However, despite the enormous populations of the Pacific and Atlantic
Oceans, and despite an opportunity of about five million years for new
(useful) variability to arise – which should have introduced differences
between the two oceanic populations – it does not appear to have done so.
We should note that five million years is approximately 1% of the time in
which higher animals are supposed to have evolved from very simple
aquatic life and 5–10% of the time for primates such as man to have
evolved from early shrew-like mammals (see Ch. 10). So, if these long-
term evolutionary scenarios, based on the supposed occurrence of constructive mutations, are valid, then surely we should see some sign of significant evolutionary development between the two oceanic populations
over the course of a few million years? But, instead, they are 'virtually
identical'.

Disjunct plant populations

One of the notable features of the worldwide distribution of plants is that
the flora of eastern Asia resembles that of eastern North America much
more closely than it does the geographically nearer western North America. In particular, there are well over a hundred plant species from 65 genera, including species of magnolia, hydrangea and orchid, which occur in

almost identical morphological forms in these two locations, but not in-between. That is, the populations of these discontinuous or 'disjunct' species, as they are known, are separated by an expanse of thousands of kilometres which includes the Rocky Mountains and Pacific Ocean. One of the first to draw attention to this odd distribution of flora was the American botanist, and keen supporter of Darwin, Asa Gray.

For most of these very similar species it is thought that the present day populations are descended from a common ancestral species which extended across a much wider geographical range, perhaps covering much of the northern hemisphere, but which, for some reason or other, has been lost from the intervening area. Based on fossil evidence, estimated times since separation of the Asian and American populations range from 2 to 25 million years, depending on the species concerned. These separation times are supported by comparisons of DNA sequences which indicate the accumulation of some genetic differences (neutral substitutions) between them. It is remarkable that so many species have remained morphologically static for so long in two different parts of the globe. What is even more remarkable is that many pairs of morphologically similar species from the different locations are interfertile, including two species of the plant *Campsis* which are thought to have been separated for more than 24 million years.

What is important for the present discussion is that these disjunct species do not seem to have acquired new variability in terms of genes that affect morphology or interspecies compatibility, even over many millions of years. As with the oceanic faunas mentioned above, we should note that 25 million years is more than 5% of the time since the first simple plants colonised the land, and approaching 20% of the time since flowering plants first appeared (see Ch. 10). Surely, if macroevolution proceeds according to the Neo-Darwinian model, then over time-scales such as these we should see clear evidence of the mutations which give rise to new, constructive and even progressive variations?

Of course, various theories are offered to explain away observations such as these. One is coevolution which I discuss further below. Another is stabilising selection, i.e. that new variations arise but they are weeded out be selection. Yet surely that is just another way of saying that advantageous mutations have not arisen – which is exactly the point I am making. Perhaps the most feeble, yet common, 'explanation' is simply that these species are in the 'stasis mode of evolution' which is invoked to describe the many cases in the fossil record where species seem unchanging for long periods. But clearly that is no real explanation at all.

Evolution within Limits

So what are we to make of all this? First, we saw in Chapters 7 and 8 that there is a *prima facie* case against a random origin of biological macromolecules due to their exceeding improbability, and that there is no known mechanism that can adequately overcome this obstacle. Then – the focus of the early part of this chapter – we saw that the well-known examples of adaptation and diversification amply demonstrate the shuffling and segregation of genes, but shed no light on their origin. Further, although the evolution of higher organisms from lower ones would require the production of new genetic material, there is no sign of this in situations where it would have been expected. It seems we are faced with a dilemma – evidence of evolution on the one hand, albeit only in terms of gene segregation, but an insurmountable hurdle to the first part of the process.

If it is true that constructive new genetic material does not arise, then it follows that a group of related species (perhaps, but not necessarily, corresponding to one of our accepted taxonomic groups such as a genus or family) is substantially limited to the genetic material of some primordial population – whatever that may have been – from which the various species have descended by segregation and selection from the initial gene pool. I realise, of course, that whilst similar views were expressed by some 19th-century biologists, it is heresy to modern biology! But, however extreme or anachronistic this proposal may seem, and whatever the wider implications (which I shall discuss in due course), it is certainly more consistent with the facts than the current orthodox evolutionary position which assumes that all forms of life have progressively developed from simple organisms by selective accumulation of constructive mutations – because there is no evidence for those mutations. Apart from the examples that I considered earlier, support for this idea of evolution from highly plastic primordial populations also comes from various other observations, in particular from some instances of coevolution.

Coevolution

Coevolution is the term used to describe situations where organisms from different ancestral lineages have striking resemblances of some form or other: they are considered to have evolved in a parallel or even convergent manner. An example of what is seen as coevolution on the large scale is how fish, sharks, ichthyosaurs (extinct marine reptiles, see Ch. 10) and dolphins (mammals), though from very different animal groups, all have a similar streamlined profile, appropriate to their aquatic lifestyle. The standard evolutionary explanation is that, in all of these groups, many variations arose (by mutation), including those which promoted a streamlined profile, and these were preferentially selected so that in time an appropriate profile emerged. This sort of coevolutionary explanation, along with the prevailing Neo-Darwinian view of mutations in general, encouraged evolutionists to believe that appropriate variations can and do arise. These views formed, of course, before there was any understanding of what was involved at the genetic or biochemical level in producing such variations.

Cichlid fish

Coevolution is also observed within more closely related taxa. A well-known example of this is the cichlid fish (which includes many of the freshwater 'tropical fish' kept in domestic aquaria) of various lakes in East Africa. In Lakes Tanganyika and Malawi there are about 200 and 400 species respectively of these fish, and in each lake there are several species (at least a dozen) which bear a fairly close morphological resemblance to species in the other. This in itself is not remarkable as it could have been that each of the similar species descended from a common ancestor which entered both lakes some time ago, and then subsequent evolution has led to the minor differences between them. However, what makes these cichlids interesting is that genetic studies have shown that all the cichlid fish in each lake are more closely related to each other than are the morphologically similar species from the different lakes. In other words, the various species in each lake appear to be descended from a more recent common stock than the common ancestors of the morphologically similar species from both lakes. So the occurrence of these morphologically similar species but from different ancestral lines is a classic case of coevolution.

We could suppose that the different stocks of fish have each somehow acquired similar variability (despite the improbability of this), and some biologists favour this sort of explanation because it fosters the idea of evolution not being too difficult, and even that we can expect appropriate variability to arise readily. However, as studies on these fish have continued, increasingly the view is that all of the present-day cichlid species – in both lakes – are derived from some very ancient stock which happened to be very variable and adaptable, populations of which entered the lakes in the distant past and became the ancestors of today's species. That is, it is now thought that the modern species are derived predominantly if not solely by genes segregating in different ways from an ancient highly variable species or possibly a few such species. Given the enormous variability of these fish and the hundreds of actual species that exist in each of the two lakes, it is not at all surprising that a dozen or so from one resemble species from the other. (It may be noted that generally they are much less similar than the disjunct plant species mentioned above.) Further, it is easy to see how, from a common stock with extensive variability, divergence and selection can favour specific gene combinations, for example to suit particular ecological niches which are similar in the two lakes. Either way – whether the similar species have arisen by chance or selection – this is a much more credible scenario than that the same variability arose independently in separate populations. And, of course, it is consistent with the idea of there being highly variable primordial species from which today's species are derived – but derived by gene segregation, not by mutation – which is my main reason for giving this example.

But there is a further reason – and that is to highlight how, once again, evolution which is clearly due to gene segregation is misused to support the notion of evolution by mutation. Because there is no doubt that coevolution occurs within some closely related groups of species by gene segregation, this is taken as evidence for the idea that fish, dolphins etc. have coevolved by the occurrence and selection of appropriate mutations. Such extrapolation is not justified, because the processes in question are totally different. In so far as various aquatic groups have acquired similar streamlined profiles, it is only because their primordial ancestors already had appropriate genes – there is no evidence that such genes have arisen subsequently. Using the coevolution of the cichlid fish as support for the coevolution of unrelated taxa originated before adequate knowledge of the relevant genetics. Now that we know the genetics, this support should be discarded – but all too often it is still cited, simply because it suits the evolutionary outlook to do so.

Loss of diversity

As mentioned above, one of the points emphasized by biologists is the enormous genetic diversity of most natural species; and, although selection reduces that variability, it is supposed to be replenished by mutation, so that there is always potential for variations to occur. However, if, as I am suggesting, species are derived by segregation and selection from a highly adaptable primordial species, and that consequent loss of variability is not restored by mutation, then perhaps we should see some evidence of this.

That is, we are well aware that domestic breeding reduces variability, and we do not know of any instances where new variability has arisen; but the evolutionary response to this is, of course, that any domestic breeding is much too recent for new mutations to have arisen to replenish variability. However, although most natural species have substantial variability, there is a significant minority – about 5% of both plant and animal species – with much lower variability than we might otherwise have expected. Given that selection has taken place in nature over much longer periods than mankind's breeding, then it is surprising when we find long-established species with relatively low variability because, according to the evolutionary model, their variability should continually be being topped up.

It is evident that the wolf has extensive genetic variability which has enabled the numerous diverse breeds of dog to have been derived from it. To a lesser extent, the same is true of the wild cat from which domestic cat breeds are descended; and, like domestic dogs, the various breeds of domestic cats are interfertile. As we would expect, domestic breeds of dogs and cats have relatively limited variability, sometimes resulting in the well-known cases of diminished fitness.

However, what is of particular interest here, is that the large cat species (lion, tiger etc.) also have generally lower than average genetic diversity.

The standard evolutionary explanation for this is the so-called 'bottleneck' effect: it has long been recognised that small populations have lower variability, and it is argued that some present day species have low variability because they had a reduced population in the not too distant past. But to produce the low variability seen in the large cats, each of the species' populations would need to have been reduced substantially, perhaps to less than 100 breeding individuals, for several generations – but there is no evidence that this has actually happened. On the contrary, a recent study of big cats concluded that low genetic variability was a natural feature of these species (Shankaranarayanan *et al.*).

An alternative explanation for their low variability is that an ancient and variable large-cat population diversified to produce modern species, each with reduced variability – analogously to the production of several different domestic breeds from a common stock. It is, of course, generally accepted that the large cats have a common ancestor, but the point in question here is whether or not they have acquired new variability since their divergence. Although they do not normally interbreed in the wild, most if not all of these are in fact interfertile, which certainly is an indication that any new genetic material is minor.

Probably the most well-known case of low genetic diversity in the big cats is the cheetah. It is so low that skin grafts from one cheetah are usually accepted by any other. (In contrast to this, consider the efforts that must be made to ensure tissue compatibility for transplanting skin or organs in humans because of our genetic differences.) This has been explained in terms of the bottleneck effect and inbreeding. However, the difficulty with this explanation is that the world cheetah populations are currently several thousands and the cheetah appears to have been much more abundant and widespread in the past – extending over Asia, Africa, Europe and North America – so there is no sign of anything like a bottleneck. Also wild cheetah populations (as opposed to those in captivity which do often inbreed) do not show signs of inbreeding such as infertility or reduced litter sizes. Current concerns about the future of the cheetah arise from its loss of habitat, not any inherent poor fitness.

A more likely explanation is that here we have a species which, through progressive natural selection, has become highly adapted to a particular way of life, and in so doing has been reduced to a very restricted set of genes. It can be compared with domestic breeding where continued selection has taken a species to the limit of selection. So far as the cheetah is concerned, it is well know to be the fastest land animal, capable of achieving well over 100 km/h for short distances. It would appear that natural selection has matched the ability of domestic breeders to obtain the best possible performance within the limits of available genetic material, and it would not be surprising if in doing so it has abandoned unnecessary features. For example, compared with other cats, it has relatively short and blunt claws which it is cannot fully retract – a clear adaptation for rapid pursuit of prey. Interestingly, in contrast to the general shortness of its claws, that on the first digit of its fore paws (the so-called dewclaw) is exceptionally long, and curved: the cheetah uses this as a hook to catch the legs of its prey (whereas species such as the lion use their weight to overpower prey). The cheetah is known as the most specialised of the living felids, and it is noteworthy in the context of the present discussion that this is accompanied by such low genetic diversity.

Also, cheetahs have been around for quite a long time. The earliest fossil cheetah is dated about 3.5 million years ago, and, as just mentioned, it used to have a much more extensive range than it does now, so there has been plenty of scope and time to acquire new variability. In fact, what makes it even more interesting, is that the earliest fossils indicate that the claws may have been retractable; so this points to the cheetah having become even more specialised in that time.

Another example of an animal with low genetic variability is the polar bear, and this too appears to be highly adapted to its environment and way of life. It is the only carnivorous bear, indeed it is the largest terrestrial carnivore. It is also the only marine bear, capable of swimming more than 50 kilometres per day! And it has a uniform white fur which is suited to its snowy habitat whereas other bears exhibit a range of colours, usually various shades of brown.

A final example of loss of genetic diversity comes from humans, in particular with regard to skin colour. The traditional view, no doubt arising from the fact that fair skin tans in sunshine and darkest skinned races are from sunny climates, is that dark-coloured skin has arisen (by mutation) in response to sunny environments. However it has been known for a long time that not all races from sunny locations have dark skin and, as Dobzhansky (1977) commented: 'Indians in the American tropics are not particularly dark. The explanation that these people have not lived long enough in their tropical habitat to become more heavily pigmented is rather unconvincing, but no better explanation is available.'

I suggest the difficulty arises from a fundamental misconception: It is not that dark-skinned races have acquired dark-coloured skin through living in a sunny environment, but fair-skinned have lost (or reduced) the ability to produce dark-coloured skin through living in non-sunny areas where the protection it affords to UV light would be of limited value but there is some benefit from fair skin in producing vitamin D in response to sunlight. And, through loss of genes due to selection for the protection that dark skin affords, black-skinned races have reduced ability to produce fair skin. That is, the early human species had genetic variability – the ability to produce a range of skin colour; but in the course of time different gene combinations have been selected in different subpopulations to suit their environment. And where they have subsequently migrated to different environments they cannot regain the genes they have lost. However, when black- and white-skinned intermarry this restores some of that diversity; notably, their offspring are not simply brown but likely to exhibit a range of skin colour.

Fixity? ... Evolution? ... or What?

Although the idea of species arising from some kind of primordial stock is more consistent with the facts than the orthodox evolutionary view, I have little doubt that many will have considerable difficulty coming to terms with this sort of concept. One of the stumbling blocks is not so much scientific as historical. It stems from the medieval belief in the fixity of species which arose out of classical biology, persisted through the scientific revolution, and is unhelpfully perpetuated in some quarters even today.

As outlined early on in this book, the perception of species being unchanging can be traced back to Plato with his essentialist view of the world. It was then adopted by Christianity, especially under the influence of Augustine who accepted much of Plato's philosophy, and the fixity of species became associated with the church's doctrine of creation, persisting right through the Middle Ages. It was reinforced by the influx of Aristotle's teaching because of his *scala naturae* in which each creature had its place and no movement was allowed. By the time of the scientific revolution the concept was so firmly established that classifiers such as John Ray and Linnaeus could scarcely contemplate the prospect of either extinction or the appearance of new species.

However, gradually it became evident that many species are not rigidly fixed: at the very least some exhibit variation, and for a few there could be traced a progressive change in form over several generations. When Darwin articulated his theory of evolution, not only did it allow for variations, but he gave them a key role as stepping stones from one species to another: he saw a complete gradation from variations to races, subspecies and new species, and then on to higher taxonomic groups. The major extrapolation which he made, based on the fact that species do vary, was to presume that there is no limit to the variation they could acquire over the course of time. Just like Buffon before him, once he was satisfied that some morphological change occurred, he could see no reason to doubt that any amount of change was possible – whether in terms of adaptation or of higher organisms developing from lower ones.

> Slow though the process of selection may be, if feeble man can do much by his powers of artificial selection, I can see no limit to the amount of change, to the beauty and infinite complexity of the coadaptations between all organic beings, one with another and with their physical conditions of life, which may be effected in the long course of time by nature's power of selection. [*Origin*, Ch. 4]

Wallace expressed a similar view in the title of his paper, *On the Tendency of Varieties to Depart Indefinitely from the Original Type*. To the 19th-century mind, once the fixity of species was broken, it seemed there need be no limit to the changes that could be possible. So, with the general acceptance of Darwin's theory, over quite a short period of time, not only was the fixity of species abandoned, but it was replaced with the whole progression from inanimate matter to higher plants and animals.

In the second half of the 19th century that was entirely believable. Whilst it was a large step to extrapolate from small scale variations to an unlimited continuum, it did not contradict the prevailing science. Voices were raised against it – based primarily on the limits of domestic breeding and the lack of intermediates in the fossil record – but these did not provide a convincing reason as to why there should be limits, and there was always the possibility that new variations or new fossils would be found. Similarly, one or two scientists at the time (such as William Hopkins (1793–1866), see Hull) recognised that the prevailing knowledge of biological structures was minimal, and argued that the issue of evolution could not be decided until much more was known, and suspected that biology would be too complicated to have a natural explanation; but that was an intuitive response, not an objective one, and few heeded it.

So the important difference between then and now is that we have considerable knowledge of the structure and workings of biology at the chemical level. We now know that substantial variations can arise through the shuffling of genetic material, but that there is a limit to what can be achieved in this way. In particular, compared with simpler organisms, higher plants and animals require many new structural proteins and many more new control proteins with their corresponding DNA control sequences in order to implement their increased number of cell types and levels of coordinated activities. Further, we now know that these macromolecules are very improbable structures and, despite our extensive knowledge of biology and chemistry, we have no satisfactory explanation for their origin. Indeed, the more we find out about the complexities of life, the more unscalable the mountain becomes. In other words, with our modern understanding of chemistry and biology we not only observe that there are limits, but we can also understand why there are limits.

Most modern texts still follow the 19th-century evolutionists and would have us believe there is a stark choice: either a rigid fixity of species – each species was created individually and cannot change at all – or the whole theory of evolution from inorganic material to the most complex organisms. It suits evolutionists to put it in these terms because it is now so easy to discredit the idea of species being fixed. This is why so much effort is made to present evidence for adaptation, divergence, ring species, and the mechanisms for speciation; and why some biologists, such as Mayr, argue that macroevolution should be equated with speciation. Evidence of species changing and for new species arising is presented as unequivocal proof for the whole gamut of evolution (although we have seen that is not) and, as a corollary, complete refutation of any sort of alternative. However, approaching the question from an unprejudiced stance –

not trying to defend a traditional creationist or traditional evolutionary position – there is no reason why the choice should be so limited. It seems to me that the evidence points to the creation of highly plastic and adaptable primordial organisms (i.e. populations with extensive genetic variability) from which present day species have evolved by gene segregation.

Although this suggestion may be new to many readers, in fact ideas along these lines have been proposed, in various guises, by quite a few authors over the last couple of centuries. You may recall from Chapter 4 that Buffon thought that today's species have been derived from a relatively small number of primordial forms, and even Linnaeus moved towards this point of view later in life. In the opening Historical Sketch to later editions of the *Origin* Darwin refers to similar views expressed by Geoffroy Saint-Hilaire that 'what we call species are various degenerations of the same type', and by the Rev. W. Herbert that 'single species of each genus were created in an originally highly plastic condition, and that these have produced, chiefly by intercrossing, but likewise by variation, all our existing species'. Even in the 1950s a number of scientists advocated this (see Mixter). But all of these preceded knowledge of molecular biology and lacked a fundamental basis for *why* there should be a limit to available variations.

One way to test this hypothesis is, of course, to examine the fossil record, so it is to this that we now turn.

10

Gaps and Gradations

The theory of evolution was based on large-scale morphological character-istics of plants and animals. Initially there was no problem in proposing that organisms could acquire and progressively accumulate unlimited morphological variations because at the time biologists just did not know what was involved in doing so. However, we have seen in Chapters 7 and 8 that molecular biochemistry has shown biology to be immensely more complex than the 19th century naturalists ever imagined, and based on exceedingly improbable macromolecules which cannot have arisen in a progressive evolutionary manner. Consequently, molecular biochemistry reveals a fundamental flaw in the theory of evolution.

Long before the biochemical difficulties became apparent, the theory had become well established because of the many examples that came to light of natural selection in action and other processes related to speci-ation, which were seen as clear evidence for evolution, and are still cited as such today. However, in Chapter 9 I have shown that such observations are entirely explicable in terms of segregating or changing the frequency of pre-existing genes. None of this ecological evidence provides any indi-cation whatever of how the intrinsic improbability of biological macro-molecules might be overcome; moreover, there is a marked lack of evidence for the constructive mutations which macroevolution requires.

No doubt many people will still dismiss these objections regarding the origin of macromolecules as somewhat theoretical, and will point to the fossil record as clear demonstration that evolution has occurred. This is because the commonly-held view is that fossils show life on the large (morphological) scale has evolved, even if we do not yet understand the mechanism of how this was achieved on the small (molecular) scale. My primary aim in this chapter is to consider this claim; and we will see that the story told by the fossils is nothing like so convincing as is generally believed. Arising from this I will look at the patterns evident in the fossil record, and consider these in the light of my earlier comments, especially regarding the important distinction between the origin and segregation of genes in the overall process of evolution.

This chapter refers to many of the main geological strata and plant and animal groups, and I draw readers' attention to the stratigraphical column and classification scheme given in the appendices. The latter is strongly biased – it is designed to guide readers with the relationships of those taxa mentioned in the text rather than to give a balanced classification of all living organisms.

THE MISSING RECORDS

In Chapter 3 we saw that Cuvier was one of the first geologists to see a temporal sequence in the fossil records. From his excavations in the Paris Basin it was clear that fossils in superficial deposits resembled modern animals, but that at progressively greater depths they were increasingly different. His researches were only in fairly recent rocks where mammals predominate, but fossils from older strata elsewhere continued the trend. By the early 19th century the main series of strata were seen as dominated by particular groups of animals: mammals in the Cenozoic, reptiles in the Mesozoic, fish were abundant in the upper Palaeozoic strata such as the Devonian, and with invertebrates below them. It was this general progression – from 'simpler' to 'higher' forms of life as one ascended the geological column – which promoted evolutionary ideas before Darwin, such as those depicted by Robert Chambers in his *Vestiges*.

When Darwin proposed his theory, he recognised that the fossils then available lacked the intermediate forms between the various groups of creatures which gradual evolution required. He ascribed this to the prevailing scanty knowledge of the fossil record, pointing out that only a small part of the globe had been explored geologically, and he maintained that in time the intermediates would be found. Even in his day there was criticism that this optimism was unfounded but, despite this omission of intermediates, the other attractions of his theory meant that most were happy to accept evolution and, like Darwin, hope that the missing records would turn up in due course. That may well have been a tenable position in the mid 19th century.

However, 150 years on, after millions of man-years of searching in which whole new areas of the world have been explored – notably Africa, Asia and Australia – opening up major new fossil beds from which, quite literally, thousands of times more fossils have been unearthed than were known in Darwin's time, the missing records have still not been found. Predominantly what has been found is more of the same: more of the same species, more species from the same genera ... more classes from the same phyla. That is, the new finds have added more representatives to the established taxonomic groupings. In a few cases, particularly the Cambrian deposits of the Burgess Shale (Canada), species have been found which do not belong to any previously known phyla and new phyla have been defined for them – new phyla, but not intermediates between the phyla that were previously known. So we have many millions of fossils, classified into hundreds of thousands of species, but the radical divisions between groups of organisms persist: the longed-for links remain as elusive as ever. Indeed some palaeontologists (e.g. Nelson) have commented that, far from resolving questions of how extant groups of organisms are related, on the whole fossil finds have increased the number of 'obscure relationships' for which we do not have a satisfactory evolutionary explanation.

It is not as if there are only a few insignificant gaps – many speak of 'the missing link' as if there were just one or two – on the contrary, there are yawning gulfs between most if not all of the main groups of organisms. Something of this can be appreciated by scanning phylogenetic trees presented in many evolutionary texts: they consist of lines of descent of various groups of organisms, the origins of the groups being joined by dotted lines to indicate the proposed ancestral links – proposed, and generally believed to have occurred, but there is no evidence that they actually existed. So, for the benefit of readers unfamiliar with the realities of the fossil record, this section gives an outline, highlighting some of the main discontinuities.

The Cambrian explosion

The overall pattern of the fossil record is typified by the abrupt appearance of a wide range of organisms, predominantly invertebrates, with hard (fossilizable) parts in Cambrian strata dating from about 540 million years ago. They were complex, well-developed organisms with many types of differentiated cells, and it is generally accepted that they could not possibly have evolved from unicellular precursors within a short space of time. It is to be expected that non-skeletonized predecessors will leave few if any fossils, so it would not be surprising for one evolving line to appear suddenly in the fossil record, once it reached the stage of being fossilizable. But what is surprising is the wide variety of fossilizable forms which appeared at more or less the same time. In fact this is an example of how, as we have learnt more, instead of solving the problem, it has become more acute: early discoveries dated the emergence of different forms in the Cambrian over a period of up to 100 million years; but more recent discoveries, especially from China, have reduced this span considerably – some would say to as little as 10 million years. (Even the Edicarian fossils which were once thought to precede the Cambrian by as much as 100 million years are now dated within 13 million years, see Valentine *et al.*) The number of totally different body types which appear in a very short period of geological time is greater than at any other, such that their emergence is often referred to as the Cambrian explosion. All phyla represented by modern day animals, certainly all those with fossilizable parts, were included, yet for none is there any clearly identifiable ancestor. Explaining their abrupt appearance is one of the leading challenges in evolutionary biology.

There are two main options to account for this sudden diversity: either the different forms arose at the start of the Cambrian from a single common stock from which the various forms radiated, or the divergence occurred much earlier. If they diverged at the beginning of the Cambrian then this goes some way to explaining the absence of preceding fossils; but then it is difficult to account for the wide divergence of characteristics occurring so suddenly, i.e. that the various phyla appear so different as to be unrelated. But if they diverged considerably earlier in the Precambrian, so as to allow time for the divergence of various forms, then it is extraordinary that all the various lines should reach a fossilizable stage at much the same time. The coincident appearance of various fossilizable forms is all the more remarkable when one considers their very different types of skeleton:

> The most well known animals of the Cambrian are the trilobites. These are a subphylum of the arthropods which are invertebrates having an exoskeleton or cuticle comprising protein and chitin which is a polysaccharide quite similar to cellulose used by plants. Present-day arthropods include insects, although they did not appear until much later. Another group of arthropods which appeared in the Cambrian are the crustaceans, in many of which the chitinous exoskeleton contains deposits of calcium carbonate and/or calcium phosphate.

> The simplest multi-celled animals are the Porifera or sponges. They have a skeletal framework comprising fibres of a protein called spongin (which gives their familiar spongy character), and/or mineral spicules of calcium carbonate or, notably in the Hexactinellida or 'glass sponges', of silica. Both calcareous and siliceous sponges are found in the Cambrian.

> Then there are the echinoderms (e.g. starfish and sea urchins) with their skeleton of ossicles, composed of a magnesium-calcium carbonate, which often protrude to give them a 'spiny skin' from which their name is derived.

> Another major phylum appearing at this time was the molluscs, including the familiar calcareous shells of gastropods (e.g. snails) and bivalves (e.g. common seashells such as clams).

> And finally, as well as the various invertebrates, also appearing in the Cambrian are the first vertebrates, possessing bone which contains hydroxyapatite, a form of calcium phosphate.

All too easily, because most invertebrates are 'simple' lower animals, we might assume that their skeletons must be rudimentary and easily formed in a primitive environment. But that would be a mistaken view. It is evident that the proteins of the various organic matrices require appropriate genes to encode them; and it should not be overlooked that the polysaccharide components require specific enzymes for their synthesis. Then, the mineral components of skeletons do not arise by simple precipitation, but this too results from biological processes requiring appropriate enzymes and control systems. In fact we still understand little of how biomineralization occurs.

Of particular note is the construction of spicules in the sponges, because their cells are usually rather independent, but several cells cooperate in spicule formation. For example, to produce a three-rayed calcareous spicule, three cells partially fuse together, each then divides, a ray of mineral starts to form between the cells of each pair, and then the rays fuse together at precisely determined angles. Each pair then continues to grow its ray: one cell extending it and the other thickening it. As one textbook puts it: 'Sponge cells are so poorly coordinated in most matters that the relatively complex cellular cooperation in skeleton formation is amazing' (Meglitsch and Schram, Ch. 3).

Equally impressive are the echinoderms because each of their ossicles is a discrete crystal which is laid down by a single cell or its daughter cells. Interestingly, echinoderms use their skeleton as a store for calcium and mobilise it as necessary in much the same way as vertebrates use their bones.

Even though the cuticle of arthropods is essentially organic rather than mineral, this does not mean it is a simple structure. On the contrary it is composed of several distinct layers which are secreted by cells of the underlying epidermis. Further, of course, because the rigid outer cuticle encases the arthropod, it restricts the animal's growth. This problem is overcome by periodic moulting in which the inner layers of the cuticle are digested, a new cuticle is formed, and then what is left of the old exoskeleton splits and is shed. At each moulting, new sensory structures in, and muscle attachments to, the new cuticle must be made. All of this is under hormonal control, the detailed mechanisms of which we are beginning to elucidate.

In summary, what is generally proposed, is the extraordinary coincidence that these diverse types of organism (and others I have not mentioned), with their radically different skeletons, all reached fossilizable stage within a relatively short period of time. A century ago it was a somewhat optimistic proposition; as we learn more of what is actually involved it looks increasingly unrealistic. The explanation usually offered is that there was a substantial increase in oxygen at about this time which stimulated biological progress; but that completely ignores the improbability of biological macromolecules, whether oxygen is plentiful or not.

The enigma is compounded because, not only do different phyla appear suddenly, but also, within most of the phyla, quite distinct classes arise, again at more or less the same time:

For example, there are four classes of sponges, and all of them appear in the Cambrian. Despite their relatively unspecialised structure, the classes are distinct and difficult to relate to each other (the Hexactinellida in particular have many peculiar features), there are no signs of intermediate fossils, and there is no consensus as to how they might have evolved from a common ancestor. Indeed, despite their primitive form, and it might have been thought that other animal groups evolved from them, the sponges are quite separate from the rest of the animal kingdom.

Similarly, so far as the arthropods are concerned, the different subphyla of trilobites, horseshoe crabs and crustaceans arise in the Cambrian.

In addition, the crustaceans are very diverse: they are as important in the aquatic environment as insects are on land. All four major classes of the crustaceans and many lower taxa are found in the Cambrian; but, despite this wealth of fossils, no trace can be found of the supposed intermediates which would link the different groups to a common ancestor.

The same is true of the molluscs, of which at least four classes arise in the Cambrian. Also, similar in external appearance to the bivalve molluscs, but with a very different internal structure so they are in a separate phylum, are the Brachiopoda (lamp shells). There are two main types of brachiopod depending on whether the shells are articulated or not; the inarticulates are considered more primitive than the articulates, but both groups appear together at the beginning of the Cambrian.

So the Cambrian explosion raises the sorts of questions that occur repeatedly regarding the fossil record. First is that major new types of organism appear suddenly, without preceding partly-developed forms. Second, many different lines, exhibiting the same sort of significant development, arise at about the same time. On the one hand they are so diverse that it is hard to believe they had a recent common ancestor, but it seems so unlikely that the same sort of advance could have arisen independently in several lines, especially simultaneously. And, repeatedly, there is the lack of fossil evidence of intermediates or identifiable predecessors which might explain how the proposed evolution occurred. We will see many more examples of this as we progress up the geological column.

Vertebrates

Turning now to the vertebrates, generally regarded as the most highly developed group of animals, and of course to which humans belong. Vertebrates are the major subdivision of the phylum of chordates which are animals having, at least in their embryo stage, a notochord which comprises a fibrous rod along the back and an accompanying hollow nerve cord. In vertebrates the notochord becomes reinforced with cartilage and/or bone to provide the spinal column, and surrounds and protects the spinal chord.

Jawless fish

The earliest vertebrates were various groups of jawless fish (Agnatha). A few have been found that resemble their eel-like modern counterparts – lamprey and hagfish – but most were of a more conventional fish shape, and many had bony dermal (skin) plates. It should be noted that the jawless fish were quite advanced organisms. Being jawless, they were filter feeders – drawing in water through structures similar to gill slits, and filtering out food particles (like lamprey); but they were genuine fish, typically 20 to 30 cm long, with fully developed spinal column, a tail which they used for propulsion (like modern fish) and various sensory organs including fully-formed eyes. But despite the obvious complexity and well developed structure of these earliest known vertebrates – a complexity which would require considerable periods of evolution – there are no fossils known of any part developed form.

The taxonomic group of chordates includes some present-day organisms which are simpler than the vertebrates – the cephalochordates – the best known being amphioxus, and there is a somewhat similar organism known as *Pikaia* in the Burgess Shale. Because it is recognised that the vertebrates are highly developed organisms, it is proposed that they evolved via creatures such as these. However, there are substantial differences between the two subphyla. Of especial note is that cephalochordates have nothing equivalent to the vertebrate brain. Also, neural crest cells, which develop into a wide range of adult cell types are unique to vertebrates, as are the hard tissues of cartilage, bone and teeth. There are no intermediates known between cephalochordates and vertebrates, so it is mere conjecture that one evolved from the other.

The problem of the origin of the first vertebrates is compounded by the fact that, as indicated above, there were many diverse groups of jawless fish. The most common had a large bony shield which extended over most of the anterior (front) half of their body, but others lacked this shield and instead were completely covered with small bony scales. Some groups had paired fins whilst others did not; some appear to have been bottom dwellers, whilst others were free-swimming. The groups are fully differentiated with no evidence of intermediates between them.

Jawed fish

Whilst the earliest fish were jawless filter feeders, the vast majority of modern fish have jaws, and the earliest fossils of jawed fish appear in Silurian rocks. The differences in skeletal structure between a jawless fish and a jawed fish are substantial, but various mechanisms have been proposed for the transformation, e.g. from a gill arch to jaws, but no intermediate forms are known. What is remarkable, is that the transition from jawless to jawed fish is widely accepted, yet, as well as there being absolutely no knowledge of any transitional form, there is not even a likely candidate for the ancestral jawless fish:

> We would expect that among the great variety of ostracoderms [main group of jawless fish] there would be some group containing annectent forms, linking the Agnatha [overall class of jawless fish] with the higher fishes. Yet the fossil record fails to show any such link. [*Colbert's Evolution of the vertebrates*, Ch. 2)

In the light of this absence of a link, these authors conclude that the divergence between jawless and jawed fish must have occurred much earlier than their appearance in the fossil record. But – a repeated theme – that merely puts off the question of their origin, it does not answer it.

There were also several groups of the early jawed fish, such as the acanthodians (spiny sharks) which are quite distinct from any other group –

there is no trace of their ancestors, or even anything to link them with modern cartilaginous fish (e.g. present-day sharks). As for the other early jawed fish, although collectively known as the placoderms for convenience, there were many disparate types. One order, the arthrodires, stands out for their peculiar arrangement of bony shields: as well as one about their head (common to many early jawed fish), they also had a thoracic shield, and the two shields were hinged – apparently allowing the head to articulate upwards as the jaw opened, enabling an even wider gape. Some of these were extraordinarily large – the combined shields reaching 3 metres in length, and the fish as a whole estimated to be about 10 metres long! In fact, the placoderms were so diverse that most palaeontologists consider them to be polyphyletic i.e. the various distinct groups having multiple origins. Which is a remarkable conclusion given that no credible ancestor can be identified for any of the jawless fish.

Bony fish

Bony fish, in which the skull and jaws, vertebrae and ribs, gill covers and scales are of a bony composition, appeared first in Devonian strata. They are divided into two main groups depending on the structure and control of their fins. In ray-finned fish (with which we are most familiar) the fins are supported by bony rays whose movements are controlled almost entirely by muscles within the body wall, whereas in lobe-finned fish the fins contain small bones and muscles which control the fins. There were also other differences between them, for example in their number of fins and type of scale. What is striking is that both groups arise fully differentiated and appear at about the same stage in the fossil record. With their bony skeleton etc. both groups have what is generally considered to be a significant advancement over their predecessors, and would be expected to have arisen from a common line, but no common bony ancestors are known. So, yet again, their evolution by way of divergence from a common line is assumed to have occurred before their appearance in the fossil record, but there is no evidence of it.

Early amphibians

Lobe-finned fish are believed to be the ancestors of land vertebrates, proposing that their bony fins developed into tetrapod limbs. The overall class of lobe-finned fish includes lungfish, both fossil and three present-day species, which has further suggested that they were predecessors to air-breathing land animals, although it is not generally thought that terrestrial animals are direct descendents of the early lungfish.

It is easy to be drawn into thinking that bony fins could readily develop into the limbs of a terrestrial animal, and the evolutionary perspective also emphasizes that some of the early amphibians resembled fish in their overall hydrodynamic shape and one or two even had fin-like structures on their tail. However, this emphasis completely overlooks the major structural differences between even the most amphibian-like fish and earliest fish-like amphibian (typically *Ichthyostega*), mainly due to the fundamentally different modes of locomotion and the need for a terrestrial animal to support its weight in a way which is quite unnecessary for an aquatic organism which is buoyed by the water.

First, the bones of the posterior paired fins of the lobe-finned fish were not attached at all to the backbone (and are not attached in modern lobe-finned fish), whereas in tetrapods the hind limbs are connected via a well-developed pelvis. Second, the anterior fins are attached directly to the skull; but such an arrangement for a land animal would transmit tremors from the impact of limbs on the ground directly to the skull and thence brain, so it is not surprising that in tetrapods (including the earliest amphibians) the forelimbs are attached instead to the backbone. This attachment is not immediately behind the skull, but there are several intervening vertebrae which provide further buffering and permit movement of the head independently of the rest of the body. This arrangement of supporting the spine near each end then permits the body organs to be suspended from it – a suitable arrangement for a land animal. Land animals also tend to have interlocking vertebrae so that the spine can perform this structural role.

Even more striking than the differences in limb attachments is the origin of the polydactyl limb in the first amphibians. It is well known that land animals have a similar overall bone plan in their limbs: starting with a single bone where it contacts the rest of the body, then a pair of bones, then several bones leading to multiple digits (for example the humerus, ulna and radius, carpals and metacarpals, then phalanges in the arm and hand of man). In most land vertebrates there are five digits and the whole structure is termed a pentadactyl limb, although the earliest amphibians had more and modern amphibians have four! The bone structure in fins is quite different from a tetrapod limb: a single bone equivalent to the humerus can be identified, and perhaps two bones comparable to the ulna and radius, but then any possible homology breaks down, and there is nothing at all comparable with the digits. So the whole polydactyl structure had to arise between the last known fish predecessor and the first amphibian.

Another marked difference between fish fins and amphibian limbs is their overall orientation: In fish the fins are swept backwards, consistent with reducing hydrodynamic resistance; whereas amphibians have the usual land-animal arrangement of the limbs being directed forwards.

So there are at least four major structural differences between lobe-finned fish and the early amphibians: the attachments of fore and hind limbs, the orientation of the limbs, and the formation of polydactyl limbs. There are no fossils between the fish and amphibians showing creatures having only some of these features, nor of any having an intermediate stage for any of these structures, e.g. in the formation of a polydactyl limb. Somewhat disingenuously, many evolutionary texts depict a transitional stage (for example with limbs beginning forwards but then swept back-wards, and with part-formed digits), no doubt trying to make the gap look more bridgeable – but there is no such intermediate form known.

All too often, the substantial modifications that are required to effect this change in lifestyle – from fish to amphibian – are just trivialised or glossed over to the point of misrepresentation. For example, here is an account from a modern textbook on evolution:

> Sometimes, by chance, an organ that works well in one function turns out to work well in another function after relatively little adjustment. Fins in both fish groups evolved for swimming. In some lobe-finned species, they probably came to be used for skuttling around near the seashore or on the bottom of rivers or lakes. From this point, only a small change was re-quired for the fish to walk on land. Whatever the details involved, it is a reasonable inference that the lobe-finned skeleton was, unlike a ray-fin, preadapted to evolve into a tetrapod limb. The term preadaptation is ap-plied when a large change in function is accomplished with little change of structure. [Ridley, Ch. 13]

This quotation also illustrates the widespread tendency on the part of bi-ologists to adopt an 'inheritance of acquired characteristics' or 'effect of use' way of thinking in evolutionary scenarios. We must not lose sight of the fact that to be of any evolutionary significance there must be a genetic basis; but any changes will occur at random, and utility can be determined only retrospectively by natural selection. A fish cannot anticipate the po-tential value of limbs or in any way promote an appropriate change. I will take up his theme in the next chapter.

The time between the latest proposed piscine predecessor and the ear-liest known amphibian is reckoned to be 20 to 30 million years. This is acknowledged to be a relatively short period in geological terms, and would require quite dramatic changes to take place. The situation is even more problematic because there are several distinct groups of the early amphibians. For example, appearing at much the same time as the ich-thyostegids were the anthracasaurs whose vertebrae structure was more suited for living on land; and another group to appear early on were the temnospondyls with a yet different vertebrae structure, even more suited for load bearing, and were generally more heavily built.

Of particular note is the subclass of lepospondyls, the first of which ap-peared early in the Carboniferous. Not only were their vertebrae very dif-ferent in terms of final structure, but it is also apparent that they formed differently – by direct deposition of bone rather than the more usual being preformed in cartilage. They are subdivided into five orders which are quite distinct morphologically – especially in terms of vertebrae, limbs and skull – and in their lifestyles. For instance the aïstopods had much reduced limbs or were even limbless and snakelike, whereas the microsaurs were very varied and predominantly aquatic. Following a recent detailed study Robert Carroll concluded that the common features of these groups suggest they are monophyletic, but there is no sign of a common ancestry in the fossils. Further, 'no member of any of these groups provides obvious evidence of relationships with other Palaeozoic tetrapods'.

In other words, not only do distinct subclasses of amphibians appear without connecting family tree, but this is also true even within a subclass. This has prompted some palaeontologists to propose that the amphibians, like the jawed fish, were polyphyletic; in other words that the transition from lobe-finned fish, even though it involves several large steps, nevertheless occurred more than once, in different lines, within a short period of time. Not surprisingly, others have rejected this option on the grounds of its improbability; but that leaves the problem of how different groups of amphibians arose from a common fish-like ancestor without leaving any trace of intermediates.

Because of the diminutive or even absent legs of the aïstopods, these are sometimes cited (in popular rather than scientific literature) as intermediates between fish and fully-limbed amphibians; but this is not a view shared by palaeontologists who recognise their substantial differences from fish in other important respects such as vertebral structure, as well as not being the earliest group of amphibians.

Modern amphibians

Although somewhat out of sequence so far as the geological column is concerned, it is appropriate to comment here on modern amphibians. There are three groups of these: anurans (frogs and toads), urodeles (newts and salamanders), and caecilians (limbless, wormlike burrowers). These three groups have in common a range of features which are quite distinct from any of the early amphibians. Most significant is that instead of scales they have a moist glandular skin which is permeable to water and gases and through which they respire i.e. their skin acts rather like lungs, exchanging oxygen and carbon dioxide with the atmosphere. They also have in common a distinctive vertebral structure, two bones in their middle ear rather than the single bone of the early tetrapods, and the anurans and urodeles have four digits on their limbs rather than the usual five. Finally, they have teeth which include fibrous tissue between the calcified root and crown – a structure which is not found in any other vertebrates. Despite these similarities, there is no known common ancestor – all three groups appear fully differentiated in Jurassic strata. Further, there is not the slightest sign of any feasible intermediates linking them to the early amphibians: virtually all of those died out by the end of the Permian period with just one or two lingering into the Triassic, and those bearing no resemblance to modern amphibians.

Reptiles

The most significant differences between amphibians and reptiles relate to their eggs. Amphibians, like fish, lay eggs in water or other moist environment and (with just one or two exceptions) the eggs must remain wet to maintain their structural integrity and to avoid desiccation. In contrast, reptiles lay eggs on land and, consistent with this, reptilian eggs have fundamental differences in structure from amphibian eggs. Most obvious is the rigid calcareous outer shell which provides structural integrity. But within that, as well as the yolk sac which is common to fish and amphibians, there is an amniotic membrane (which gives the name of amniotic egg for reptiles, birds and mammals) which is fluid-filled and envelops the developing embryo, an allantoic sac for the accumulation of metabolic waste, and chorion for the exchange of oxygen and carbon dioxide. The requirement for an amniotic egg to 'breathe' means that it must be laid in air not water, which is an important difference from amphibian eggs. Also, arising from the use of an amniotic egg, whereas fish and amphibian eggs are fertilized after being laid, reptilian eggs must be fertilized first. Clearly this requires coital mating and consequent substantial differences in the reproductive organs of both sexes. Also, corresponding with how the different types of egg are laid, amphibian young have gills, whereas reptiles do not.

Although these features of the egg, reproductive organs and breathing constitute fundamental differences between amphibians and reptiles, they do not result in such marked contrasts of skeletal structure that would be distinguishable from fossils as, for example, those between fish and the first land animals. Consequently it is not surprising that some of the early reptiles resemble some of the contemporaneous amphibians, just as, in a general sense, modern-day lizards and salamanders are similar. The amphibians most like reptiles are called Seymouria, a suborder of anthracasaurs, which, although having distinctively amphibian skull and teeth, in the rest of the skeleton had a number of features which resemble those of reptiles. Seymouria cannot have been actual ancestors of the reptiles because they lived much later than their first appearance; but it is widely held that reptiles evolved from a related group of amphibians. However, others have concluded that reptiles evolved as a sister group of the lepospondyls (Laurin and Reisz). This difference of opinion among palaeontologists illustrates clearly that the fossil record does not show an actual transition from amphibians to reptiles.

The earliest reptiles are identified as captorrhinomorphs, which were small land carnivores. But what is striking about the fossil record is the way in which diverse groups of reptiles arose at about the same time. There were at least two other distinct types of very early reptiles: one was the mesosaurs which, right from the outset, were highly adapted to an aquatic life style, such as having webbed, paddle-like feet; and then there was the earliest group of reptiles to have some mammal-like features such as differentiated teeth (see below). The differences between the various groups are so substantial that, as with the different types of bony fish or amphibians, some palaeontologists have considered whether these groups had a polyphyletic origin. But, as we have already noted, whereas this suggestion may go some way to explain away the lack of intermediates, it requires the highly improbable scenario that the essential reptilian features (notably the amniotic egg and appropriate reproductive organs) arose independently multiple times.

The first reptiles are identified in late Carboniferous strata, with some new groups appearing in the subsequent Permian period at the end of which there was a major extinction, especially of marine organisms. Following this, many new reptile groups arose in the Triassic and Jurassic periods which, together with the subsequent Cretaceous period, form the Mesozoic era which is known as the age of reptiles. But throughout this long period, with diverse groups of reptiles arising, including the various types of dinosaur, the occurrence of new forms is abrupt. Although family trees, linking various groups, are proposed, they are only speculative because clearly identifiable intermediates are not known.

The most remarkable examples of radically new reptile groups appearing abruptly are those specifically adapted for aquatic and aerial life, the ichthyosaurs and pterosaurs respectively. These creatures were quite distinct from the terrestrial reptiles – they appear on the scene fully developed and specialised, with no trace of any progressive evolution from earlier land-dwellers.

Although there were other groups of aquatic reptile, none is remotely like the ichthyosaurs, and there is no sign of any link between them. As recently as 2001, Colbert *et al.* (Ch. 12) commented that 'The basic problem of ichthyosaur relationships is that no conclusive evidence can be found for linking these reptiles with any other reptilian order.' The ichthyosaurs were remarkably dolphin-like in their overall shape (although with a vertical rather than horizontal tail fin), including having nostrils on the top of the head. Consistent with their exclusively aquatic lifestyle (so their spine did not have to support the weight of the body organs), they did not have the interlocking vertebrae which are typical of land-animals. Also, as amniotic eggs cannot be laid in water, the ichthyosaurs were viviparous – they gave birth to live young, just as some modern reptile species do.

The pterosaurs were equally well adapted to their aerial lifestyle. Their wings consisted of skin stretched between the limbs, including a very elongated fourth digit, and body, somewhat like present-day bats; and, like birds, they had hollow bones for lightness. Some of the pterosaurs were huge – with a wing span of more than 10 metres.

The dinosaurs arose towards the end of the Triassic period. Right from the outset there were two completely distinct groups, which differed in various ways, but have as a primary distinguishing feature the shape of the pelvis. One group, known as the Ornithiscans because they had a bird-like pelvis, were mainly herbivorous and included among their number the well-known Stegosaurs and Triceratops. The other group, known as the

Saurischians because they had a lizard-like pelvis, were mainly carnivorous and included the Velociraptors and infamous Tyrannosaurus.

Birds

It was during the time of the dinosaurs that the first birds appeared, notably *Archaeopteryx* from Jurassic strata in Bavaria. Because it had fully developed feathers, including primaries and secondaries which were arranged on its wings in the same way as modern birds, *Archaeopteryx* is recognised, and classified, as a bird. However, its skeleton had a number of reptile-like features, notably a long bony tail, claws on the digits of its fore- as well as hind limbs, and in having teeth; so it is seen as a link – showing that birds evolved from reptiles. In particular it is widely thought that birds arose from a group of dinosaurs known as the theropods. Features of the latter that are seen as anticipating the avian form are their overall modest size (unlike many of the giant dinosaurs), they were clearly bipedal (rather than walking on four legs as did some dinosaurs), their hind legs were long and clawed, and, like birds, they had only three digits on their forelimb. Further, some of the theropods had hollow bones, and some later ones did not have teeth, with their mouth resembling a beak.

However, despite similarities between theropods and birds, there are substantial differences which clearly militate against an ancestral relationship. First, contrary to what might have been expected, the theropods were Saurischian not Ornithischian, i.e. they had the distinctive lizard-like pelvis rather than a bird-like one. Then, although both theropods and birds have only three digits in their forelimbs, they are different ones: theropods have the first to third of the usual five digits, whereas birds have the second to fourth. A particularly important difference is that the theropods, whilst having long and well-developed hind limbs, had very small fore limbs (comparable with Tyrannosaurus). Such short limbs were hardly credible precursors for the extended and well developed fore limbs of birds, including those of *Archaeopteryx*, which must support their wings. Also, typical of running animals, they lacked clavicles which are present in *Archaeopteryx* and fused to form the familiar avian wishbone (furcula) which contributes to support of the wing. Finally, *Archaeopteryx* had a significantly larger braincase, reflecting the need for additional sensory input and motor control required for flying.

Also worth mentioning is that birds have a unique arrangement for distributing air around their bodies – to support the very high metabolism associated with flying. There is nothing to indicate that any reptile group had such a system or even anything that could be considered a precursor to it. Related to this, one group of scientists (Ruben *et al.*) concluded that the theropods had a crocodilian mode of breathing, and that to convert from this to the bird system would mean going through a non-viable intermediate stage:

The modern bird system requires a single thoracic cavity, whereas the crocodile and theropod system require the thorax to be divided into two separate airtight chambers. Ruben and colleagues argue that the earliest stages in the evolution of avian respiration from the theropod system would have required selection for a diaphragmatic hernia in the intermediates. This would have prevented the animal from breathing... [Thomas and Garner]

Importantly, the most bird-like theropods did not arise until late in the Cretaceous period – about 70 million years after *Archaeopteryx*, so could not have been its ancestor. In fact, they were even later than various groups of Cretaceous birds which were more similar to modern ones than was *Archaeopteryx*.

This is a convenient point to comment on the recent finds of 'feathered dinosaurs' in China. These too are from the Cretaceous period, so cannot have been ancestors of *Archaeopteryx* but might be from sister groups. Also, their feathers are much less well formed than on *Archaeopteryx*, and some scientists doubt that they really were feathers (e.g. Ruben and Jones). Nevertheless, many biologists see these as primitive (evolving) feathers, assuming that they illustrate the sort of creatures that were ancestral to *Archaeopteryx*. Others (e.g. Geist and Feduccia) believe it is more likely the 'feathers' were vestigial; that is, they consider the fossils to be of flightless birds, descended from *Archaeopteryx* or similar – just as modern flightless (ratite) birds are believed to have descended from earlier flying ones rather than vice versa (de Beer, 1975).

Finally there is the general problem of how powered flight started. Was it by gliding down from trees, or by running fast along the ground? It is a long-standing debate because there is no convincing evidence, and some contrary evidence, either way. The point is that *Archaeopteryx* was a fully-fledged bird, capable of proper flapping flight, not an intermediate runner or glider, and there is no convincing running or gliding predecessor.

So it may be fair to conclude that *Archaeopteryx* was of intermediate structure – in the sense that it was a bird with some features shared by some reptiles; but that does not make it transitional in terms of phylogeny. As with so many other classes of animal, the birds arise in the fossil record without any clear link to earlier creatures.

Mammals

It is generally assumed that, like birds, mammals evolved from reptiles. Comparable with the main differences between amphibians and reptiles, the primary feature that characterizes mammals from other vertebrates – the mammary glands – are clearly soft tissues and hence not discernible from fossils. This is also true of other important distinguishing characteristics such as a four-chambered heart, diaphragm and being warm blooded. However, there are some skeletal features that are specific to mammals, of which the most prominent, especially compared with reptiles, are the following:

Ear ossicles: Mammals use three bones (malleus, incus and stapes) in the middle ear to transmit and amplify vibrations from the eardrum to the inner ear; in reptiles, there is only one, the stapes.

Jaw: The mammalian jaw is formed from a single bone, the dentary; whereas several bones are used in reptiles, notably including the quadrate and articular which form the jaw joint.

Teeth: More than any other vertebrates, in most mammals teeth are differentiated into incisors at the front, then canine, with premolar and molar behind. As well as this differentiation, the molar and premolar teeth of upper and lower jaw are closely matched to enable effective maceration of food.

Skull: In addition, like birds, mammals generally have a relatively much larger braincase than other land animals, consistent with their expanded and more developed cerebral cortex. And they have a double occipital condyle forming the articulation for the skull with the first vertebra of the neck.

Spine: Unlike reptiles, in mammals the ribs on the cervical vertebrae (neck) are much reduced and fused with the vertebrae, and ribs are absent from the lumbar region, and the bones forming the pelvis are also fused.

Right from their first appearance in the late Carboniferous period, and at various stages throughout the subsequent Permian and Triassic periods, various groups of reptiles appeared, collectively classified as synapsids[1], which are known as the mammal-like reptiles because they exhibited some features which are similar to those of mammals. For instance, from early on they showed some differentiation of their teeth, and all of them remained quadrupeds as distinct from many of the dinosaurs which were bipedal. In view of their similarities, it is of course assumed that the various synapsid groups are connected in an phylogenetic sense, but there were in fact substantial differences between them, especially between the early pelycosaurs and later therapsids: fossils do not show a transition from one to the other.

Some of the therapsid species had various mammal-like features such as a secondary palate, which is a bony plate that separates the nasal passage from the mouth and thus enables breathing to continue while the mouth is full of food. And some therapsids called the cynodonts were doglike in their overall size and structure, in particular with legs below their body, and knees pointing forwards rather than outwards, enabling a fairly upright posture rather than the more typical squat stance of many reptiles. Of particular interest is a progressive change in structure of the jaw bone and jaw joint that can be traced through various therapsid genera: a grad-

---

[1] The term synapsid is used to classify those reptiles with a single lower temporal opening behind the eye; to be compared with, for example, the anapsida which have none and the diapsida which have two.

ual reduction in the size of the quadrate and articular bones – tending towards the mammalian arrangement. Further, from an embryological or developmental point of view, these two bones are widely considered equivalent (homologous) to the malleus and incus in mammals. This is probably the closest there is in the fossil record to a transition from one class of vertebrates to another. Evolutionary biologists see it as conclusive.

However, the equivalence of the jaw and ear bones (and the evolution of mammals from synapsid reptiles which relies on this presumed homology) has been challenged for a long time by the palaeozoologist Erik Jarvik (1907–98), for instance in his *Basic Structure and Evolution of Vertebrates* (1980), and more recently by Hans Bjerring (1998). In the light of which it is not surprising that 'The fossil record does not record the details of the final transformation of jaw joint to middle ear bones' (Eldredge, 1991, Ch. 6). One cannot help thinking that their views are marginalized – despite the quality of their anatomical work – simply because they do not conform with the accepted evolutionary scenario.

In addition, it should be noted that different reptile species had different mammal-like features – it is not as if a particular line became increasingly mammal-like – and none had the specifically mammalian features identified above. There remain, therefore, substantial differences in skeletal structure and dentition between the most mammal-like reptiles and the first fossils that are recognised as truly mammalian – morganucodonts, which were small mouse-like creatures and very different from the doglike cynodonts.

Further, whilst it is to be expected that evolutionary biologists will emphasize the similarities between cynodonts and morganucodonts, that is only one side of the coin. Because, in a pattern we have seen several times already, almost as soon as the first mammals appear, there were five different orders of them. They were so distinct that they are categorized into two separate subclasses, at least two of which are not even thought to be derived from the morganucodonts. In effect, a polyphyletic origin is proposed, i.e. that different therapsid lines evolved to give different mammalian orders; but the difficulties with that sort of scenario have already been adequately mentioned. All of these orders petered out during the Cretaceous, except for the Multituberculata which flourished in the Cretaceous and persisted into the Tertiary, and they were replaced by modern mammalian orders – replaced by, but there is no fossil record to link them to the early ones.

There are three divisions of modern mammals. Because the monotremes lay eggs they are seen as primitive, not far removed from the egg-laying reptiles. The best known is probably the 'duck-billed' platypus, and the only other modern species are anteaters. It seems they have always been a minor group. Marsupials, well known from the Australian species, are those in which the young are born in a very immature form and development continues in the mother's pouch. Then there are the placental mammals which are considered more advanced than the marsupials: the young are not born until highly developed, and in general they have relatively larger brains. The dominance of the placentals is seen as evidence of this, and it is thought they evolved through an intermediate marsupial stage. However there is no evidence of this progression – the fossil record shows both groups arising virtually simultaneously.

Finally, although placental mammals share various defining characteristics in fact they are remarkably diverse. Cuvier struggled with classifying them, and modern taxonomists fare little better – the more we find out, the more we realise how different these groups are from each other. It is assumed that the distinct orders (of which there are about thirty) diverged from the early placental mammals, but they first appear in the fossil record at various stages through the Tertiary, fully differentiated, with no trace of a family tree. As with the reptiles, especially prominent examples of this are the mammals specialised for aerial or aquatic life – the bats and cetaceans.

Bats' wings consist of a membrane stretching from their bodies and arms, and supported by very extended digits. The earliest bats date from the Eocene epoch, about 50 million years ago, and are fully specialised for flying. From the evolutionary viewpoint, there is no doubt that bats (some of which are insectivorous, whilst others eat fruit) evolved from the early mammalian insectivores. But there is absolutely no species known which can in any way be considered intermediate between them. Further, as is well known, most bats use echolocation rather than sight for navigation. There is no necessary connection between flight and echolocation but, so far as we can tell from the early fossils, the first bats not only flew but used echolocation too. Modern bat species that use sight belong to genera which appeared only later in the fossil record.

And then there are the cetaceans – the whales and dolphins – which, like the ichthyosaurs, are presumed to have evolved from land animals but appear fully specialised for marine life without any sign of intermediate creatures:

> Like the bats, the whales appear suddenly in early Tertiary times, fully adapted by profound modifications of the basic mammalian structure for a highly specialized mode of life. Indeed, the whales are even more isolated with relation to other mammals than the bats; they stand quite alone.
> [Colbert, Ch. 26]

These specialisations include front limbs modified as flippers, including a greatly increased number of phalanges, and fundamental changes to the ear which enable sophisticated echolocation.

Until now I have focussed on the vertebrates, but it is worth mentioning a few other groups to illustrate that the gaps pervade all types of organism in the fossil record.

Insects

In discussing the Cambrian explosion I mentioned that some of the major groups of organisms arising then were arthropods – the trilobites, horseshoe crabs and crustaceans – and that the insects appeared later. All of these share the chitinous exoskeleton and segmented bodies which are characteristic of the arthropods, and it was initially assumed that the later arthropods arose from the earlier, or at least that they must share a common arthropod-like ancestor. However, as we have become more aware of the diversity of the different groups (and, indeed, within them) and that, despite extensive searching, appropriate transitional forms have not been found, an increasing number of biologists believe they probably had independent origins. That is, many no longer regard the arthropods as a monophyletic group, but that at least the crustaceans and insects should be regarded as separate phyla. All of the arthropod groups appear abruptly in the fossil record without any evident links to each other or any other earlier forms, and for none is this more pronounced than for the insects.

The insects are by far the most abundant modern group of arthropods, comprising about 75% of the million or so known animal species, and it is thought there may be several times more insect species yet to be identified. However, despite their abundance and diversity, past and present, the origin of insects – at least in an evolutionary sense – is a complete mystery.

Wings

The earliest known fossil insects are wingless springtails and bristletails. They appear abruptly in Devonian strata as fully developed insects, having the typical trisegmented body, and are not very different from some modern wingless insects. Because they are highly developed, it is assumed that insects must have been evolving for a long time previously – perhaps since the Silurian period when land plants appeared – but there is no sign whatever of this development in the fossil record.

The first winged insects appear much later, towards the end of the Carboniferous period. Quite naturally, it is presumed that winged insects evolved from earlier wingless ones, but there are no known intermediates – the winged insects appear as abruptly as the wingless ones. Their wings are so specialised and highly developed that, as for the insects as a whole, it is assumed there must have been a considerable period of prior development – but there is no sign of this. Indeed, the origin of insect wings is one of the intriguing evolutionary questions.

But that is not all. Just as we have seen so many times already, the enigma is compounded because two different types of wing appear at more or less the same time. Some of the early groups of winged insects, such as dragonflies, had wings which extended from the insect's body, and could not be folded up. Other early groups, such as the cockroaches which were similar to and equally abundant as modern ones, were able to fold their wings over their abdomen.

Because the most popular theory of the evolution of insect wings is that they developed as projections of the insect's skin, it is thought that the unfolded wings were the first to arise; and, because the ability to fold wings is seen as an advance (it allows the insect to access confined spaces, and the wings are less vulnerable to damage), it is thought that folding wings evolved from rigid ones. However, both groups of insect appear with no sign of development from one to the other or of a common ancestor. In fact, in a pattern we have seen many times before, even within these two broad groups separate orders of insects arise: for example, mayflies (Ephemeroptera) along with the dragonflies (Odonata), and grasshoppers (Orthoptera) with the cockroaches (Blattodea). And there are not even any intermediates known to link these morphologically related orders.

Metamorphosis

Wingless[2] insects do not undergo metamorphosis: the young insects are but small versions of the adult into which they grow by a series of moults (recall this general feature of arthropods). The earliest orders of winged insects undergo partial metamorphosis: for example the young insects may have no or incompletely formed wings or reproductive organs; these structures develop over the course of successive moults. Then there are the insects which hatch as a larva and mature into the adult insect via a dormant pupa stage, for example a caterpillar which forms a chrysalis from which it emerges as a butterfly. Insects which undergo this sort of full metamorphosis appear in the Permian period, with groups which include the flies and exceedingly diverse beetles (comprising about one third of all insect species). The various orders are morphologically distinct from the earlier insects, and there is nothing in the fossil record to link them.

In summary, although the insects are the most diverse animal group, and have been abundant at least since the Carboniferous period, we cannot trace an evolutionary origin of insects as a whole or of their principal groups.

Plants

The sort of pattern we have seen with animal fossils – that major developments appear suddenly without identifiable precursors, and often in multiple distinct lines at about the same time – is also seen in the various plant phyla.

Vascular plants

Algae are believed to have been in existence since long before the Cambrian, and the first 'higher' plants appear from the late Silurian period onwards. These early 'higher' plants had complete vascular systems, comprising xylem and phloem for the transport of water and nutrients respec-

---

[2] That is, what are considered 'true' wingless insects, the apterygota; some insects are classed as 'pterygota' even though their wings are vestigial or totally unformed.

tively. They are highly developed structures, involving the interaction of many differentiated cells; but there are no known predecessors with part-formed structures, and these developments appear within a short space of time in several distinct subphyla – the ferns, horsetails and club mosses.

A further significant feature of these vascular plants is the first appearance of leaves. As with a few modern plants, some fossil vascular plants did not have leaves, but there are no examples of plants with partly formed leaves. Theories have of course been put forward as to how leaves evolved, for example by modification of thin branches or development of flaps of tissue from the stem; but the matter remains conjectural and disputed simply because no intermediate forms are known. What is especially confounding is that two totally different types of leaf appear at this time – the familiar 'simple' leaf, and the more frond-like fern leaf; and we cannot account for the evolution of either type.

It is interesting to note that some of the early species were very large: in particular the Devonian forests were dominated by a variety of large woody trees of these early vascular plants – with trunks over half a metre wide and growing to more than 30 m tall. They formed a substantial part of what are today's major coal seams. By contrast, almost all their modern counterparts are herbaceous and of modest size, such as the horsetail.

Bryophytes

In discussing the appearance of the vascular plants, I commented that their vascular tissues arise abruptly without any earlier simpler form. Simpler plants do exist, the bryophytes, which is a collective term for the mosses, liverworts and hornworts. These have tissues which enable some transport of water and nutrients, but lack the differentiated xylem and phloem that characterize the higher plants, and their systems are not as effective. Consequently, the bryophytes are of only limited size and live predominantly in wet habitats.

Bryophytes do not appear in the fossil record until the Devonian for liverworts and the Carboniferous for mosses. Clearly the occurrence of the highly developed vascular plants *before* the bryophytes is not consistent with the usual evolutionary perception of increasing complexity. Given that they inhabit moist environments, which are generally those most amenable to fossilization occurring, it is difficult to believe that they could have arisen before the vascular plants but escaped fossilization until much later – but this is widely presumed to be the case in order to account for evolution of the vascular plants.

Secondly, readers will probably no longer be surprised that the origin of the bryophytes is just as much an enigma as for the first higher plants. It is generally assumed that they evolved from single-celled algae that preceded them but, although the bryophytes are much more complex than any of the algae, there are no known fossils showing part-formed bryophytes or organisms which could be considered as intermediate between them. Also, the different groups of bryophytes are completely distinct right from their first appearance.

Seed-bearing plants

Like the preceding algae and the subsequent mosses, the early vascular plants propagated by means of spores, which means their range is restricted to moist areas. What is seen as the next advance, especially that it enables colonisation of drier land, is the use of pollen and seeds. There are two main subphyla of seed-bearing plants, those without flowers which give rise to seeds without an outer covering (gymnosperms, such as conifers), and those with flowers which produce covered seeds (angiosperms). The gymnosperms appear first in the fossil record, being frequent by the Carboniferous period, but there is no trace of their ancestry, no indication of how the pollen and seed-producing mechanisms arose.

Flowering plants

Most striking of all, so far as the origins of plants are concerned, is the appearance of the angiosperms, which are better known as the flowering plants. These arise without any apparent predecessor at the beginning of the Cretaceous period, and by the end of the Cretaceous they exhibit a diversity comparable with the modern day. Although some have tried to draw similarities between the 'simpler' angiosperms such as magnolia and some gymnosperms, there is a fundamental difference in the structure of the reproductive organs – a gap which is totally unbridged by anything in the fossil record. Also, right from the earliest forms, as well as the novelty of the flowers, there is substantial differentiation of the angiosperm leaves which have a distinctive branching veinous structure which is absent from any of the earlier plants.

Because of these complex structures, some have inferred that the appearance of the flowering plants must have been preceded by a substantial period of evolution, perhaps in upland areas where conditions for fossilization did not occur. The difficulty with this suggestion is that the distinctive angiosperm pollen grains (which clearly could be transmitted far from the plants themselves e.g. to locations where fossilization was occurring) are also found only from the Cretaceous onwards.

The flowering plants are divided into two main groups, the monocotyledons and the dicotyledons depending on whether their seeds have one or two seed leaves (the gymnosperm seeds usually have several). This distinction is associated with many other different characteristics between the two groups. Both of these plant groups appear independently, there are no indications of any intermediate forms having shared characteristics. So, once again, we have profound developments (flowers etc.) arising in markedly distinct groups of organisms.

Even in his day, Darwin considered the origin of the angiosperms to be a grave problem because of their very sudden appearance and diversification. Nearly 150 years later the situation is just the same. It is worth commenting that the flowering plants (including flower-bearing trees) are by far the most numerous of modern plant species (about 80%); yet their origin – in evolutionary terms – is a complete mystery.

PATTERNS IN THE FOSSIL RECORD

I trust the preceding section has gone some way towards rectifying the widely held view that evolution must be a fact because it is clearly demonstrated in the fossil record. Whilst it can be argued that in general there is a progression from simpler to higher organisms, it should be noted that there are significant exceptions to this such as the bryophytes after the vascular plants. But the key issue that challenges an evolutionary interpretation of the fossil record is its profound discontinuities. Right from Darwin's conception of the theory, even before there was any idea of how complex living organisms are, it was recognised that any development in an evolutionary sense must be gradual; but the required intermediates for a gradual progression are not there. The abrupt appearance of distinct groups is the most conspicuous feature of the fossil record and I will expand on this a little here. I will then discuss the other important aspects of diversification and stasis.

The abrupt appearance of distinct groups

This is the pattern right from the start with the Cambrian explosion when all animal phyla, past and present, appear within a fairly limited period of time: they have completely different body plans, there are no intermediates, and for none is there a recognisable precursor. And a similar pattern is seen with the main plant groups – the various phyla of the bryophytes and early vascular plants, the gymnosperms and angiosperms – with all appearing suddenly, without any sign of prior development. So when we talk about gaps in the fossil record, we are not referring to minor pieces missing from an otherwise complete picture; they are not just trivial vacancies in the family tree of some small or obscure group. On the contrary, it is the major taxa, notably including the dominant animals and plants – the insects and the angiosperms – those that are most abundant in terms of number and diversity of species, number of individuals, and widespread occurrence. Far from perceiving a progressive evolutionary origin for such as these, their origin is a complete mystery: they appear suddenly in highly developed and diversified forms, without any trace of a predecessor – usually there is not even a remote candidate.

This unconnectedness of the fossil record is emphasized by the fact that it is not just the phyla, but in most cases distinct classes within these also arise abruptly, usually with no intermediate forms or recognisable common predecessor. Indeed, as we have seen, it is often the case that when a new phylum or other major taxonomic group appears, it does so as multiple distinct lines – such as the different orders of winged insects and the monocotyledons and dicotyledons of the flowering plants. As I have already commented, these situations raise a particular difficulty for an evolutionary origin of phyla. New classes within a phylum do not always arise simultaneously: sometimes other classes arise well after the initial appearance of the phylum, and when they do so, this too is abruptly, without discernable link to an earlier group. This is especially true of the various types of arthropod. Another example is, of course, the different classes of vertebrates that appear at various stages through the fossil record – such as the jawed fish, bony fish, amphibians and birds.

The abrupt appearance of new groups of organisms is also seen at lower taxonomic levels. Notably, for each of the vertebrate classes – from jawless fish to mammals – distinct orders arise without any sign of intermediate forms. Often they occur when the class first appears, but sometimes later. It is especially noteworthy that this is true even for the orders of modern mammals which have arisen relatively recently, during the Cenozoic, from which fossils are abundant. Despite their common features as mammals, the various orders are notoriously difficult to relate to each other because they have such distinctive features; from an evolutionary perspective, this isolation is paralleled by a lack of discernible links between them in the fossil record.

I have already mentioned the bats and cetaceans; another prominent mammalian example is the rodents. In terms of their number of species (there are about 1700 which is more than all the other mammalian groups taken together), number of individuals, and world-wide distribution, they are the most abundant modern group of mammal, and it is thought to have been similar throughout their history. It is assumed that, like most other placental orders, ultimately they are derived from the earliest insectivores. But there is no trace of any ancestry linking them to non-rodent predecessors: the first rodent-like mammals, typically *Paramys* which appears in the Palaeocene, already have fully developed rodent features, notably the chisel-like incisors, one pair in each jaw, giving the tooth arrangement of just four incisors which is unique to the rodents. Further, these teeth are specialised – they grow continually throughout the animal's life, and are harder at the front than the back so that differential wear keeps them sharp; and rodents have a particular jaw articulation to enable an effective gnawing action. Also, rodent molars are suitably shaped for grinding plant material, quite different from those of the early insectivores. The earliest rodent-like mammals had all of these specialised features, and a long low skull which is also typical of rodents. As the rodents are abundant and widespread, and many, including the earliest, are squirrel-sized, their fossils are reasonably abundant and we would expect to see something of their early development in the fossil record – but there is nothing.

Diversification

Until now I have emphasized the abruptness of the fossil record by show-ing it is not just the phyla which appear suddenly, but substantial groups within these such as classes and orders, or even lower taxa; and several distinct groups may appear more or less together. However, once a new group has appeared, it is often the case that subsequent species differ only moderately from the earliest members, and sometimes progressive mor-phological change is evident. In such cases it seems very likely that the later species have descended from the earlier ones in a conventional evolu-tionary sense. Also, it is not unusual to find a line appearing to diverge and diversify – to provide a kind of family tree.

For instance, in discussing the reptiles, I mentioned how the ichthyo-saurs appear in the early Triassic, fully specialised for marine life, without apparent link to any preceding reptile group. Following this appearance there were progressive small changes in their overall shape, with the later ichthyosaurs generally being fatter and having longer skulls and flippers. Also, in mid Triassic the ichthyosaur line appears to have split into two, with the distinguishing feature being that one line had wide flippers whilst the other had long ones, and there were other differences in the shapes of their dorsal fin and tail. The former line became extinct in the Jurassic, whereas the latter persisted to about the end of the Cretaceous.

Evolution of the horse

The best known example of progressive morphological change in the fos-sil record is the ancestry of the horse (see Fig 11). Horse-like mammals were abundant through most of the Cenozoic, especially in North America where many fossils have been found. Their ancestry was first identified in the 1860s, when it helped to persuade many of the truth of evolution, and it has been further elaborated since then.

The first horse-like mammal, called *Hyracotherium* (originally known as *Eohippus*), was about the size of a fox, and appeared in the early Eo-cene, about 55 million years ago. It had long bones in its legs and toes, and the latter were held almost vertical so that it walked on the tips of its toes; all of which made it well suited for running. It had three and four toes re-spectively on its hind- and fore limbs. Its teeth were relatively small, but clearly those of a herbivore. During the Eocene and Oligocene it appears to have developed, through intermediate genera known as *Orohippus, Epi-hippus* and *Mesohippus* to *Miohippus* by about 25 million years ago. Com-pared with *Hyracotherium, Miohippus* was somewhat larger and had longer legs and feet. Of particular note, the forelegs had lost their outer-most toe, so there were three toes on each foot; and in each case the mid-dle toe was now clearly larger than the outer ones. Also, its premolars were more like molars, and all the cheek teeth had higher crests.

During the Miocene, the horse line diversified. One line, *Archaeohippus*, remained little changed, and lasted until about half-way through the Mio-

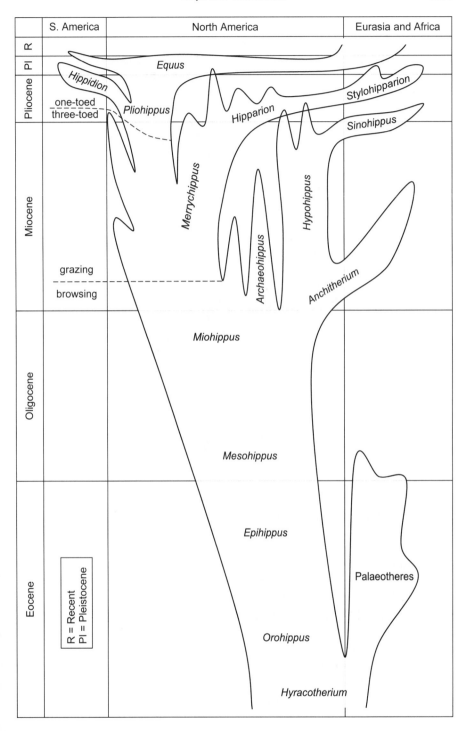

FIGURE 11. Principal genera in the ancestry of the horse.

cene. Another, *Anchitherium*, generally remained like *Miohippus* in both feet and teeth, but became as large as a modern horse, and extended its range into the Old World, before it too died out. Of particular importance was *Merychippus*, which was about the size of a pony, and in which all the side toes were so small that it walked on only its middle toes, which now had a significant hoof. The teeth were more convoluted, with cement – providing improved abrasive surfaces for grinding plant material. *Merychippus* also diverged, with one line known as *Hipparion* having more developed teeth, although with relatively unchanged feet. The generic term *Hipparion* includes various related groups which became very wide-spread in both America and Europe where it survived until the Pleistocene. The other important line was *Pliohippus* in which not only were the teeth more developed, but the side toes were so reduced that these animals effectively were single-toed. It was probably from *Pliohippus* that the modern horse, *Equus*, is derived.

All of this horse evolution took place in North America, with occasional migrations of some genera to the Old World; but all of these became extinct there. *Equus* also migrated to the Old World, including Africa where it founded species such as the zebras, and to South America. Interestingly, after North America having been the birth-place of the horse, *Equus* became extinct there at about the end of the Pleistocene, but was reintroduced from the Old World by man in the 16th century.

Apart from the overall increase in size, there are two major trends in passing from *Hyracotherium* to the modern horse. First, although with its tip-toed gait, *Hyracotherium* was already well suited to running, this was accentuated by the reduction to a single toe and an accompanying modification of the foot ligaments (making them more elastic). Second were the molarization of the premolar teeth, prolonged growth of all the cheek teeth, and various changes in structure which meant that, when worn, the upper surfaces provided a better surface for grinding grass which contains silicates in addition to the cellulose of plant material generally. These latter changes in particular are seen as accompanying a change in feeding habit from browsing (e.g. leaves from trees) in *Hyracotherium* to grazing of the modern horse.

Anomalous rates of evolution

So it is at this level – diversification and adaptation within a fairly low taxonomic group such as an order or family – that we can reasonably infer evolutionary relationships between species or even genera, and sometimes we can perceive evolutionary change of a species. However, there are two points to note about this evolution: it is of limited morphological extent, and it is slow.

These are well illustrated by the ancestry of the horse. Compared with *Hyracotherium*, modern horses are larger, with longer legs, longer and more convoluted teeth, and fewer toes. Yet these relatively minor changes took place over 50 million years or more. That is about one third of the time between the first mouse-like mammals and the earliest horses; but the morphological differences between *Morganucodon* and *Hyracotherium* are much greater than three times the difference between *Hyracotherium* and *Equus*.

A similar conclusion comes from other mammalian groups, especially the bats and whales. These, too, emerged about 50 million years ago and we can trace some morphological change in them since then. But if, based on the rate of morphological change that we infer from their fossils, we try to retropolate to a common mammalian ancestor, then such an ancestor would need to be much more remote than the earliest known mammal. In fact it would probably precede even the most primitive vertebrates of the Cambrian period. Put another way, the 50 million years or so since the appearance of modern mammalian forms, such as of *Hyracotherium*, is about 10% of the time over which modern mammals are supposed to have evolved from amphioxus-like organisms, but require many more than ten times the degree of morphological change. Similarly, although we can see some morphological change in the ichthyosaurs, the time over which this occurred – from the early Triassic through to the late Cretaceous – was at least 100 million years.

Whilst I have focused on the vertebrates, the same sort of pattern can be seen elsewhere. For example, many insects are clearly adapted to pollinate flowers, and flowers adapted for insect pollination; but this co-adaptation arose only gradually through the Cretaceous:

> It is actually not until the late Early Cretaceous that there appear many of the first definitive representatives of insects belonging to present-day anthophilic [pollinating] groups, and all these fossil forms are clearly generalized in morphology. Not until the Cenozoic do insects consistently appear with structures specialised for flower-feeding, most notably the repeated appearance of elongate proboscides. [Grimaldi]

Clearly, these rates of evolution are inconsistent: There are major discrepancies between the slow evolution that we can reasonably infer from morphologically related species, and the rates that are required to account for large scale evolution. In other words, whatever mechanisms effect evolution within a new group after it has appeared certainly do not account for its initial appearance; the latter requires a completely different sort of explanation. So the observations of evolutionary change serve to reinforce the point I have made already – that new groups appear abruptly.

It is worth noting that this sort of pattern is the same as that evident from molecular data. Recall from Chapter 7 that the degree of difference between amino acid or nucleotide sequences is seen as support for evolution: whilst there are notable discrepancies, in general, closely related morphological species have the most similar sequences, and in unrelated species they are dissimilar. Further, the inferred rates of nucleotide substitution are seen as a clock that can be used to estimate the time since divergence from a common ancestor. When relativity few sequences were known, a reasonably consistent picture could be maintained; but as more sequences have become available, these have challenged the conventional evolutionary explanation. Noteworthy for the present context is that, whilst closely related species may have very similar sequences, the differences between higher taxonomic groups are larger than would be expected by extrapolation from the differences between closely related species. What this means is that when attempts are made to estimate the times of divergence between various lines based on molecular data, all too often the estimate places the presumed divergence much earlier than is consistent with the fossil record.

Two especially prominent examples of this are the mammals and flowering plants. These are significant because they are the most recently emerged groups of the animal and plant kingdoms, so we may reasonably expect both the fossil record and molecular clocks to be most reliable. However, in each case, the estimated divergence between the various groups of flowering plants or mammalian orders, is much further back than is consistent with the fossil record. That is, the extrapolation implies that evolution of the flowering plants and modern mammalian orders was taking place long before the first appearance of the relevant fossils. However, palaeontologists specialising in the appropriate areas have commented that, in view of how good they believe the relevant fossil record to be, it is unlikely that such evolution could have occurred without leaving some fossil evidence. For example:

> The most recent molecular time scales for mammals place the origin of many modern orders before the Cretaceous/Tertiary boundary, indeed as far back as the early Cretaceous. This implies a missing fossil history spanning 64 million years. According to Foote *et al.* (1999), this hiatus is difficult to accept because estimated preservation rates for Cretaceous mammals are too high to explain such a gap. [Allard *et al.*]

> In our view, the absence of distinctive triaperaturate pollen grains [i.e. distinctively angiosperm] in numerous, rich, Triassic and Jurassic palynofloras from both hemispheres precludes the long cryptic period of evolution implied by some estimates of rates of molecular evolution. [Crane *et al.*]

So, yet again, the evidence points to substantial discontinuities in the fossil record; in particular that, whatever subsequent evolution may be discernible, major new groups arise abruptly, without evidence of prior evolution. Before pursing this issue further, we need to consider stasis.

Stasis

The other notable feature of the fossil record that needs to be recognised is that, although we can identify a series of morphologically related species, on the whole the species themselves tend to be static. That is, whereas Darwin envisaged that species would change gradually over the course of time, at least so far as fossil species are concerned, we rarely see this. What we usually see in the fossil record is a sequence of species, but not a gradation from one to the other. For instance, in the ancestry of the horse we can trace a progression of characteristics through the series of genera from *Hyracotherium* to *Equus*, but we do not see a gradation of one to the other; i.e., we see a progression of species but not a merging of species. It was Eldredge and Gould, of punctuated equilibria (see below) fame, who stressed that, in contrast to the classical Darwinian expectation of seeing gradual change, a pervasive feature of the fossil record is species stability or, to use their word, stasis. Hence, the usual pattern is for species to appear on the scene, persist for a while during which they are fairly static, and the vast majority of species then simply disappear. Put another way, rather than one species evolving into another, most species become extinct and do not seem to lead anywhere; but often they are replaced by another from within the same family.

However, whilst highlighting the stability of species, it is important to recognise that 'stasis' does not mean morphological uniformity or a total lack of change. In his book *Fossils* Eldredge stresses that members of fossil species were not carbon copies of each other: just as modern species show some temporal and spatial variation, the same was true of fossil species. Over a period of time some feature may show a definite trend, but subsequently it is just as likely to return to its starting point as to continue its trend. And any change was of a minor degree – comparable with the normal range of variation for modern species.

EXPLAINING THE GAPS

Although the popular perception is that fossils tell a convincing evolutionary story, most fairly recent texts on palaeontology, and even many on evolution, now accept that this is not so. There are gaps at the large scale – between major groups of organisms, and at the small scale – between species. Nevertheless, the prevailing view is not to question that evolution has occurred, but that there must be an explanation for why there is little evidence of it in the fossil record.

Some, following Darwin, still hold out hope that fossils of the proposed intermediates will be found. Others accept they are not there, but argue that this is not because the intermediates never existed, but that fossilization is a rare occurrence and they were never fossilized and/or that the relevant fossils formed but have been lost by subsequent erosion.

On the other hand, several palaeontologists have carried out investigations to assess the reliability of the fossil record and have concluded that it does give a fair account of past life. For example, Foote *et al.* (see above) estimated that the fossil record would need to be 10 to 100 times worse than it is for modern orders of mammals to have lived in the Cretaceous but not be known from the fossil record. And Paul (1985) has argued that even if fossils occurred at random then the order in which they recorded the first appearance of species would be right 50% of the time, so the order in which they are actually found will almost always be right. This latter point is especially significant for instances such as the bryophytes which do not appear until after the more advanced vascular plants.

Punctuated equilibria

It was primarily Niles Eldredge and Stephen Jay Gould who, in the 1970s, challenged the scientific community to face up to the fact that the fossil record does not demonstrate evolution – that what it shows is stasis. They contended that for too long evolutionists had excused the absence of intermediates and the failure to demonstrate clear gradation of characteristics on the vagaries of fossilization, and that it was time take on board the realities of the fossil record. They emphasized that any viable theory of evolution must take account of the fact that species are substantially constant: that species tend to be replaced one by another, rather than changing significantly over time. As Eldredge put it in *Reinventing Darwin*, 'Evolution cannot forever be going on someplace else' and 'consistency of species through a long period of time is evidence of stasis, no matter how gappy the record' (Ch. 4).

What they proposed is that new species generally arise in small populations, somewhat isolated from the main stream of things. This follows Ernst Mayr's ideas about geographic separation being an important factor for speciation to occur. Random effects within these small populations produce a range of variations, some of which are successful, e.g. perhaps being more suited to a changing environment. These proliferate and disperse beyond the local population sufficiently that they have a good chance of being fossilized and subsequently found by us. But only the successful species proliferate in this way; the intermediates which arise transiently between them are limited in number, few if any get fossilized, and even those that do are so rare that it is not surprising we have not found them.

Clearly this theory takes on board what the population geneticists learned early in the 20th century (see Ch. 5). In particular that, whereas large populations tend to remain genetically stable, in small populations significant random fluctuations in gene frequencies can arise more easily. In this way, within a small population, a wider range of genotypes (and corresponding phenotypes) can be generated, so it is more likely that individuals can be produced which are better adapted to the prevailing or changing environment. Hence punctuated equilibria provides a plausible explanation for the lack of intermediates between species, based on the well established principles of population genetics, and without resorting to macromutations or the likes of Goldschmidt's 'hopeful monsters'. It also explains the instances where parent and daughter species appear to have lived concurrently, even side by side, whereas traditional evolutionary theory would have expected them to be sequential.

Limits of punctuated equilibria

However, a little thought shows that there is a fundamental limitation to the explanatory scope of punctuated equilibria. Whilst it makes sense for variations to arise more readily in small populations due to random fluctuations of genes, it is obvious that those genes are an essential prerequisite; that is, for this means of speciation to work, the appropriate genes must already be available, they cannot be generated there. Let me expand on this.

So far as the possible production of genes is concerned, it would be a little easier for new mutations to gain a foothold by chance if the population is very small (again, see Ch. 5). But this is true whether the mutations are good, bad or indifferent: it is just as likely that a beneficial gene will, by chance, be ousted by a detrimental one, as vice versa. However, the primary problem with new useful mutations is that they are so rare – so rare that the only ones we know of are those conferring resistance to antibiotics and the like (discussed in Ch. 9); and the only hope for finding even these is to have very large populations – of the order of a billion, even when only a single nucleotide change is required. As shown in Chapter 7, to find a completely new useful gene is well beyond the capability of any possible resources. By restricting the origin of new species to isolated populations, this limits the scope for generating new mutations even more; so it is totally out of the question for new genetic material to arise in these small populations.[3]

In other words, the production of new species in small populations can only realistically be based on sorting genes: it provides a mechanism for enhancing the segregation of genes, but not the production of new ones. The crucial, though unspoken, assumption of punctuated equilibria is that the relevant small populations have the necessary genetic variability to generate a range of genotypes, and consequent phenotypes. To use the vocabulary of my preceding chapter: it is realistic for small populations to be a source of new variations, but not of new variability. In fact, the concept of punctuated equilibria is a clear pointer that the morphological evolution we can see in the fossils results primarily from gene segregation rather than the acquisition of new genetic material.

Hence, whilst punctuated equilibria can explain the gaps between related species, it cannot be invoked to explain the gaps between high taxonomic groups such as phyla, classes and probably orders or even lower. Punctuated equilibria offer no help with the major riddle posed by the fossil record – the abrupt appearance of new groups of organisms. The latter entail a considerably greater scale of morphological change, crucially re-

---

[3] Some readers may wonder if it is possible for a beneficial mutation to arise in the population at large, and then be selected in the isolated population. However, this is not valid. Quite apart from the intrinsic improbability of macromolecules so that no population is large enough to find them, we cannot keep relying on chance that the desired mutation will fortuitously be in the key small population when we need it.

quiring the production of new genes rather than merely their segregation. And – especially confounding from an evolutionary point of view – these more profound changes take place extremely rapidly, much more rapidly than the subsequent diversification, as indicated by the anomalous rates of evolution mentioned above.[4]

Primordial variability

As I have indicated, although we rarely see a species evolve into another, there are instances in the fossil record where species clearly appear to be related in an evolutionary sense, with the ancestry of the horse being a particularly good example. Because the theory of evolution in its widest sense envisages that overall there has been a progressive increase in complexity from the earliest organisms to the present higher plants and animals – a progression which necessarily involves the production of new genetic material – it is generally assumed that when we see evolutionary change in the fossil record, especially if this includes some form of specialisation, then this is due at least in part, if not primarily, to the acquisition of new genes.

However, we have seen in Chapter 9 that significant morphological change, including specialisation, can be achieved simply by segregating genes. The prime example of this is domestic breeding for specific traits; a more natural example would be the yarrow plant which has specialised to different habitats. So what I am suggesting here is that the evolutionary changes we can perceive in the fossil record – these, too, are attributable to gene segregation rather than to the production of new genetic material. The obvious implication from this is that, when a new group of organisms arises, its founding members have extensive genetic variability – sufficient to fuel subsequent morphological diversification, which may include specialisation. The question is, of course, What evidence is there for such a radical hypothesis?

A closer look at horse evolution

Interestingly, one of the clearest indications that diversification, even specialisation, is based on the segregation of genes present in a primordial group rather than subsequent acquisition of new genes by mutation comes from the ancestry of the horse.

Before looking in more detail, it is worth noting first that *Hyracotherium* was a thoroughgoing perissodactyl: It was fully herbivorous, with appropriate teeth and its intestines included a caecum (cf. appendix) which accommodated bacteria for digesting cellulose and other plant ma-

---

[4] Eldredge says the big gaps are just an extension of the small ones. He sees a major role for extinctions in causing ecological gaps that are readily filled by new species; and suggests that bigger extinctions lead to bigger evolutionary events. But he does not recognise that speciation in small populations relies on gene shuffling rather than gene production, and fails to consider the improbability of new useful genetic material.

terial. Also, it was lightly built, with slender legs and tip-toed gait, all of which made it well-suited for running. All of the subsequent morphological changes were relatively minor modifications of characters already present in *Hyracotherium*. Even changes of the skull were only quantitative rather than qualitative, i.e. can be accounted for in terms of differential growth of various parts of it (termed allometry, see e.g. de Beer, 1971), rather than requiring novel features. An indication that the differences between the various horse-like mammals were rather limited is that all of the species are classified within a single family, Equidea. Also, whilst different characteristics are involved, the scale of the changes is comparable with those between the various breeds of dog (which we even regard as the same species – if it were not for the time barrier, could *Hyracotherium* have bred with a modern horse?). So it is reasonable to suggest that changes of this magnitude could be achieved by gene segregation alone.

When the horse ancestry was first identified, it was perceived as a linear progression from *Hyracotherium*, through various intervening genera such as *Miohippus* and *Merychippus*, to *Equus*, along which there was a steady development of features such as increased overall size, reduction to a single toe on each foot, along with the specialised ligament, and enhanced tooth development. However, as more fossils have come to light over the past century, it has become clear that the actual ancestry was far more complex than this. Whilst the horse evolution in North America was a fairly linear succession of genera through the Eocene and Oligocene to *Miohippus*, in the Miocene there was substantial diversification, with some genera not changing much thereafter, others developing new features, and some having a mixture of old and new.

Several diverse characteristics are used to distinguish the different genera. They include the single-toed ligament and various structural features of the teeth, such as having high crowns. Two significant developments in the skull were the retraction and deepening of the nasal opening, and gradual disappearance of the preorbital fossa (a depression). In addition, whereas the general view is of a gradual increase in the size of the horse, some lines became even larger than modern horses, and others became very small. Similarly, although the usual trend was for lengthening of feet bones, some became shorter.

The variations in both directions of general body size and feet bones are good indications by themselves of selection from a generalised ancestor. But the most remarkable finding, which is especially relevant to our discussion, is that all of the features that I have just listed 'evolved time and again in very different lineages of horses' (Forsten).

Now we could choose to believe that the same mutations arose independently in diverse lineages, perhaps persuading ourselves that the horse-like mammals had some sort of propensity for these particular mutations. But surely a much more credible explanation is that the genetic bases for these features were already available within *Hyracotherium*.

Then, through a combination of random fluctuations and selection, in various lines appropriate gene combinations arose, and the various traits emerged – analagous to the races of North American house sparrow or the species of Galapagos finches (Ch. 9). One of the striking observations which supports this latter view is that in various species of *Hipparion*, although they still had the 'primitive' three toes, nevertheless they also had the 'single-toe' ligament system. Given that the latter is supposed to be an adaptation to the single-toed status, it would be very odd if it 'evolved' by mutation in three-toed animals.

Accounts of horse evolution tend to focus on what took place in North America, because that is where it reached the modern horse. However, at the start of the Eocene America was still joined to Europe and *Hyracotherium* also lived there; but the two continents split soon afterwards and the two founding populations of *Hyracotherium* evolved separately. (Occasional land bridges occurred in the region of the Bering Strait from the Miocene onwards, allowing the migrations to the Old World mentioned above.) The European line diversified almost immediately into what are known as the Palaeotheres, reaching levels of diversity in the Eocene comparable with that occurring in North America in the Miocene (about 30 million years later), but became extinct early in the Oligocene. What is relevant to the present discussion is that this line developed some of the same characteristics as those in North America, notably large body size and advanced teeth. These similar developments arose quite independently of those in North America, being separated by the nascent Atlantic, and the timing was quite different anyway. So, yet again, we have to propose that essentially the same mutations arose in completely different populations of *Hyracotherium* – or, surely more likely, that the necessary genes were already present in the original *Hyracotherium* population. In addition, there are some more general observations which point to this diversification arising from gene segregation.

First, the much earlier diversification would be all the more difficult to explain in terms of accumulating appropriate mutations, whereas there is no difficulty in achieving rapid diversification by gene segregation. In fact, the diversification is accounted for by the fact that, during the Eocene, Europe was an archipelago, and we are well aware that geographical separation, such as on islands, promotes gene segregation.

Second, all the early European horses became extinct by the end of the Oligocene. Ann Forsten, no doubt aware that diversity generally favours adaptability, comments that 'At the time of their demise in Europe, in the Eocene, horses were generically diverse and flourishing, whereas at the same time in North America they survived despite being little differentiated.' However, I think this is confusing morphological diversity among related groups with genetic diversity within each of those groups, and I suggest an alternative view: The European horses were morphologically diverse because the original genetic diversity had been segregated; although there was a variety of genera, the individual genera had limited genetic variability. Consequently they were less adaptable, less tolerant of changing conditions, and so more susceptible to extinction. In contrast, because the North American horses changed little during the Eocene and Oligocene, they retained their genetic variability, and survived. But then, having diversified from the beginning of the Miocene (possibly because their habitat, whilst still continental, also diversified), most were extinct by the end of the Miocene. Could this have been because, like their European cousins long before, in diversifying, individually they had lost their genetic variability and adaptability?

In summary, the ancestry of the horse provides a clear indication that there is a limit to the variability of founding members of a taxonomic group. First, because exactly the same specialisations arise in separate lines – even in rather different environments (an archipelago and a continent). Second, because, as the group diversified, the separated genera had reduced variability, making them less adaptable to meet future challenges. Importantly, this latter reason gives further supporting evidence for the point I made in the preceding chapter – that there is no evidence for the loss of variability arising from diversification being restored by mutation, even over 50 million years.

General and specialised species

There are other features of the fossil record which many palaeontologists find hard to explain in terms of traditional evolutionary theory based on the assimilation of mutations, but which are readily interpreted and make good sense in terms of segregating genes from an original highly variable (and adaptable) primordial group.

A common pattern is for the early members of a group to be generalised – in terms of morphology and environmental tolerance e.g. they may inhabit a wide range of environments. Along with this, many survive a very long time, and some are included among species which are known as living fossils i.e. modern species whose ancestors appeared on the scene perhaps hundreds of millions of years ago, but there has been relatively little morphological change over that time. Interestingly, it has been noted that even long-standing generalised species do not seem to accumulate more adaptive change in the course of their existence (Eldredge, 1995, Ch. 5) – a further indication that they do not acquire new genetic material. A prime example of all this is the horseshoe crab which appeared very early, probably in the Cambrian. They are still in existence today, and are a favourite example in evolutionary texts of only minor morphological change having taken place over the last 500 million years.

Often, it appears that an initial general species may split into two or more somewhat less generalised species. This seems to have been the case for *Hyracotherium* in Europe at the start of the Eocene, and *Miohippus* in North America at the start of the Miocene; and the daughter genera continued to diversify and specialise through the ensuing geological periods. This is often called adaptive radiation, and I shall comment further on it below. Sometimes, it appears that a series of specialised daughter species branch off from the parent general species, leaving the latter substantially unchanged. Whichever way the more specialised species arise, the dominant pattern in the fossil record is that they survive much shorter times than general ones. Indeed, the average life of a species is only about four million years, and most become extinct without leaving daughter species.

It is appropriate to repeat here the point I made a little earlier: Because the popular perception is that evolution proceeds by the acquisition of new genetic material, almost automatically it has been assumed that specialisation, because it is seen as evolutionary 'advance', must be due to such acquisition. However, an equally valid interpretation is that specialisation is not due to the gain of genetic material, but to its loss. Though this may sound odd, it is in fact analogous with conventional breeding programmes (i.e. excluding genetic engineering) where considerable specialisation is achieved, and no one doubts it is due to selecting subsets from the available gene pool, i.e. through loss, not gain, of genetic material.

Further, this interpretation (of specialisation being due to loss rather than gain of genetic material) makes more sense of some observations than does the more traditional explanation. In particular, it is quite common for a line which is evolving and specialising – being 'successful' – suddenly to become extinct. If evolution were through the acquisition of new material then it would seem anomalous that a successfully evolving species should suddenly fail. But if that evolutionary success had been through intense selection of a very limited set of genes, then it is not surprising that it be more susceptible to changes in circumstances and more prone to extinction. Loss of adaptability is irreversible – because such a species cannot recover the genes it has lost by prior selection. In other words, the mistake has been to equate evolutionary progress with the acquisition of new genes.

Finally, selection of genes from a genetically versatile progenitor also explains the existence of vestigial organs. Vestigial organs have long been seen as evidence of evolution, especially because evolution emerged in contrast to the traditional idea of the fixity of species, and it did not seem to make sense that unused organs be part of a designed special creation (though some have argued that we can never prove an organ has no use). However, it is quite feasible that a highly adaptable primordial group had the genetic capability to produce a well-developed organ, but which, in some descendant lines which have little or no use for it, the organ has been progressively reduced to minimal or no function, but not actually lost.

Adaptive radiation

As mentioned above, after a new group has arisen abruptly, we may well see diversification from its founding members. Because the diversification is from a common origin, it is fair to describe it as a radiation.

However, the term 'adaptive radiation' is also used freely to refer to the approximately simultaneous appearance of various morphologically related groups, such as groups of jawed fish, reptiles, modern mammals, winged insects, flowering plants etc., even when there is no indication of common predecessors. In other words, to refer to such instances as adaptive radiation is nothing more than imposing an evolutionary interpretation on the observations. That is, it is recognised that it is too unlikely for the same sort of major development (e.g. amniotic egg or insect wings) to have evolved more or less simultaneously in multiple lines, so it is argued that the first group to make the advance was at such an advantage that it was able to proliferate and diversify rapidly into the new ecological niches that were open to it. This initial diversification was so rapid that, although intermediates existed, they escaped the fossil record. The fact that adaptive radiation can be seen at lower taxonomic levels is seen as adequate support for presuming that the same has happened at higher levels. But such retropolation is not supported by the facts, and to refer to the sudden appearance of diverse groups, albeit with some common feature(s), as a radiation is to misrepresent the facts.

CONCLUSION

A popular view is that the fossil record provides key evidence for evolution. But no phylum can be traced from a preceding one in the fossil record, in fact we cannot account for the origin of a single phylum: they all appear abruptly. This is also true of lower taxonomic groups such as classes and orders, and possibly lower still. Especially confounding from an evolutionary perspective is that taxonomic groups often appear more or less simultaneously, with no indication of common ancestry.

We find evidence of an ancestral line only for genera and species. The fact that some evolution can be discerned at this level, notably of the horses, has persuaded many that fossils provide proof for the whole gamut of evolution – but it does not. What is especially relevant is that the level of evolutionary changes that can be discerned in the fossil record can be accounted for wholly in terms of gene segregation. So, as in the preceding chapter, although we find evidence of evolution in terms of gene shuffling, yet again we find no explanation for the origin of genetic material. The fossil record offers no evidence at all that the fundamental problem of the improbability of biological macromolecules has been overcome. In fact, it has strengthened the case that we do not see evidence for the traditional view of evolution – of variability arising by the progressive accumulation of mutations – even over many millions of years.

11

New Wine and Old Wineskins

At the heart of the theory of evolution is the idea that substantial advancement can be made through the progressive assimilation of small improvements. Each improvement is small enough to arise naturally, it can then be retained by natural selection, and in this way an apparently unlikely end product can arise through a series of practicable steps. In particular, the complexity of higher organisms could have built up incrementally. There was so much evidence for evolution at the morphological level that, once a satisfactory genetic basis had been found, the theory was firmly established by mid 20th century. So much so that there was no doubt in most scientist's minds that as we unravelled the biochemical nature of organisms this would surely be consistent with the theory and even add further support to it. However, our knowledge of biochemistry has not done this. Rather, it has raised some awkward questions and presented serious challenges to the overall evolutionary account.

Even though it became apparent that proteins are highly improbable sequences of amino acids, such was the demonstrable power of natural selection that it was confidently assumed a natural selection sort of process could account for their formation (and of genes). This is the prevailing explanation, but Chapter 7 shows that in fact it cannot work: natural selection operates on differential fitnesses, so for it to operate there has to be some fitness. But so much of a protein's sequence has to be right for it to have any utility (fitness) that it is too improbable to occur. Hence natural selection cannot account for the origin of useful genetic material because the first step is much too big. That is, if one could somehow get enough of a sequence right to have some meaningful activity then it may well be possible to improve on that activity by random mutation and selection; but currently we have no viable theory for how that first step might be achieved; relying on chance will not do.

Importantly, the challenge of biochemistry goes beyond the improbability of individual macromolecules. That may be the fundamental problem; but as we uncover more of the biochemical structure of tissues and the mechanisms of their formation, these present serious obstacles to macroevolution. So I now return to the point I made at the start of Chapter 7 – that the theory of evolution as a whole must be examined in the light of what we now know about biochemistry. In particular, any proposed morphological evolution must fully take on board the genetic and biochemical implications.

Biochemical implications of evolution

By way of introduction I want to draw attention to the sort of biochemical differences that exist between groups of organisms. You see, the most striking feature of early biochemical discoveries was of how remarkably similar much of biology is at that level. Most fundamental is the universality of the genetic code (although we now know it is nothing like so uniform as first thought) and the basic mechanisms of transcription and translation to produce various RNAs and proteins. Then, not only are there common core metabolic pathways, but many of the relevant enzymes have similar amino acid sequences, as we saw, for example, with cytochrome c. And structural compounds, such as the proteins of the cytoskeleton or the lipids of cell membranes are substantially the same. Understandably, discoveries such as these conveyed the distinct impression that most of the biochemistry of life had been sorted out by the time of the earliest fossils, and that, no matter how substantial the morphological evolution since then, biochemical novelties were minor, prompting comments such as:

> It is now clear that novelties in coding sequences, i.e. in protein structures, can have hardly been a main driving force in the diversification of multicellular organisms. What distinguishes a butterfly from a lion, a hen from a fly, or a worm from a whale is much less a difference in chemical constituents than in organization and distribution of those constituents. ... As has been often emphasised, differences between vertebrates are a matter of regulation rather than structure. [Jaçob]

This is a pervasive view, but only one side of the coin, and is somewhat misleading.

Specific structural proteins

First, it is clear that alongside the common biochemical components there are many structural proteins that are absent in 'lower' organisms but present in 'higher' ones, and there are various specific proteins that are exclusive to particular groups of organisms, such as different classes of vertebrates.

I shall discuss the supposed evolution of eukaryotes from prokaryotes in Chapter 13. We shall see there is an enormous increase in cellular organization which requires many new structural proteins. But it is worth pointing out here that, for example, although filaments and microtubules are common structural elements of the eukaryotic cytoskeleton, the cytoskeleton may take many different forms depending on the particular organism and cell type, and these variations are achieved through the diversity of other structural proteins which interact with the filaments and microtubules, bundling and crosslinking them in different ways to form a variety of larger structures.

Also, the preceding chapter mentioned various invertebrate groups with distinctive exoskeletons. In some, such as sponges and arthropods, specific proteins are an integral part of their hard parts; in others, although proteins may be only a minor component, specific enzymes are required to synthesize the hard parts, e.g. chitin, and notably for biomineralization.

Then there are the vertebrates with their tissues of cartilage, bone and teeth, each requiring specific proteins for their structure and/or synthesis; and clearly there are many structural proteins specific to the amniotic egg of reptiles, birds and mammals. Further, only mammals possess the biochemical means for milk production; and we should note that this is not just a matter of synthesizing specific milk components such as α-lactalbumin, but also formation of the mammary gland and hormonal control which ensures milk production occurs only when required. And 'decoupling protein' occurs only in the brown fat of placental mammals, where it is important for heat production, especially in the newborn.

The 'higher' animals also have much greater levels of inter- and intracellular organization. This requires many proteins that have a regulatory role in cell metabolism (this is different from regulating genes). To give just one example: the most widespread way of controlling the activity of eukaryotic enzymes is by the addition or removal of one or more phosphate groups at specific sites on the enzyme. This is effected by enzymes called kinases and phosphatases respectively, which may themselves be subject to de/activation, resulting in large and complex regulatory networks. In higher organisms there are many of these enzymes: in the order of 2000 different kinases and 1000 different phosphatases are encoded in the human genome.

Regulatory genes

However, whilst there are many structural proteins specific to various groups of organisms (and we will meet a few more examples later in this chapter), what I want to emphasize is that, even if most of the building blocks are common, the application of these to produce a range of tissues requires many different regulatory genes – notably for transcription factors and DNA control sequences. In other words, regulating or organizing pre-existing structural genes in different ways requires the production of new ones; or, without forcing an evolutionary interpretation – different applications of similar genes require other genes that are different.

For example, what is perhaps most remarkable about neural crest cells which are peculiar to vertebrates is the diverse range of cell types they produce. Whilst these cell types have many common biochemical features (though there are specific structural proteins too), it is clear that reliable production of these different cells – the right types in the right places – involves sophisticated gene regulation, requiring a complex array of control genes.

Similarly, it is not just that spiders' web silk consists of specific proteins, but we also need to consider formation of the spinneret (the organ in which the silk is produced) and the control of silk production; and even web-spinning behaviour must have a biochemical basis, ultimately based on regulatory proteins and control sequences.

MORPHOGENESIS

It is especially important to take account of the biochemical nature of tissues and the biochemical basis of their formation when we come to consider the supposed evolution of novel morphological structures.

Morphological plasticity

Darwin, being aware of morphological variations that arise naturally, but knowing nothing of the underlying genetic mechanisms or biochemical implications, believed that biological tissues are innately plastic. Also, from the success of domestic breeding it was evident that variations are heritable and that significant morphological change can be achieved by repeated (artificial) selection. In the light of which, Darwin extrapolated that any amount of morphological change could be achieved through the selection and accumulation of small beneficial variations. In effect he believed there was no fundamental limit to the plasticity of biological tissues: morphological change was limited only by the viability of the resulting structure.

Now, of course, we realise that this 'plasticity' of tissues is a consequence of underlying genetic and biochemical mechanisms. In particular, morphological variations arise through different combinations of genes, so there is a limit to the extent of possible variations because of the finite pool of genetic material available. (Some variations may be due to the organism's prevailing environment or way of life (e.g. specific growing conditions) rather than to the expression of genes, such as in Johannsen's beans (see Ch. 5); but such variations are not heritable and hence, even if beneficial to the individual in question, are not relevant to evolution.)

Further, over the last decade or two we have started to unravel the biochemical processes of how a fertilized egg develops into a mature adult. As yet we are only scratching the surface, but it is evident that all tissues – comprising the right types of cells, each correctly differentiated and associated with its neighbours – are produced, not by some sort of vague plasticity, but through complex biochemical machinery, involving the expression of an array of interacting genes which require many control proteins and regulatory DNA sequences as well as specific structural proteins.

Unfortunately, many contemporary biologists, although they are aware of the genetic and biochemical aspects of variations, still follow the uninformed 19th-century view when it comes to considering the origin of new morphological structures. That is, they totally ignore the genetic or biochemical implications, and assume that variations can arise and accumulate indefinitely through the imagined plasticity of biological tissues. It is as if they consider morphological changes to tissues and the biochemistry of tissues in entirely separate compartments; whereas in reality they are intimately connected – importantly, the former are entirely dependent on the latter.

In so far as any consideration is given to the fact that novel morphological features require new genes, it is uncritically assumed that the necessary genes will arise through mutation, and have arisen in the past to fuel evolution so far. This is only in a general or vague sense, one that is no better than that of the early Neo-Darwinians based on deleterious mutations, and certainly does not take into account what we now know of the specific nature of biological macromolecules. And, as I have pointed out in preceding chapters, not only are new genes *prima facie* improbable, there is also a marked lack of evidence for constructive mutations. Evolutionary literature is full of simplistic scenarios of how morphological structures might have arisen – scenarios which, even at the start of the 21st century, are no better than the naive ideas of the 19th. Here I discuss only a couple of them – relating to the eye and feather.

Blinkered vision

The eye is the classic example of a highly specialised organ, considered by Ray and Paley as incontrovertible evidence of design in nature. Darwin recognised that the eye was a challenge to his theory, but in the *Origin* indicated how he believed it could have arisen progressively from a simple light-sensitive tissue through a series of variations, each of which offered some minor improvement over its predecessor.[1] The eye remains a favourite example used by evolutionists to show how they believe a complex, well-adapted organ can arise opportunistically – of how design is but apparent design.

Optimistic optics

A prominent example of this comes from just a few years ago when a couple of biologists, Dan-E. Nilsson and Susanne Pelger, produced a scientific paper in which they not only described how an eye could arise readily in a progressive evolutionary manner, but also estimated how long the process might take. Despite always taking what they considered to be pessimistic options, they concluded that the process would require only a few hundred

---

[1] Darwin saw support for this sort of scenario in the occurrence of various shapes, sizes and optical capabilities of eyes in nature – arguing that these illustrated steps in the evolution of a simple eye into a more complex one. Whilst it is true that a range of eye types can be found in some invertebrate groups, this is no proof whatever of the supposed evolution. Even more relevant is that we do not find a gradation for the vertebrate eye – which is arguably the most complex, and it appears fully-formed in the fossil record. Unfortunately, some biologists uses instances of vestigial or even absent eyes – such as in moles and some cave-dwelling fish – as evidence for progressive eye evolution, when it is clearly nothing of the sort.

thousand generations; so geological time would be more than ample for it to occur, in fact would allow an eye to evolve many times. Not surprisingly, this study, with its seemingly scientific approach to the evolution of an eye, and published in a respected scientific journal, has been referred to many times by later authors as demonstrating the efficacy of the evolutionary process; so it warrants some attention here.

On even cursory perusal of their analysis it is evident that, rather than proving an eye could evolve readily, in fact it illustrates clearly how many contemporary biologists completely ignore modern knowledge of genetics and biochemistry in their proposed evolutionary scenarios.

This failing in their approach is evident right from the outset when they comment that 'Taking a patch of pigmented light-sensitive epithelium as the starting point, we avoid the more inaccessible problem of photoreceptor cell evolution.' That was Darwin's starting point too: he did not speculate as to how a photoreceptor might arise, dismissing the problem by expressing his belief that any nerve could become sensitive to light or sound (*Origin*, Ch. 6). Darwin can be excused for having such a simplistic view as he was unaware of the complex biochemistry of sensory cells. Now, however, we know a good deal of the workings of a photoreceptor cell, that it requires many specialised macromolecules – Box 11.1 outlines just the core processes. In the light of which, the position adopted by Nilsson and Pelger is tacit admission that the biochemistry of a photoreceptor poses a formidable if not insurmountable hurdle to an evolutionary origin. Nevertheless, whilst I think this is a major omission and a revealing comment on their part, my main criticism of their study relates to how they depict substantial morphological change arising through evolutionary pressures, without any regard to the biochemical implications.

Nilsson and Pelger assessed eye performance in terms of visual acuity, i.e. the ability to discriminate between parts of the image, which depends on different photocells having different fields of view; and they used this as the basis for selection and hence for directing morphological change.

In terms of making minor changes to an existing structure, visual acuity can be improved first by the initially flat patch of photoreceptors becoming concave. That is, the authors assume that the patch of photoreceptors is not always flat, but varies in shape, and in at least some individuals it is somewhat concave; because the latter improves performance slightly, organisms having a concave photocell patch survive better and reproduce preferentially. Importantly, the authors make the crucial assumption that change in shape is heritable (i.e. based on genes) and hence that the progeny also have concave photosensitive patches.[2] Then, some of the progeny have even more indented patches, which are selected for in preference to the rest, and so on – no matter how much morphologi-

---

[2] As a point of detail, they assume that the morphological variations are not wholly heritable, only 50%; but the crucial assumption is that they are heritable at all.

Box 11.1   Biochemistry of light detection

The photosensitive compound in a photocell is rhodopsin which consists of a light-absorbing pigment called retinal and a protein called opsin. The retinal is an a particular configuration, described as 11-*cis* retinal, but on absorbing a photon of light this transforms into a different configuration known as all-*trans* retinal. This change of configuration activates the opsin, enabling it to convert a protein called transducin from an inactive to an active state, which in turn activates another enzyme, cyclic GMP phosphodiesterase. This enzyme converts cyclic GMP to the non-cyclic form of GMP, thus reducing the concentration of cGMP in the cell. Because high concentrations of cGMP are required to keep open certain sodium channels in the cell membrane, reducing its concentration results in some of these channels closing. This sequence of reactions is summarised here.

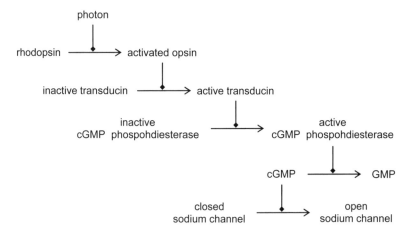

This cascade of reactions means that although one photon activates only one opsin molecule, the signal is multiplied such that many hundreds of sodium channels are closed, reducing the flow of sodium ions into the cell by about 10 million and causing a significant change in voltage across the cell membrane. It is this change of voltage which alters the release of neurotransmitters from the photocell, which is detected by an adjacent nerve cell, and an impulse is relayed to the brain. And all of this happens in a fraction of a second.

Much more could be said about how each of the components operates, and there are mechanisms for regenerating their inactive forms, which of course involve yet more macromolecules; but for the sake of brevity I have omitted these details.

cal change and selection has occurred in the past, the authors assume that more change is possible. That is the second crucial assumption; one which clearly indicates – though the authors are not explicit – that their scenario must be based on new genetic material arising, as we know that there is a limit to the change that is possible through the recombination of existing genes. Indeed, if they considered their model of eye evolution to be solely through segregating pre-existing genes then that would completely undermine the relevance of their study.

Once the indented patch of photocells is as deep as it is wide then, rather than becoming deeper, visual acuity is improved more effectively by constricting the orifice of the depression; and the authors assume that appropriate variations to effect this will arise as required, be inherited, and lead to substantial morphological change. This continues until an optimum is reached: although further constriction would improve visual acuity by narrowing the angle of incident light to each photocell, this is offset by the reduction in light admitted to the photocells.

When this optimum has been attained, further improvement can be achieved only by addition of a lens, and the authors boldly assert that 'Even the weakest lens is better than no lens at all, so we can be confident that selection for increased resolution [i.e. improved optical acuity] will favour such a development all the way from no lens at all to a lens powerful enough to focus a sharp image on the retina.' And they assume that this occurs through progressive thickening of an epithelium in front of the nascent eye.

From this account it is plain to see that the authors simply follow Darwin's view that biological tissues are innately plastic. The essence of their evolutionary model is that variations appropriate for an organism's evolutionary progress are sure to arise and, without any consideration given to the basis of those variations, it is assumed they can be inherited and enhanced; such that in principle there is no limit to the morphological change that can be accumulated in this way. All that we have learned in the last 50 years of genetic mechanisms and the biochemistry of tissues is totally ignored. Regrettably, this sort of blinkered reasoning is widespread among evolutionary biologists; and one cannot help thinking that the biochemistry is ignored not just as an oversight, but because it presents such a formidable hurdle – because to give it its proper weight would completely undermine their supposed evolutionary scenarios.

Before leaving Nilsson and Pelger it is worth commenting on their calculations relating to the rate at which they believe an eye could evolve.

First, the size of their evolutionary steps – of each minor improvement effected by variation – is measured in terms of performance, and they adopt a value of 1% enhancement of visual acuity. That may sound reasonable; but, as they point out, as visual acuity improves, the magnitude of each improvement in absolute terms increases, and the size of morphological change causing the improved performance will generally increase too. That is surely an optimistic scenario, and challenges their claim to have always adopted pessimistic options. Given that the fundamental change taking place is one of structure, it would seem much more appropriate to base the scenario on sensible step changes of structure rather than performance.

A much more serious criticism relates to the way Nilsson and Pelger estimate the time (number of generations) over which the proposed changes could take place. Their approach seems to be entirely inappropriate because they base their estimate on calculations that relate to the change in characteristics of a species due to domestic breeding; that is, on artificial selection where a clear cut-off is made by a breeder based on the extent of a specific characteristic (such as size of body or yield of crop). Hence, although Nilsson and Pelger adopt 'pessimistic' 1% incremental improvements in visual acuity, their calculations imply the scenario that all organisms having that level of improvement (or greater) survive to reproduce, whilst all those that fail to meet this standard (i.e. not just fail to improve, but improve by a smaller margin, even by as much as 0.9%) are rejected. In effect, this is conferring 100% fitness on the former and zero fitness on the latter, which is not merely optimistic but clearly unrealistic and inapplicable to the proposed natural evolution of an eye; and this criticism is scarcely mitigated by the authors acknowledging that in reality natural selection will not be based on eye characteristics alone.

Further, it is clear from the source for their calculations (Falconer) that, consistent with the context of breeding, the calculated rate of change of characteristics is based on the recombination and selection of existing genes – not on the production of new ones; whereas new genes must surely be required for the sort of evolution that Nilsson and Pelger envisage. Falconer recognises that selection reduces the variability on which morphological change can be made, and assumes that for longer term evolutionary progress variability is topped up by mutation. However – and this brings us back to my main point – this role for mutation is understood only in a vague sense which is no better than that of the early Neo-Darwinians, based on deleterious mutations, which I discussed fully in Chapter 9. Absolutely no consideration is given to what must be required for constructive mutations at the biochemical level. This is true even in a later edition of Falconer's work, published only a few years ago – which underlines the point I am making that many of today's biologists are simply ignoring these biochemical implications even though they are crucial to the origin of substantial evolutionary novelties.

Embryological plasticity

We have seen how Darwin's misinterpretation of morphological variations as arising from a supposed innate plasticity of biological tissues has cast a long shadow over subsequent biological thinking. Unfortunately, there is a similar problem with the perception of embryological plasticity. Biologists in the 19th century thought of embryonic tissue as plastic, because they could see that a 'simple' fertilized cell developed into a fully formed individual with specialised limbs and organs. Also, because adult structures arise through developmental processes, there is a sense in which evolution must occur through the modification of those processes.

Understandably, today's embryologists also refer to embryological tissues as being plastic in their formative stages, before their ultimate fate is determined. A good example of which is formation of the lens in the vertebrate eye. The eye originates embryologically as an outgrowth (called an optic vesicle, one on either side) from the forebrain. As an optic vesicle expands, it contacts the surface epithelium (which covers the body and normally develops into skin), and where it does so the latter responds by forming a lens over the front of the developing eye. Of particular interest is that if an optic vesicle is transplanted to almost any other part of the body's epithelium, then that epithelium will produce a lens, even though normally it would not do so. Phenomena such as this have been known for a long time, and have reinforced early ideas of embryological plasticity. No doubt they have also encouraged the sort of scenarios envisaged by Nilsson and Pelger where it is thought that a lens can arise spontaneously.

However, the truth of the matter is that lens formation is far from spontaneous. It is prompted by the release of one or more chemicals (known as inducers) from the optic vesicle, which in epithelium activate the genes that implement development of the lens. Also, formation of the lens is far from just a thickening of the epithelium as envisaged by Nilsson and Pelger. That is the initial response – to form what is called a lens placode – but that is not a lens. The thickened epithelium bulges towards the optic vesicle and then pinches off to form a hollow lens vesicle, with the epithelium healing over behind it. It is somewhat like blowing a bubble from a soap film, with the film re-forming afterwards – except that here we have living tissues which develop this way through internal biochemical mechanisms.

It is the lens vesicle which then develops into a lens. The cells of the posterior wall of the vesicle (i.e. towards the optic vesicle) transform into lens fibres, elongating to 50 or more times their original length such that they fill the cavity of the vesicle, and occupy almost all of the volume of the lens. (The cells on the front of the vesicle remain a thin layer.) This transformation entails various changes that are essential for their light transmission properties. One important change is that they lose their internal organelles, even including the nucleus, which is beneficial because these structures would scatter light passing through the fibres, rather than allowing uninterrupted transmission.

They also produce large quantities of proteins known as crystallins. In fact, about 40% of the mass of a lens fibre is protein, which is several times more concentrated than in typical cells – giving them a high density and hence high refractive index which confers the lens' light-bending properties. Most proteins, if as concentrated as this, would tend to agglomerate and/or denature (unfold), as a result of which the lens would lose its transparency and become cloudy or even opaque (like cooking an egg white). However, the lens crystallins are generally stable proteins and, in particular, the largest group, the α-crystallins, not only are extremely stable, but have an especially important role in stabilising the other crystallins. They are known as molecular chaperones. Clusters of about 40 α-crystallin molecules associate to form hollow balls, and these are connected by other proteins, called CP49 and filensin, to form structures known as beaded filaments. These beaded filaments dominate the lens fibres and are crucial for enabling such high protein concentrations to adopt a transparent almost crystalline state.

They also protect other proteins from degrading with the passage of time. Unlike almost every other biological tissue, proteins in the lens are not recycled – you still have the same crystallin molecules in your eyes as when you were born! Over time, most proteins normally gradually degenerate, and this would lead to the lens becoming cloudy, but the beaded filaments prevent this happening in the lèns. Cataracts can occur in old age if the beaded filament structure is faulty or has become saturated with denaturing proteins.

Finally, the lens fibres are packed tightly together, hexagonal in cross-section, and aligned parallel to the axis of the eye: they are very much like a regular array of crystals, transmitting light clearly through the lens. So we can see that the lens is a unique biological structure in which both the molecular architecture of the fibres and their overall regular arrangement contribute to its transparency.

All of this is from cells which, without the appropriate stimulus, would have developed into the outer layer of skin; so it is no wonder that embryological tissues are regarded as plastic – capable of transforming into a range of cell types. However, we need to be clear about what this embryological plasticity is.

We now know that all of the information required to construct an individual is present in the DNA of the fertilized egg, and that the organism grows – with all that that means in terms of the appropriate multiplication and differentiation of cells and tissues – through the properly regulated expression of that information. Indeed, the marvel of development is that from a fertilized egg, solely through the coordinated activation and suppression of genes, the whole adult organism is fashioned. Development is a process in which cells progressively become committed to a particular fate. And embryological plasticity is simply that, in the course of such development, some nascent tissues are not fully committed to their fate but retain flexibility and – up to a point – can even be redirected. Almost all cells retain a copy of the organism's DNA, but with various parts of it inactivated; and we know that some of this information can be switched back on again. Importantly, comparable with morphological plasticity discussed above, embryological plasticity is entirely dependent on existing genes: it reflects a measure of flexibility in the expression of those genes rather than the production of new ones.

Molecular morphogenesis

Let's now look a little closer at the sort of biochemical mechanisms that are involved in morphogenesis, which we are starting to uncover.

As mentioned earlier, lens formation is prompted by induction from the optic vesicle. At present, we do not know the precise nature of the inducer, but it causes the epithelial cells to synthesize a transcription factor known as Pax6 which then activates the genes that cause the epithelium to form a lens placode and then lens vesicle. Subsequently Pax6 is directly involved in activating the crystallin genes. Pax6 also has a role in the initial formation of the optic vesicle, and in differentiation of the cells in the retina (see below). In view of its key roles, Pax6 is known as the master control gene of eye development, and we will meet it again in the next chapter.

But how can the same factor have these separate roles in different cells? At least part of the explanation is that its action is modulated by or even dependent on other factors which are specific to the relevant target tissues. For example, recent research indicates that the action of Pax6 in epithelium is necessarily in conjunction with another protein known as Sox2: the binding together of these two transcription factors on a specific DNA sequence (identified as DC5) appears to be the genetic switch for initiating lens differentiation. Further, the action of Pax6 to activate the crystallin genes at a later stage in lens development is not only another example of how gene regulation is necessarily cell-specific, but we can also see that the timing of gene expression is crucial for correct development, and this is under accurate molecular control too.

Once the optic vesicle has contacted the epithelium, it spreads outwards and folds in on itself to form the hollow eyeball. It is the inner layer of this, the neuroepithelium, that develops into the retina. (Interestingly, as well as the optic vesicle being essential for the induction of lens formation, later development of the eye is dependent on factors secreted by the developing lens. Not only is this a further indication of the complexity and cooperativity of development, it also casts even more doubt on the sort of simplistic evolutionary scenarios depicted by Nilsson and Pelger where the lens is a late addition after the eye as a whole has evolved.) In the retina, as well as photoreceptors (of which there are rods and up to three types of cones), there are four major types of neuronal cell (horizontal, bipolar, amocrine and ganglion), and with subdivisions of these. They differ in terms of morphology, molecular composition, and notably in position within the retina where they form characteristic layers, with the photoreceptors outermost and ganglion cells innermost. The neuronal cells connect with the photoreceptors and each other – collecting the raw visual input and effecting some processing of this data – and the ganglion cells communicate with the brain via the optic nerve. In addition there are glial cells which provide structural support: they pass right through the retina, holding the layers of cells together, somewhat like rivets. All of these very different cells are derived from the neuroepithelium. Indeed, a question attracting much current interest in developmental biology is how, from a common source of proliferating cells – which appear to be homogenous – this array of appropriately differentiated cells arise, arranged in an orderly laminar pattern, and selectively interconnected with each other.

From what we know so far, as the neuroepithelium grows, whilst some of its cells keep multiplying to produce more undifferentiated neuroepithelial cells (Pax6 has a role in maintaining this), others stop dividing and start migrating inwards. They then differentiate according to their position within the thickness of the retina, that position being determined through the use of inductive signals. We have already seen how the optic vesicle induces epithelial cells to produce a lens. One of the ways in which different cell types are produced according to their position, in many aspects of embryogenesis, is to use a concentration gradient of an inducer, i.e. it is more concentrated near its source and less further away. It is easy to see that with just one inducer, a gradient can determine distance from the source, and not too difficult to appreciate that, by using multiple inducers from different sources, location can be defined more precisely and/or in different directions. It can be compared with global positioning systems where location is determined by comparing the signals from different satellites. It is thought that the inducers in the nascent retina include a number of growth factors.

As a result of their exposure to the inductive signals, cells become committed to a pathway of differentiation that matches their position within the retina. However, the initial signal is not in itself sufficient to ensure complete differentiation of the cells: this requires continued regulatory signals between neighbouring cells as they mature. Literally dozens of different factors have been identified in the developing retina, and many of them have been shown to affect various retinal cell types. The factors and their receptors are distributed in specific patterns corresponding with the different types of cell that are developing.

Early signs that the cells are beginning to specialise is production of various transcription factors (including Pax6 in some cells), the combination of which depends on what type of cell is being produced. This is a first step in the sequence of events which lead eventually to the cell synthesizing macromolecules and assembling structures appropriate to its function, such as layers of membranes full of rhodopsin in the photoreceptors.

We are only just beginning to work out how development – such as of the eye – is implemented, but as we discover more and more of what is involved, of one thing we can be sure: embryological tissues are not plastic like a piece of modelling clay, that varies or can be moulded into all sorts of forms. Rather, tissues are produced through explicit and precisely controlled molecular mechanisms, under the direction of an array of genes. It really is time biologists abandoned naive notions of morphogenesis, and fully took on board what we now know of molecular biology.

Some may still argue that embryological development is where mutations can have a substantial effect, possibly with long-term, i.e. evolutionary, consequences. And it is true that mutations affecting genes that control development can have profound and heritable effects, such as the deleterious mutations observed in *Drosophila*. But it is clear that such mutations are merely disrupting the normal array of developmental genes, and they cannot be used to support the notion of mutations producing constructive new genes, or useful new morphological structures.

Finally, because some of the lens crystallins are the same proteins as enzymes used in metabolic processes, this is a favourite example used by evolutionary biologists to illustrate the opportunism of evolution and show how the lens is but a gradual development of 'ordinary' cells. However, the most abundant molecular component of the lens, α-crystallin A[3], along with CP49 and filensin – all of which are essential for the beaded filaments and hence for the lens' light transmission properties – are unique to the lens. Further, although CP49 and filensin are regarded as part of a large group of compounds known as intermediate filaments, in fact the mechanism by which they form filaments is quite different from that of other intermediate filaments. Also, in terms of the novel biochemistry required for novel morphological structures, every bit as important as the genes for these specific structural proteins are, of course, those that implement their synthesis in the lens and, indeed, of all the other lens proteins in the right amounts, and to effect other changes such as loss of organelles.

Flight of fancy

One of the key problems relating to the evolution of birds is the origin of feathers which are their primary distinguishing feature. The problem is acute because, as with so many other biological novelties, feathers appear in the fossil record abruptly and fully formed (on *Archaeopteryx*), without any trace of part-evolved structures: 'we lack completely fossils of all intermediate stages between reptilian scales and the most primitive feather' (Bock).

From quite early on it was thought that birds evolved from reptiles, so it was obvious to compare feathers with reptile scales and, as both are made of the protein keratin, the idea arose that feathers evolved from frayed reptile scales. That may have been a perfectly reasonable theory when it was first proposed, but quite frankly it is ridiculous that it is still promulgated today, knowing what we now do of genetic mechanisms and the structure and functioning of feathers.

The most fundamental objection to this sort of idea is that it stems from and still follows the pre-Darwinian notion of the inheritance of acquired characteristics and ignores the obligatory role of genetic mechanisms. That is, even if a reptile that had damaged its scales gained some benefit (e.g. thermal insulation) from their frayed state, even if it were fitter and

---

[3] There is also an α-crystallin B which is not exclusive to the lens.

survived better than it would have otherwise – this would be of no evolutionary value because to be inherited the characteristic must have a genetic basis. To have any credibility at all, the idea that feathers evolved from frayed scales must take on board the genetic basis of heritable characteristics. In view of which, it would seem the best evolutionists might realistically hope for is that there was a genetically determined propensity for the scales to fray. And, of course, there would then need to be a genetic basis for every subsequent step from frayed scale to fully sculptured feather with its many specialised features.

Specialised structure of feathers

Which brings me to what we now know of that structure. The central shaft (rachis) of the feather is tubular with a sponge-like pith which gives a light but stiff structure. Branching out from the rachis at regular intervals are barbs which in turn have smaller branches called barbules. Typically, a flight feather from a pigeon has 600 pairs of barbs and each barb has 500 barbules. The fine structure of the barbules depends on the type of feather, of which there are two main types – contour (which includes flight feathers) and down.

As is quite widely known, in the vane of a contour feather the barbules on one barb interlock with those on the adjacent barb to give the familiar zip-like action that enables a split vane to be realigned. To implement this, the barbules on either side of a barb are shaped differently. On one side the barbules have several tiny hooks, and on the other they have grooves or channels which can be engaged by hooks from barbules on the adjacent barb. One of the remarkable features of this fine structure of flight feathers is that, although the vane is not solid, the holes in it (spaces between the barbules) are so small that there is very high frictional resistance to air passing through them, and the feather presents a surface to the air that is almost as if it were impervious – yet with a lighter, and flexible, structure.

Then, bearing in mind the cellular nature of all biological tissues, including every part of a feather, each barbule is derived from a stalk of cells which are serially differentiated along its length, i.e. have a different structure depending on their position within the barbule. Those near the barb are compressed and fused, providing a strong base and attachment to it, whilst those further away are joined end to end, and have appropriately shaped projections to give hooks or grooves depending on which side of the barb they are on.

Clearly, this is a long way from a frayed scale! Feathers consist of differently specialised cells from the base of the quill right to the tip of each barbule. And we must not forget that every feature of specialisation must have a genetic basis – they are not going to just happen that way.

At first sight a down feather does not seem to have such a specialised structure as a contour feather as it lacks the interconnecting barbules and its barbs are not arranged in a specific way, such as to form a vane. Having what appears to be a simpler structure it was thought that down feathers are more primitive, i.e. must have evolved first, and that contour feathers were a subsequent specialisation. Also, at least superficially, it seems less of a transition to get from a frayed scale to a down feather than to a contour feather. However, we now realise that down feathers are in fact specialised for their downy structure. Their barbules resemble bamboo stalks with spiny outgrowths at the nodes, and these spines seem to interfere with each other, keeping the barbules separated, which maintains their downy texture and ability to trap air for insulation. In other words we can no longer consider down feathers to be primitive: they are as specialised as contour feathers. In fact it is interesting to note that, whereas contour feathers are specialised to permit interlocking of the barbules, down feathers are specialised to keep them apart!

As each type is so specialised – and in opposite ways – it is no longer realistic to think that contour feathers have evolved from downy feathers, or vice versa. Consequently, current ideas consider both types to have evolved from a common less-specialised ancestral feather. But one of the problems with this is that a non-specialised feather is much less functional: it offers much less advantage to favour its selection. In particular, without something comparable to the specialisation that keeps barbules apart in down feathers, it even casts doubt on the supposed insulative value of a frayed scale.

Molecular composition

The morphological structure of a feather is built up by proliferation and specialisation of cells within the feather follicle (see below), and an important part of the precise sculpting of the feather is selective cell death. As cells that are to provide the final substance of the feather reach their final stages of development, they synthesize large amounts of keratin, lying it down within the cell which subsequently dies, leaving the keratin structure. About 90% of the completed feather is keratin.

Keratins are a major class of proteins used in the construction of various tissues, especially the integument or outer covering (e.g. skin) of vertebrates. There are two main groups distinguished by whether their folded structure contains α-helices or β-sheets: α-keratins occur in fish and amphibians and pliable skin of other vertebrates; β-keratins occur in hard parts of reptiles, birds and mammals, such as scales, claws, feathers and hair. A subgroup of the latter are the ϕ-keratins which occur only in birds; and within the ϕ-keratins there are two main types which can be distinguished by molecular size and function: the larger ϕ-keratins are used in beak, claws and scales, and the smaller occur exclusively in feathers.

There are about 20 different feather-keratins, and most if not all are synthesized in each of the cells that contribute to the feather, but the relative proportions vary depending on which part of the feather is being produced. When feather-keratins are synthesized, they spontaneously fold into their 3D shape and associate to form filaments which in turn combine to form keratin fibres which are the main structural elements. All of the different types of feather-keratin fold up individually into very similar 3D shapes – with variable regions of the amino acid sequences being on the outside of the folded protein – and one might have thought that the differences between them were unimportant. However, when the folded polypeptides aggregate, the different polypeptides interact differently, and the different combinations within the various feather cells form differently shaped filaments and fibres. As the different fibres are formed within the various cells, these confer shape and mechanical properties appropriate to the part of feather being produced. Hence, although the differences between the various feather-keratins are relatively small in terms of their amino acid sequences, it is clear that these differences are important to provide the range of building blocks required for a feather. Also, of at least equal importance to these differences in the structural genes are the differences in gene regulation which ensure that each type of feather-keratin is synthesized in appropriate amounts depending on the cell's position in the feather.

So – the point I have been emphasizing throughout – it is not only structural proteins that matter, but also the genes that ensure appropriate expression of those structural genes. It is not yet clear to what extent the feather keratin fibres automatically adopt correct orientations and large scale organization within the cell, and to what extent other organizing mechanisms are involved. Clearly, additional organizing mechanisms will require appropriate genetic control too. Because all parts of a feather seem to require most if not all types of feather-keratin (and appropriate control mechanisms), and given that the earliest feathers (contour and down) appear to be modern in structure, this means that a great deal of the genetic basis would need to have arisen in a very short space of time, if not simultaneously. Yet another of those incredible coincidences that macroevolution requires. We are also beginning to elucidate the genetic mechanisms that control the early development of the feather, comparable to those I mentioned earlier in connection with development of the retina, and they involve many inducers and transcription factors.

Feather follicles and tracts

Not only are feathers novel structures, but so are the follicles in which they are fabricated, which penetrate through the upper layer of skin (epidermis) down to the underlying dermis (resembling mammalian hair follicles). A particular puzzle from an evolutionary point of view is that, although both reptile scales and feathers are substantially flat, feathers are synthesized in a rolled up form inside a cylindrical sheath which originates within the follicle, and unroll to adopt their flat profile only when their synthesis is complete and they emerge from the sheath. It is difficult to envisage why or how evolution should go by such a roundabout route, and certainly undermines further the very idea that feathers have evolved from reptile scales.

Mention should also be made of the fact that feather follicles do not occur uniformly or randomly over a bird's body, but in distinct groups called tracts. Feathers in a tract are linked by an elaborate system of muscles and elastic fibres to control the erection of the feathers from the skin. Of particular note is the presence of muscles to depress the feather, which are not required to balance the erector muscles, but which counteract external forces, such as from air currents. These muscles are unique to birds – there is no equivalent in reptile skin or for mammalian hair follicles – and they are essential for controlling feathers in flight. Further, useful control of feather position requires not only the means to move them but also nerves to sense the position of and forces on the feathers. In fact, not only are the feather muscles richly innervated, but the sensory apparatus of avian skin, especially near the follicles of contour feathers, is very rich and diverse.

All of which serves to emphasize that feathers are not just a modest development of reptile scales. They are specialised structures with equally specialised tissues intimately associated with them. And, so far as we can tell, all of these features are necessary for feathers to be useful:

> The sprouting of feathers on the surface of the skin without a machinery to control and regulate their position is unlikely to have been selectively advantageous for a reptilian organism. Feathers are also unlikely to have evolved as individual structures and must have evolved as a coat of feathers from their very inception.

> any scenario that reconstructs the evolutionary history of feathers must deal with the entire integument and its subcutaneous structures.
> [Homberger and de Silva]

Alan Brush, who has researched this subject for many years and recognises the substantial novelties associated with feathers has commented (1996) that 'reptilian scales and feathers are related only by the fact that their origin is in epidermal tissue. Every feature from gene structure and organization, to development, morphogenesis and tissue organization is different.' Further, unlike so many contemporary biologists, Brush recognises that novel features such as feathers ultimately depend on novel genes and that proposed evolutionary scenarios must take these genetic implications into consideration:

> Few have attempted to deal with the mechanisms involved in the evolution of feathers, especially at the molecular or cellular level. The existing morphological approach tends to focus on *why* feathers evolved or *where* feathers came from. At this juncture neither is as illuminating as to ask *how* they arose. [1996, emphasis in original]

However, in view of various factors such as the complexity of even the earliest feathers, he has concluded that the earliest feather follicle must have had substantial genetic variability. In a recent symposium he commented that 'the most primitive feather, whatever its morphology, had an innate potential for structural variation' and 'The existence of follicular capacity to produce morphological diversity existed prior to the complex morphology.' (Brush, 2000)

But surely that is just begging the question. How do we explain the evolution by natural selection of the capacity to produce morphological structures which have never previously occurred so could not have been subject to selection? Is it not tantamount to saying that feathers appeared suddenly and did not *evolve* at all?

QUANTUM BIOLOGY

Before Darwin, there were many morphological structures, such as eyes and feathers, which, because of their clear adaptation of form to function, interdependence of parts, and fine structure of some of those parts, seemed to preclude any possibility that they could be the outcome of undirected natural processes. Darwin, however, with the concept of selection operating on naturally occurring variations, proposed how complexity and/or a high degree of adaptation could arise opportunistically through the progressive assimilation of small advantageous modifications. With the successful amalgamation of Darwinism and Mendelian genetics, this evolutionary concept was wholeheartedly embraced by the scientific community and continues to be the leading explanation for form and function in biology.

However, it is my contention that this key explanatory principle of modern biology has not been properly reviewed in the light of what we now know of biochemistry and genetics. In fact there is a widespread tendency in evolutionary thinking to consider morphological and molecular aspects entirely separately. As we have seen, many evolutionary scenarios are proposed, occurring through the assimilation of novel morphological variations, without any thought whatever given to the genetic basis of those variations, or even recognition that variations of evolutionary significance must have a genetic basis. In this respect they are no better than a century ago, and hence lack any serious credibility. We cannot legitimately divorce morphological change from the underlying genetic and molecular mechanisms; and, when we take them into account, it is clear that these mechanisms challenge the Neo-Darwinian thesis in a number of important respects.

First – a point I discussed fully in Chapter 9 – is that the variations we observe in nature are due to the shuffling and segregation of existing genes, so there is a limit to the variation that can be achieved in this way. Because of the abundance of natural variability, Darwin assumed that appropriate variations would always be available, and modern biologists still follow this: where evolutionary scenarios are concerned, there is an underlying assumption that if a small modification will confer some advantage then we can be confident it will, in due course, arise. However, this assumption completely ignores the fact that for substantial evolutionary advance (macroevolution) new genes are required. In so far as any thought is given to the fact that new genes must be generated, it is uncritically assumed that they will arise by mutation. The variability introduced by deleterious mutations is still used as a basis for believing that new constructive genetic material can be produced by random mutation. But meaningful genes are very improbable, and there is no evidence that new ones have ever evolved.

Second, as we unravel the biochemical basis of biological tissues – what they constitute and how they are formed – increasingly we realise that they involve a large number of interdependent macromolecules. Complexity at the molecular level far outstrips that evident from morphology alone. And from this we are beginning to appreciate what sorts of genes must be required to implement novel morphological structures: new structural proteins, control proteins and control sequences in DNA. It is no longer realistic to think that new structures can arise through a single fortuitous mutation or even a few; and, as emphasized in Chapter 8, it is their interdependence that makes the evolution of any of the genes so very unlikely.

To some extent biologists are beginning to appreciate that macromolecules, at least in terms of structural proteins, are improbable, and recognise that their evolution poses a problem; but in general they have not yet recognised the complexity and hence improbability of effective gene regulation, and still think that this can be implemented readily. This can be seen in, for example, proposed scenarios for the evolution of an eye, which, even today, start with some sort of light sensitive cell, assigning its origin to 'stochastic' processes (i.e. by chance); but then think that substantial morphological change – development of the anatomical structure of the eye – can occur gradually and readily. Whereas, in fact, this development entails a comparable number, or even more, of control proteins to the structural proteins of a photoreceptor, and of comparable or even greater improbability.

Finally – and the most important point I want to make here – our modern knowledge of biochemistry undermines the fundamental evolutionary concept of progress through small increments.

Following Darwin, no matter what complexity or sophistication we discover in biology, the evolutionary answer is to break it down into small steps; believing that if a sequence of gradations, with progressive improvement of function, can be identified, then it proves an evolutionary origin is possible. From an evolutionary point of view, all that is required is to identify a gradual change in structure that confers a progressive change in performance (there does not need to be a direct relationship between them) so that the development can proceed in small steps, hence avoiding the need for 'macromutations' or any other sort of unlikely jump.

Nilsson and Pelger used 1% incremental improvements which they believed to be reasonable, even pessimistic. If critics feel it is too large a step, then the evolutionary response is simply that the steps can be made smaller, and that the important point – that a reasonable evolutionary route can be identified – still stands. Richard Dawkins expressed it like this:

> Is there a continuous series of Xs [intermediates] connecting the modern human eye to a state with no eye at all?

> It seems to me clear that the answer has to be yes, provided only that we allow ourselves a sufficiently large series of Xs. You might feel that 1000 Xs is ample, but if you need more steps to make the total transition plausible in your mind, simply allow yourself to assume 10,000 Xs. And if 10,000 is not enough allow yourself 100,000, and so on. [*Blind Watchmaker*, Ch. 4]

At the morphological level, there are debates about various structures as to whether or not a progressive incremental evolutionary route can be identified for their development.

The eye, of course, has been the subject of such debate for a long time, with the scheme by Nilsson and Pelger, where development is directed by improving visual acuity, being but one of many that have been proposed. The essence of their scheme is typical of this sort of evolutionary scenario for producing morphological structures: a series of stages, with only a small change of structure between them, but each offering at least some functional improvement over its predecessor.

The development of feathers from frayed reptile scales is another well-worked example. Typically, to begin with increased fraying is selected for because of improved insulative value, and gradually the developing feather acquires aerodynamic properties which direct its further development. In the early stages this could be something as unspecialised as increasing air resistance – which could help slow the descent from a jump, then it could be to assist gliding or improve streamlining, and finally to be useful for powered flight. The related structures of feather follicle, tracts, and the sensory and motor mechanisms for controlling feathers which seem to be inextricably associated with feathers do pose substantial difficulties to such evolutionary accounts, but biologists can be very ingenious in devising possible scenarios.

However, although this sort of thinking – that evolutionary progress can be broken down into infinitesimal steps – is widespread among biologists, modern knowledge of genetic mechanisms and the biochemistry of tissues poses a serious challenge to this.

Even at an elementary level, once we realise that gene shuffling cannot provide an indefinite amount of morphological change, it is evident that at some stage or other (and probably at many stages) we need new genes. Some might argue that continuous (or very nearly continuous) gradation of characters does occur in nature. As recounted in Chapter 5, for a time this was seen as a challenge to particulate (Mendelian) genetics. But it became apparent that even continuously varying characters are based on discrete genes: gradations of phenotype emerge from complex interactions of genes. There is now no doubt that, ultimately, for substantial new structures, new genes are required.

This conclusion is reinforced when we look at how tissues are formed:

To begin with, some organs are clearly not just gradual developments of some earlier tissue. For example, Nilsson and Pelger envisaged a lens evolving as a progressive thickening of the epithelium, whereas in fact it is a discrete body which develops from a lens vesicle. It might be argued that the lens placode could have some lens-like properties, but once it starts to invaginate, it is more likely to impair visual acuity than enhance it.[4] Vision is enhanced by the complete structure, not a partly-formed one.

Even more relevant is that we are now aware development proceeds not through some vague plasticity of embryonic tissue, but through precisely engineered biochemical processes, orchestrated by batteries of interdependent genes. It is now evident that any new structure is likely to require not just one but many new genes. Even though we are aware of some genes, such as Pax6, having many roles, it is clear that correct implementation of the several roles depends on other genes that are tissue-specific.

We now know without any doubt that all of biology is dependent on the existence of genes. And, crucially, genes are distinct entities which can arise only on a discrete basis – either a gene works or it doesn't. Ever since Darwin, there has been much argument about whether part-formed tissues – such as an incomplete eye, or a frayed scale – might have sufficient utility to be favoured by natural selection. But there can be no argument about macromolecules: either enough of the sequence is right for a macromolecule to have at least some of a particular function – or it isn't. And this applies to structural proteins, control proteins, and DNA control sequences. So, the important conclusion from all of this is that, ultimately, biology is not a gradation but comprises discrete units, and the current belief in finely graded evolution fails at the biochemical level.

The discrete nature of biology would not present a difficulty to evolution if the quanta – the individual genes – were reasonably attainable; but the fact is that they are not. Each of these discrete units is much too improbable to have arisen opportunistically. In particular, the evolutionary rationale that we can achieve an improbable end result through a series of realisable intermediates – that we can climb mount improbable in manageable steps – does not work with macromolecules. The step from random sequence of nucleotides to something with any sort of meaningful biological activity is inconceivably improbable and we know of no way in

---

[4] Whilst ontogeny is not phylogeny, even if a lens could develop first by progressive thickening of an epithelium, this would simply lead to the question of why different developmental processes should evolve for essentially the same adult structure. I comment further on this in the next chapter.

which it might have occurred.

In conclusion, the smallest step by which substantial increase in biological complexity could arise entails at least one new functioning macromolecule, but there is no way in which it might arise. Even more confounding from an evolutionary perspective is that any realistic advance almost certainly entails multiple genes – and that makes the whole idea totally untenable. In short, the fundamental evolutionary principle of incremental progress fails completely at the level of molecular biology.

12

Homology and Phylogeny

Although the word homology was not used in its now familiar biological sense until the 19th century, notably by Richard Owen, the concept has a long history. Aristotle recognised that in even quite different types of animals there may be many corresponding structures – of soft internal organs as well as bones; and a few others after him such as Galen, Leonardo da Vinci and Pierre Belon (1517–1564) compared the human skeleton with that of a monkey, horse and bird respectively. Then, homologies in the sense of corresponding parts of organisms with a common body plan, were the basis of the classification schemes developed from the 16th century onwards. Homology was therefore a pre-Darwinian concept, initially linked with the idea of archetypes, somewhat like Plato's Forms, without evolutionary connotations. However, Erasmus Darwin cited the similarities of quadrupeds as indicating their common origin and, of course, these similarities were included by Charles among the facts his theory explained.

In his book *Classification, evolution and the nature of biology* (1992) Alec Panchen argues that if evolution is proposed in order to explain homologies and the hierarchical classification of organisms, then, logically, homology and classification cannot be seen as evidence for evolution. Nevertheless, most present-day biologists do regard homology as important evidence of past evolution. A prime example is the tetrapod limb. It is believed that the tetrapod leg, from its first appearance with the early amphibians, has been progressively modified to produce structures as diverse and specialised as a bird's wing, seal's flipper and human arm, all based on a common bone arrangement which is clearly discernible despite the very different overall shapes and functions. That is, in just the same way as it is thought impossible that similar amino acid sequences could arise independently in 'homologous' proteins such as the globins (Ch. 7), so it is thought too improbable that such similar morphological structures could have arisen separately. Hence, it is concluded, they must have evolved from a common source. It will however be evident from Chapter 10 that we do not know of gradations between these different types of fore-limb, showing how they actually evolved from a common ancestor: the different animal groups, notably those with highly specialised limbs – such as pterosaurs, birds and bats for flying, and ichthyosaurs and cetaceans for swimming – appear abruptly, not progressively.

Further, the very idea of homology fits in well with one of the key concepts of evolution. Evolution has no foresight – it cannot anticipate what

319

would be the best form of a wing or flipper; but, it is thought, appropriate structures can arise by progressive modification (through variation and selection, and ignoring the biochemical implications) of what is already available. Put another way, in evolutionary terms it is much easier to adapt an existing structure to meet new needs than to acquire a completely new structure from scratch. Then, because an ancestral structure, in different lineages, has been modified in different ways, this has produced the diverse structures but with some retained common features – i.e. homologues – of today's species. Indeed, the essence of homology is that the same structure has, through the evolutionary process, been modified in different ways, adapted to different ends.

With this in mind, embryology acquired an important role in identifying and interpreting homologies. The point being that, even if adult structures look rather different, if they are homologous then they may be traced to a common embryological source. For all species of a given phylum, their early embryological development passes through what is called a phylotypic stage in which there is considerable similarity throughout the phylum. Indeed, this similar stage is an important criterion in defining the members of a phylum. It is thought that, over the course of evolution, modifications of the developmental processes have resulted in divergence from this common embryological form to give the range of modern day species (within a phylum).

Conversely, even if structures from two different species look similar, if they have developed from different embryological tissues then they would not normally be regarded as homologous, but due to convergent evolution. A good example of this is the vertebrate eye and that of the cuttlefish (a mollusc related to squid). In overall structure they closely resemble each other, notably in having a lens and iris, are equally specialised and with comparable performance. But they are not considered to be homologous, as there is no doubt they have arisen quite independently in separate phyla (chordates and molluscs) which have completely different body plans (and phylotypic stages). There was no common primitive eye from which they have both evolved.

Although most evolutionary texts convey a consistent and hence persuasive picture of homology, there are in fact many substantial anomalies. In particular, as we discover more of how tissues are formed embryologically, it casts doubt on much of the homology that has been perceived for so long at the morphological level. It has even raised fundamental questions about what is meant by homology – but more of this later.

Vertebrate non-homologies

Because the vertebrate skeleton is the prime example used to illustrate homology in support of evolution I will focus on this. It is not at all surprising that the general similarity of vertebrate skeletons, especially of tetrapods, should prompt ideas of a common origin and be seen as at least circumstantial evidence of evolution. However, closer inspection reveals significant non-homologies – differences which discriminate between every class of vertebrate, and even between orders within some classes.

Skull bones

Take, for example, the earliest bony fish. You may recall from Chapter 10 that they appear as two main groups, distinguished primarily by the structure of their fins, being lobe-finned or ray-finned. Although there are no known intermediates linking these two groups, in view of their similarities as bony fish it is confidently assumed that they must have evolved from a common ancestor.

However, whilst it might have appeared at first sight that their skeletons are directly homologous, in fact there is a prominent exception in the arrangement of their skull bones. The skulls of the two groups are so distinct that some biologists have concluded 'Evidently the two major patterns evolved separately and so differently that they are not directly homologous' (Webster and Webster, Ch. 4). In fact, these authors pointed out that the actinopterygian (ray-finned) skull is markedly different from any other type of vertebrate, past or present, commenting that 'Although many of these bones are given the same names as bones of other vertebrate skulls, the homologies are dubious at best.'

A major subdivision of the actinopterygians is the teleosts which is the dominant group of modern fish. So, although the vertebrate skeleton is seen as a key example of homology, as supporting a common evolutionary origin, in fact there is a marked non-homology which separates most present-day fish species from all other vertebrates.

While we're on the subject, it is worth noting also that lungfish (dipnoi), too, have a distinctive skull structure and, despite being classified along with other lobe-finned fish as sarcopterygians, are significantly non-homologous in this respect.

Tetrapod limbs

In Chapter 10 I drew attention to key novel features of the earliest amphibians, novelties which pose the greatest challenge to their supposed evolution from preceding lobe-finned fish. In particular, embryology has confirmed that feet – the carpals or tarsals (wrist or ankle bones) and digits – are completely new structures, not in any way homologous to fin rays. This is evident from the pattern of expression of the relevant developmental genes as well as in the way the bones appear to form.

Even the limbs of different tetrapod classes are non-homologous. Notably, despite the common possession of carpals, tarsals and digits, in each class these bones have a distinctive arrangement: they do not appear to be derived from a common underlying pattern by which they might be considered to be homologous (e.g. Hinchcliffe and Griffiths, 1983). And it was surprising to find that the digits of urodeles (e.g. salamanders) are not homologous with those of anurans (e.g. frogs) or other tetrapods. It is not just a question of being a non-homologous combination from the usual five – such as between theropods and birds: salamander digits are formed embryologically in a quite different way from that of other tetrapods (e.g. Wagner, 1999).

Ribs

A further skeletal feature which at first might have appeared homologous, but in fact distinguishes fish from tetrapods, and between different classes of the latter, relates to their ribs. We are familiar with the idea that tetrapods, such as ourselves, have only one set of ribs – i.e. there is just one pair of ribs to each vertebra, with one rib being on each side of the body. In contrast to this, many fish, including teleosts and lobe-finned fish, have two sets: ventral and dorsal. Ventral ribs surround the body cavity (rather like our own), and then the dorsal ribs lie between the dorsal and ventral muscles which run along the back and sides of the fish. (It is the dorsal ribs that are not readily removed by filleting, and can be a nuisance when we eat fish.) Surprisingly, although amphibian ribs surround the body cavity – and hence it might have been thought they would be ventral ribs – it is evident that their embryological source is equivalent to that of dorsal ribs rather than ventral. In other words, contrary to evolutionary expectations, the ribs of amphibians are not homologous to those of the lobe-finned fish from which they are supposed to have descended.

The supposed transition from amphibians to reptiles presents a similar discontinuity because, somewhat intriguingly, the ribs of amniotes, including reptiles, are not clearly homologous to either dorsal or ventral ribs. So, as well as the key differences between amphibians and reptiles relating to the latter's amniotic egg and all that that entails in terms of reproductive organs and behaviour, and despite the general similarity of amphibian and reptile skeletons, here too we have evidence that they are not actually homologous.

Scales

Another interesting difference between amphibians and reptiles relates to their scales which superficially might have seemed to be obvious homologues.

The outer layer of animals, which for humans is our skin and associated structures such as hair, is called the integument. In vertebrates it comprises two layers, an outer epidermis and underlying dermis. These are completely distinct, having different embryological origins, and are separated by what is called a basement membrane.

Fish scales are bony, and arise from the dermis. This applies from the earliest jawless fish right through to modern teleosts, notably including lobe-finned fish. (Whilst it might seem odd for bone to arise from layers of the skin, in fact it is not at all unusual. For example, dermal bones are important components of the skull, jaw bone and pectoral girdle (e.g. our scapula or shoulder blade), and of turtle shells.) The early amphibians also had bony dermal scales. Almost all modern amphibians are without scales, though caecilians have vestigial dermal ones.

In contrast, reptile scales do not arise from the dermis but from the epidermis, and are keratinised (i.e. primarily protein) and not bone at all. Essentially, reptile scales are thick stiff plates of epidermis, separated by more flexible sections which act as hinges, allowing relative movement between the scales. The flexible epidermis uses an α-keratin similar to that of amphibian skin, whereas the scales use a harder β-keratin which does not occur in amphibians (see Ch. 11 for more about keratins).

Hence, here we have another exception to the supposed homology between different classes of vertebrates, and another necessary change in the presumed evolution of amphibians to reptiles.

Vertebrae

In view of the importance attached to the apparent homology of the vertebrate skeleton, and the weight given to embryology for interpreting homology, it is especially relevant that vertebrae – which are distinguishing characteristics of vertebrates – form embryologically in significantly different ways for different classes of vertebrate, and even use different source materials.

Vertebrate embryos have a neural tube extending along their back, with a notochord running alongside and beneath it. Early in development, temporary structures called somites form along the neural tube, in pairs with one on either side of it, and it is usually these that develop into vertebrae.

In most tetrapods, including mammals, the cells of the somites differentiate into three discrete layers: on the outside is dermatome which will develop into connective tissue, then myotome which will develop into muscle, and innermost is sclerotome. The sclerotome grows towards the notochord and surrounds it, to form what is called the perichordal tube, which then develops into the vertebrae.

There is now good evidence that in birds the vertebrae do not develop from somite pairs on a one-for-one basis, but that each vertebra forms from the rear halves of one somite pair and the forward halves of the next – a process known as resegmentation. This is of interest to us here because it seems resegmentation does not occur in mammals, which points to another distinction between vertebrate classes.

In at least some amphibians – notably including *Xenopus* (a type of toad) which is a popular species for embryological research – somites do not differentiate into three types of cell, but consist entirely of myotomal cells from which the vertebrae develop.

In cartilaginous fish such as sharks and many 'primitive' bony fish such as sturgeons, the sclerotome grows somewhat towards the notochord, but transforms into blocks of cartilaginous tissue called arcualia. Typically, each pair of somites produces four pairs of arcualia, with one of each pair on either side of the neural cord / notochord. The arcualia then transform into a vertebra, with each arcualia destined to become a specific part of it.

In teleost fish, the process of vertebrae formation is quite different from other vertebrates, involving three distinct steps. Firstly, the sheath of the notochord (not derived from somites) differentiates into cartilaginous elements, called chordal centres, which subsequently become the main body of the vertebrae, known as the centra. Secondly, cells near the chordal centres, but again not derived from somites, form other cartilaginous elements which will become what are known as the dorsal and ventral arches of the vertebrae. And finally some sclerotome cells migrate towards the chordal centres, possibly fuse with them, and contribute to parts of the vertebrae, but transform into bone directly not via cartilage. Of particular importance here, of course, is not only that the vertebrae form by a very different route from those in other vertebrate classes, but especially that much of the vertebrae clearly derives from different source materials – notochord and other non-somite cells, and not just for some peripheral part of it, but for the centra.

Anomalies in early embryogenesis

We have now seen that there are significant differences in the formation of apparently homologous structures from the phylotypic stage. Contrary to (evolutionary) expectations there are differences in the embryological sources and developmental mechanisms by which apparently homologous structures arise. What is even more confounding is that, even though the phylotypic stage is similar within a phylum, and we would have expected it to be formed from a fertilized egg in substantially the same way, in fact there is remarkable diversity and some fundamental anomalies.

Soon after an egg is fertilized, the resulting zygote divides repeatedly to form a clump of cells, which in all vertebrates forms into a hollow ball known as a blastula (which can be very flattened for groups where there is a large yolk, especially in reptiles and birds).

The blastula then undergoes what is known as gastrulation (meaning gut formation) in which it transforms into what is at least notionally something like a sack, although its actual shape depends on the type of organism. But there is considerable – and surprising – diversity in the ways in which gastrulation occurs, even between what are generally regarded as closely related groups of organisms. In many cases the sack arises through deep invagination of one side of the blastula, or sheets of cells may move over the surface of the blastula to the inside, or cells may migrate individually from the outside through the intervening cells to the inside, or a combination of these processes.

Then, not only are there different ways in which the gastrula forms, but there is also remarkable variety in which parts of the gastrula develop into apparently homologous tissues. Notably, Sir Gavin de Beer in his small but much-cited work, *Homology, an unsolved problem* (1971), commented as follows:

> Structures as obviously homologous as the alimentary canal in all verte-brates can be formed from the roof of the embryonic gut cavity (sharks), floor (lampreys, newts), roof and floor (frogs), or from the lower layer of the embryonic disc, the blastoderm, that floats on the top of heavily yolked eggs (reptiles, birds).

It is at the gastrula stage that the three key germ layers arise: the layer of cells on the outside of the gastrula is called the ectoderm, that lining the inner cavity is the endoderm, and in-between is the mesoderm. And there is a general pattern in which the germ layers develop into characteristic tissues of the adult body: in many cases the ectoderm gives rise to nervous tissue and the epidermis, the mesoderm to skeletal, muscular and circula-tory systems, and the endoderm to the lining of the digestive and respira-tory tubes. Because at first this pattern appeared to be consistent, it led to what is known as the germ layer theory of embryological development. But we now know of many exceptions to this, again even within seemingly closely related organisms.

For example, despite the fact that anurans and urodeles are both mod-ern amphibian groups and we would generally consider them to be closely related, there is a marked difference in the source of their primordial germ-cells. In urodeles, they arise from unspecific ectodermal cells at the blastula stage; but in anurans they arise from specific cells of endodermal origin, the cells having cytoplasmic granules that originated in the unfertil-ized egg. The difference relates to such important organs – the germ-cells – especially so far as evolution is concerned, is so substantially different, and occurs so early in development, that some biologists believe it points to a very early bifurcation in the phylogeny of amphibians, even suggest-ing they might have evolved from different groups of lobe-finned fish (Nieuwkoop and Sutasurya). Clearly it would be a remarkable coincidence for two different groups to have evolved separately for so long, and yet share so many distinctive features as modern amphibians do. But the al-ternative is to propose that somehow there must have been a radical change in the relevant developmental mechanisms. Which is the sort of conclusion biologists are driven to if they insist on an evolutionary origin for the non-homologous development of 'homologous' organs.

Or consider formation of the neural tube. Whilst it always develops from the neural plate, which is part of the ectoderm, it does so in different ways. In most vertebrates the neural tube forms by the neural plate in-vaginating to form a deep furrow, the edges of which seal over to enclose a hollow tube – somewhat similar to the formation of a lens vesicle (see Ch. 11) but here to form a long thin tube. However, in lampreys and teleost fish, the neural plate does not invaginate; instead it thickens along the centre line to form what is called the neural keel, and the neural tube

forms by a cavity arising within the thickening. This difference in formation of the neural tube is especially significant because the neural tube is
one of the defining features of vertebrates, in fact of all chordates, and yet
here we have it forming in two very different ways, and very early in development – even before the phylotypic stage.

If, as the evolutionary scenario envisages, all vertebrates are derived
from a common ancestor, then, to account for the different ways in which
'homologous' organs are formed, it is necessary to postulate that substantial changes in development have occurred – and at very early stages of
development in some cases. This is true of blastula formation, development of the neural tube and alimentary canal, production of the primordial
germ-cells, and construction of vertebrae. Yet these changes in very early
development would need to have occurred in organisms which were very
advanced, i.e. where there was a lot of development following the point at
which development changed. For example, so far as formation of the neural tube is concerned, it would seem the change would need to have occurred in early bony fish in order to account for the distribution of the
method of its formation in present day organisms. But it is very difficult to
conceive of what might have led to such changes in early development,
especially when it is the final form that is the primary object of natural
selection. And, more importantly, there is general agreement that the disruption caused by substantial changes of early development, far from being advantageous, is likely to be seriously detrimental if not lethal. One
contemporary biologist summed up the dilemma like this:

> Most intriguingly, features that we regard as homologous from morpho-
> logical and phylogenetic criteria can arise in different ways in develop-
> ment. As first pointed out by von Baer in the 1820s, animals within a
> phylum, such as the vertebrates, share a common body plan, and in their
> development share a phylotypic stage in which the body plan elements
> characteristic of the phylum appear. The process of early development
> from the egg to the phylotypic stage should be at least as conserved as the
> pattern of the phylotypic stage. One might reasonably expect mechanisms
> of early development to be especially resistant to modification because all
> subsequent development derives from early processes. Traditionally, fea-
> tures of early development and conserved larval stages, even between
> phyla, have been regarded as strong homologous characters for the infer-
> ence of phylogeny. The division of animals into protostome and deu-
> terostome superphyla [see below] is based on the idea that embryonic
> similarities are homologous and have been largely immutable over hun-
> dreds of millions of years. [Raff, 1999]

The pattern of development within a phylum is often described as an hour-
glass – with the constriction corresponding to the phylotypic stage which
has many common features, but divergence in developmental processes
before and after. Unfortunately, having assigned an appropriate term to
describe the phenomenon, some biologists now write as if it were also an
adequate *explanation* – which clearly it isn't.

Biochemical homology

If all forms of life have evolved from a common source then it is to be expected that basic biochemical machinery – such as genetic mechanisms and metabolic pathways – which would have been present in the simple (single-celled) common ancestor, should have been inherited by successor organisms, leading up to higher plants and animals. Consequently, the fact that many biochemical features are common to all forms of life is seen as evidence of evolution.

One of the intriguing discoveries of recent years is that there is a marked similarity, not only in terms of within-cell biochemistry, but also in many of the genes used to implement the developmental processes of multicellular life in very different animal groups. A notable example of this is a series of genes which occur in both invertebrates and vertebrates, that have become known as Hox genes. These genes exercise high level control over the formation of different parts of the animal's body during early stages of embryological development. Although we are only just beginning to learn how they work, it is evident that the genes code for proteins that act as transcription factors, controlling the expression of batteries of genes which implement particular features of the animal's body. Not surprisingly, many see these equivalent developmental genes as further evidence of a common evolutionary origin of all life. It is assumed that the Hox genes originated in some early multi-cellular animal from which all animal phyla have subsequently evolved. On the other hand, others have pointed out that it is in fact unexpected that evolution – without any foresight – should have utilized the same developmental genes for the production of very different morphological structures.

Even more surprising are where equivalent developmental genes control the development of organs which have a similar function but have arisen independently, i.e. the organs are analogous rather than homologous. An example of this is the gene known as *distalless*. It is involved in the development of projections or appendages in various animal phyla, notably vertebrate fins and limbs, but also arthropod legs, echinoderm tube feet, ascidian syphons and ampullae, and annelid parapods. It is thought that divergence of animal phyla occurred so early that the last common ancestor did not have appendages. So, although it might be reasonable to suggest that the *distalless* gene arose before divergence and was present in a very early common ancestor – such that the genes in the various phyla are homologues of the early gene, it cannot be maintained that *distalless* already had a function in the development of appendages. Hence, it seems to be a remarkable coincidence that essentially the same gene is used for structures which are analogous, but not homologous. An even more remarkable example of this is the eye.

*Conflict between morphological and biochemical evidence
for eye evolution*

In the late 1970s L v Salvini-Plawen and Ernst Mayr undertook a comprehensive study to ascertain how many times the eye had evolved. By categorizing the eyes of different groups of organisms according to embryonic origin and final structure, they determined where it was reasonable to infer that an eye had been inherited from one group to another, and where it was not. They concluded that eyes must have evolved independently in at least 40 different phyletic lines, possibly up to 65. Hence, they reasoned, despite the evident complexity of some types of eye – notably vertebrate eyes but also of some invertebrates – a complexity which seems to challenge the possibility of an evolutionary origin, not only is it clear that nature must have been be able to find a way, but it must not be as difficult to evolve an eye as we might have thought. This multiple origin of eyes became accepted evolutionary doctrine, and was affirmed recently by Jerome Wolken (1995) from his consideration of diverse eye morphology.

> In reviewing the evolutionary development and structure of invertebrate eyes, no clear patterns could be discerned; most likely, different species evolved their eyes independently. For among invertebrate eyes, every known optical system device for forming an image has evolved, that is, from pinhole eyes to simple eyes to camera-type eyes to compound eyes and to eyes with refracting optics. [*Light detectors*, Ch. 12]

However, various biochemical facts about eyes seem to be inconsistent with this. Even at the time of Salvini-Plawen and Mayr's study, it was known that in all phyla the key molecule involved in absorbing light is rhodopsin. That is, not only is it common to all vertebrate and invertebrate eyes (i.e. in eukaryotes) but it is also used by photosensitive prokaryotes such as some bacteria and algae. This would not be surprising if there were a common photosensitive ancestor from which all of these different groups had evolved, but they concluded quite definitely that that was not the case:

> All the evidence however indicates that the earliest invertebrates, or at least those that gave rise to the more advanced phyletic lines, had no photoreceptors.

Indeed this conclusion was evident from a consideration of the molluscs alone:

> The molluscs display greatest diversity in the differentiation of eyes among all groups of animals and 7–11 different lines can be distinguished; the ancestral stock was obviously devoid of photoreceptors and neither [various mollusc groups] nor most original larvae [...] possess photoreceptors.'

Over the succeeding years, the problem has been compounded by discoveries relating to the biochemistry of eye formation. On the one hand, Salvini-Plawen and Mayr confidently asserted that 'It requires little persuasion to become convinced that the lens eye of a vertebrate and the compound eye of an insect are independent evolutionary developments.' (Bear in mind that vertebrates and insects are separated by hundreds of millions of years; and both groups appear suddenly, with fully formed eyes.) Yet, on the other, there are some striking parallels in the genetic control of their embryological development.

First to emerge was that there is a master control gene of eye development in *Drosophila* – called *eyeless* because it was first identified through deleterious mutations of its gene which resulted in no eyes forming at all – which has a very similar amino acid sequence (is considered to be homologous) to Pax6, the master control gene of eye development in vertebrates (see Ch. 11). It is a remarkable coincidence that eyes which are so very different morphologically and have arisen independently should have such similar underlying genetic mechanisms, and hard to see why or how this should have arisen through an opportunistic evolutionary process.

One possibility is that a Pax6-like control protein has some sort of propensity for promoting the development of visual systems. But, in the absence of some clear mechanism as to how this might be implemented, it has few supporters.

A more popular alternative is to postulate that, despite the lack of morphological evidence for it, there must have been a rudimentary photoreceptor in the putative common ancestor of vertebrates and insects, and it is supposed that Pax6 was involved in its formation. This primitive structure, with its role for Pax6, has then persisted and developed through the various lineages.

However, this sort of explanation becomes less tenable as we discover that not only Pax6 but several other transcription factors are common to the development of vertebrate and invertebrate eyes. Notable among these are the mammalian so-called '*Six*' genes and their equivalent in *Drosophila* known as *sine oculis*. As the action of these genes is further along the developmental pathway to their respective eyes than Pax6, it becomes increasingly difficult to envisage a common ancestral role for them from which the modern day diverse roles have evolved. Especially significant is the gene known as *Dach* in vertebrates and *dac* in *Drosophila*. Not only are they similar ('homologous') in structure, but each has a somewhat restricted role which occurs late in eye development. It is also worth noting the significant difference that fly eyes develop from a single embryonic tissue (imaginal disc) whereas vertebrate eyes develop from two (optic vesicle and epithelium), which makes a common primitive source even more difficult to believe.

Despite such anomalies this remains the evolutionary explanation as there does not seem to be a viable alternative; because the coincidental 'recruitment' of homologous genes for comparable roles in analogous (non-homologous) structures seems so improbable.

In conclusion, although some biochemical similarities support the idea of a common evolutionary origin, it is increasingly evident that other similarities are very difficult to accommodate within that sort of explanation, and may even militate against it.

Homology and evolution

Homology is a fundamental feature of biology: similarities between species – for a long time recognised as homologies – are the basis of biological classification. The theory of evolution seemed to provide a convincing explanation for homologies in terms of phylogeny: homologous organs in different species and groups of species are similar because they have been inherited – with some degree of modification – from a common ancestor. Indeed, commonality of evolutionary descent came to be the defining factor of homology.

Clearly, in the common ancestor, there would have been just one organ, with the current homologues being modifications of it. So if the evolutionary explanation of homology were correct, then derivation from a common ancestral organ should be reflected in the homologous organs being derived from a comparable embryological source – equivalent to that of the organ in the common ancestor. A corollary of this is that if two organs derive from different embryological sources then it would be possible for them both to appear in the same organism (conjunction); but conjunction is seen as a cardinal negation of homology (Paterson, 1982).

Along with this understanding that homologous structures should be derived from comparable embryological tissues, it was thought that they should also be formed by similar developmental processes (at least where homologues have a similar adult form). For example, before much was known of genetic mechanisms of development, Gareth Nelson wrote that 'the mode of development itself is the most important criterion of homology', and Louise Roth (1984) considered it essential:

> A *necessary* component of homology is the sharing of a common developmental pathway. Homologues must, to some extent, follow similar processes of differentiation which, one infers, depends on the same batteries of (regulatory and structural) genes [...] homology is based on the sharing of pathways of development which are controlled by genealogically related genes. [emphasis added]

In other words, if evolution is to explain homology and, reciprocally, for homology to be evidence or support for evolution, it is necessary to show that there is a viable route by which today's homologues could have arisen from a common predecessor. Most obviously this implies a common embryological source, and secondly the use of comparable developmental processes. The problem is, as we have uncovered more of embryological development, there are many cases which do not conform with these expectations. Most damning was that some apparently 'homologous' adult structures do not arise from comparable embryological sources, leading de Beer (1971) to write:

Therefore, *correspondence between homologous structures cannot be pressed back to similarity of position of the cells of the embryo or the parts of the eggs out of which these structures are ultimately differentiated.* [*Homology*, original emphasis]

The importance of this – at least so far as an evolutionary interpretation of homology is concerned – can hardly be overstated. Evolution had provided *the* explanation for homology – as arising by diversification from a common ancestor. A common embryological source clearly lent support to this because it could be seen how, from a common source material, divergence of developmental processes could produce a range of adult structures with some shared, but some different, features. But where 'homologous' structures arise from different embryological sources (and we have seen that the difference can be very deep), it completely undermines the evolutionary explanation. The most straightforward conclusion from the use of non-comparable embryological sources is that the adult organs have not been derived from a common ancestor, and are not homologous in the modern (post-Darwin) sense.

A similar conclusion is indicated where 'homologous' structures, even if derived from a comparable source, arise by substantially different developmental routes. There are different embryological processes and different genetic mechanisms to apparently homologous organs. And not only are there non-homologous processes to 'homologous' organs, but there is the conundrum of homologous genetic mechanisms for analogous organs.

It is clear that the expected harmony between homology and evolution is lacking. Evolution no longer explains the facts of morphological homology and, as a consequence, homology is no longer the evidence for evolution that it was once thought to be. Needless to say, there is profound reluctance to accept these conclusions, and the incongruence of homology and presumed phylogeny has provoked earnest discussion among relevant scientists over the last decade or two to try to come to terms with the facts within an evolutionary framework.

Having said that, one approach is simply to ignore the anomalies, and try to retain the evolutionary interpretation of homology despite the contradictions. For example, Darwin knew that embryological development was not always in line with supposed phylogeny, and wrote that 'community in embryonic structures reveals community of descent; but dissimilarity in embryonic development does not prove discommunity of descent' (*Origin*, 4th edn, Ch. 13). And in modern times we have:

> Although common development processes may aid in the identification of homologous structures (ontogeny as a criterion, not as a mechanism), lack of common development, be it developmental origin, process, or constraint, tells us nothing about lack of homology. [Hall, 1992, Ch. 10]

In other words, some scientists will happily use those instances where embryological development and adult similarities are consistent as evidence of common descent (i.e. evolution), and set aside those instances where they are inconsistent. But it is hardly valid to employ only the evidence that suits your hypothesis, brush the rest under the carpet, and claim the evidence supports your theory. Unfortunately, this is typical of the information presented in evolutionary textbooks: the 'good' examples are presented, and the anomalies are kept out of sight – which is why homology is still commonly regarded as evidence of evolution.

Others have grappled with the problem, trying to face up to the facts but recognising the lack of harmony between homology and the theory of evolution. One approach has been simply to remove common source and/or developmental process as criteria of homology. For example, in a later publication Roth (1988) recognised that homology can exist at different levels – morphological, developmental processes and genetic mechanisms – but accepted that these different homologies need not be congruent; even though we would have expected them to be if homologies arose from common ancestry.

It is difficult to envisage why or how developmental processes should evolve dramatically yet still lead to substantially the same morphology which is the principal object of natural selection. Nevertheless, some see the different developmental mechanisms to 'homologous' structures not as a challenge to evolution but simply as evidence that developmental mechanisms have evolved while retaining similar end structure. But surely that is circular reasoning – based on assuming that the various groups with similar morphologies but different developmental mechanisms have evolved from a common ancestor.

Some biologists propose we redefine homology, for example as structures with common adaptive constraints rather than an emphasis on their origin, to try to circumvent the non-homologous developmental mechanisms (e.g. Wagner, 1989). And some have argued that we should revert to the earlier definition of homology – which was based exclusively on a common pattern in adult organisms without reference to how the adult structures formed (e.g. Panchen, 1999). That is all very well provided homology is only a matter of comparative morphology. But as soon as we link homology with phylogeny, then we do expect homologues to be formed from comparable embryological sources, and most probably by comparable developmental processes. So those who would drop any developmental criteria of homology must recognise that in doing so they are breaking the link between homology and phylogeny.

It is not my intention to enter into the debate as to what homology is or how it should be defined, but merely to emphasize the inconsistency between what is generally considered to be homology and supposed phylogeny. Importantly, homology could be evidence of evolution only if homology can be explained in terms of common descent, but it is now abundantly clear that it cannot. There are substantial exceptions to the supposed link between homology and evolution. Consequently, as the evolutionary interpretation of homology fails, homology is no longer the evidence for evolution that it was thought to be.

Indeed, not only is homology no longer the support for evolution that it was once thought to be, but the inconsistency between morphological homology and supposed phylogenetic relationships has become a further challenge to the theory of evolution: If evolution does not explain homology, then the facts of homology are at odds with the theory of evolution and cast doubt on it. In particular, if similar morphological structures (structures that had appeared homologous and had initially been seen as evidence of common ancestry) actually develop from different embryological sources (*prima facie* indicating that they are not derived from a common ancestor), then this is strong evidence pointing away from an evolutionary relationship for the organisms concerned.

RECAPITULATION

Like homology, the idea of recapitulation has a long history, and exactly what is meant by the term has changed somewhat over the years. Here I mention only the key developments.

You may recall that from very early on nature was perceived as a Chain of Being or *scala naturae*, ranging from inanimate objects through simple organisms and plants to higher animals; and this perception persisted well into the 19th century. It was early in that century that the stratigraphical ordering of fossils became known, notably with the succession from fish to reptiles and mammals, which seemed to parallel the Chain of Being. On the one hand this stimulated theories of historical or phylogenetic evolution; and on the other – but as a quite separate issue – it promoted the idea that in the course of development the embryo of a higher animal passes through the stages of lower ones in the *scala naturae*.[1] Particular weight was given to the 'gill slits' which appear in tetrapod embryos (including mammals), and in 1824 the French embryologist Serres (1786–1868) stated that 'Man only becomes a man after traversing transitional organizatory states which assimilate him first to fish, then to reptiles, then to birds and mammals.' Serres saw this solely in terms of embryological development, with no bearing on phylogenetic evolution which he openly opposed.

Darwin, however, saw embryological development as important evidence for phylogenetic evolution – in terms of recapitulation and homology. And after publication of the *Origin*, this theme was taken up by the German biologist Ernst Haeckel who was an ardent proponent of evolution and published various phylogenetic trees to show how he believed different groups of organisms had arisen. (It is interesting to note that in his phylogenetic tree of vertebrate evolution each geological period (Cam-

---

[1] The original meaning of the word 'evolution' was unfolding, as of the embryo in development, rather than its modern usage.

brian etc.) is preceded by an ante-period in which major evolutionary events such as the emergence of distinct vertebrate classes occurred, reflecting the absence of transitional forms in the fossil record; in fact most of the ante-periods are depicted as lasting longer than the corresponding periods.) He held strongly to the theory of recapitulation, and in support of this he produced various drawings which emphasized the similarities of embryos from different vertebrate classes, which became popular and are still reproduced in many modern evolutionary textbooks. Taking this further, he depicted how man had evolved not only through 'lower' vertebrates but also how they evolved through various invertebrate forms which in turn emerged from single-celled animals such as amoeba. This scheme was nothing more than Haeckel imposing his interpretation of phylogenetic recapitulation on embryological observations. And Haeckel went even further. Not only did he see embryological development as a recapitulation of the animal's evolutionary history, he also believed that phylogeny in some way causes ontogeny. Haeckel's 'biogenetic law' was the most extreme form of the theory of recapitulation, and is reminiscent of Lamarck's inner drive for progress.

It is remarkable that the theory of recapitulation became and remained so popular for so long, because evidence showing it to be false was available from almost as soon as it was clearly formulated. In 1828 the embryologist Karl Ernst von Baer (1792–1876) published a German work in which he strongly criticised Serres' view and sought to give a more accurate account of embryological development. It is clear that a fertilized egg is (at least morphologically) completely general and unspecialised, and von Baer showed that development is a process of progressive specialisation towards the adult form. It is true that early embryos of different vertebrates resemble each other, but that is largely because of their unspecialised form, and the embryos progressively diverge as they specialise. In particular, so far as recapitulation is concerned, the young stages in the development of an animal are not like the adult stages of other animals lower down on the scale, but may resemble the young stages of those animals. Similarity of embryological form is not the same as recapitulation, and von Bear insisted that there is no repetition of ancestral forms in any embryonic period, there is only parallel development of different types up to a certain stage. Not surprisingly, in general, the closer organisms are taxonomically, the more similar is their embryological development, and the longer does the similarity persist. So, for example, in the development of a chick there is a stage at which von Baer could recognise it as a vertebrate but could not say what kind, at a later stage he could recognise it as a bird, and only later what kind of bird. Also, whilst recognising the similarities of early vertebrate embryos, they are distinctively vertebrate and do not pass through a form that resembles an invertebrate.

As knowledge of biological mechanisms increased, the extreme form of Haeckel's biogenetic law lost favour due to the absence of a mechanism for effecting it. In fact the biologist Walter Garstang (1868–1949) pointed out that, quite the opposite of changes in ontogeny resulting from phylogeny, it is changes in ontogeny that cause phylogeny. Indeed it became apparent even within the 19th century that Haeckel had misrepresented the actual appearance of vertebrate embryos in order to make them conform to his hypothesis, and this has been re-affirmed recently (Richardson *et al.*). By the early decades of the 20th century there was overwhelming evidence to show that ontogeny did not recapitulate phylogeny. Gavin de Beer wrote *Embryos and evolution* (1930) and *Embryos and ancestors* (1940, 1951, 1958) in which he marshalled various lines of evidence to show convincingly that recapitulation does not occur.

In effect, recapitulation implies that higher organisms have evolved by additional developmental stages being added to the end of the development of lower organisms. De Beer pointed out that no single case was known in which evolutionary modification of ontogeny has taken the form of addition of a new terminal phase to the final phase of an old ontogeny. Rather, not only the final result, but all stages of ontogeny are modified in evolution, and evolutionary novelties can appear at any stage of development. And some structures arise early in the development of higher animals which are completely absent from embryos of lower animals, such as the membranes of amniotes and placenta of mammals. De Beer also gave examples of where the adult descendent resembles the young stage of the ancestor, which is precisely the opposite of that required by the theory of recapitulation.

Further, some tissues arise in development in the opposite order from which they are presumed to have evolved, such as teeth forming after the tongue whereas it is thought that teeth evolved first. And various vertebrate organs such as liver and lung develop embryologically in quite different ways from how it is thought they evolved.

Then there are species which have similar adult forms but with different immature forms, which could conform with recapitulation only if the species had evolved convergently. And in a similar vein, as mentioned when discussing homology, we have realised that similar phylotypic stages and/or adult morphologies may be attained by very different developmental routes. Observations such as these show that even von Baer's view of development being an exclusively divergent process of increased specialisation does not hold.

De Beer discussed various observations that had been misrepresented as supporting recapitulation when in fact they do not.

One of the most interesting cases is that of the gill slits, or rather of the visceral pouches, which appear in the embryonic stages of reptiles, birds and mammals. These structures resemble the visceral pouches which appear in the embryonic stages of fish. The visceral pouches of embryo reptiles, birds, and mammals bear little resemblance to the gill-slits of the adult fish. Any one who can see can convince himself of the truth of this.

All that can be said is that the fish preserves its visceral pouches and elaborates them into gill slits, while reptiles, birds and mammals do not preserve them as such, but convert them into other structures such as the Eustachian tube, tonsils, thymus glands. There *is* similarity between the embryos of fish and of reptiles, birds, mammals, but the later stages of the ontogeny have diverged. [1958, Ch. 7; original emphasis]

De Beer concluded that recapitulation was an outworn theory which had seriously hindered progress in embryology. Whilst examples can be found where an embryological form resembles a presumed ancestor, the general principle of recapitulation does not apply and offers no support whatever for the theory of evolution. Despite the weight of contrary evidence, reca-pitulation continues to be mentioned in popular presentations of evolution, and even in some textbooks. The only reason why I have included this short section on the subject is for the benefit of those who have come across such accounts and are unaware of the facts. Some may still argue that, even if recapitulation as such does not occur, nevertheless the simi-larities of embryological development for similar groups of organisms is evidence of evolution. But that is the realm of homology which, as dis-cussed above, has its own problems.

CLADISTICS

The principal idea behind classical classification was simply that of order-ing species according to degree of similarity. Species with many similar characteristics were grouped into a genus, similar genera into families, and so on through orders and classes up to phyla and kingdoms. It was non-evolutionary in concept, primarily a way of organizing knowledge about types of organisms. However, the resultant hierarchical system promoted ideas of evolution, with species in the same genus being closely related (having a recent common ancestor), and being progressively more distantly related and with earlier common ancestors for higher taxonomic groups. Indeed, evolution, or phylogeny, is seen as the explanation for the hierarchical nature of classification, and the justification for perceiving it as a natural or real arrangement. In time, as evolution became widely ac-cepted, it was no longer sufficient to know only that species within a ge-nus are related, biologists wanted to know more precisely how they are related: specifically, how they had diverged or branched off from an an-cestral stock; and how genera etc. are related. Cladistics is the term used to describe the determination of these relationships: the word being de-rived from the Greek *klados* which means branch.

Important progress in cladistic methodology was made in the 1960s with publication of *Phylogenetic systematics* by W. Hennig. Consistent with the evolutionary interpretation of hierarchical classification, he reasoned that the various species with their array of characteristics must have arisen by a series of speciation events (i.e. when a species splits into two or possibly more species), and that characteristics had emerged progressively as species differentiated. That is, Hennig proposed that at any speciation event the daughter species would have been discriminated by one or other acquiring (or perhaps losing) at least one distinguishing characteristic. Subsequently, one or both of the daughter species may diverge on the basis of other characteristics, and so the branching process of speciation goes on. In almost any real situation – dealing with anything more than just a handful of species and a few defining characteristics – it is not possible to construct a phylogeny in which characteristics emerge uniquely; it is usually necessary to assume that at least some characteristics emerge more than once, and that some are lost (and may later re-emerge). That is, several trees are possible linking the species under consideration, and the choice of which is best is usually based on the principle of parsimony: on assuming that the tree which most closely represents what actually happened involves the minimum number of acquisitions and/or losses of characteristics. Another way of expressing this is that the aim is to maximise homology (similar characteristics that can be attributed to common ancestry) and minimise homoplasy (similar characteristics which must be attributed to independent origins).

Cladistics is widely seen as an objective approach for determining phylogenies, and has been employed extensively in the last couple of decades, using morphological and molecular data. The phylogenetic trees based on primary sequences of proteins or nucleic acids, mentioned in Chapter 7, are an implementation of cladistics, with a comparable application of the principle of parsimony. Various methods are now available for implementing and refining cladistic methodology; and, as mathematical methods and computer technology have developed to handle large amounts of data, phylogenetic trees may be based on many wide-ranging characteristics. Consequently, the results are generally considered reliable and are given a great deal of weight. However, at least some scientists have become aware of a number problems with cladistics – either its methodology or results – and I will consider some of these now.

The biological significance of morphological characteristics

Perhaps the chief criticism levelled at cladistics is that it considers characteristics in isolation, without regard to their biological significance; and probably the leading example of this relates to the evolution of flight. Because birds have some reptile-like features, various biologists sought to identify their reptilian ancestors, and understandably this was by comparing bird and reptile characteristics. It was through cladistic analysis of such characteristics that the theropod dinosaurs were identified as the group of reptiles most closely related to early birds and, therefore, hailed as the group from which birds evolved. However, even if theropods are the reptile group most like birds, several scientists researching the origin of flight (notably Alan Feduccia) point out that there are serious anomalies with theropods as bird ancestors. For example, as mentioned in Chapter 10, theropods had well developed hind legs but diminutive forelimbs, and they had very different modes of breathing without a viable intermediate stage. In other words, such biologists argue that overall morphological similarity is not enough: research into the origin of birds must take into account the physiology and biomechanics of flight, and to be taken seriously any putative ancestors must be suitable in respect of these factors. Advocates of cladism respond to such criticism that cladistics does not necessarily identify the ancestral group of birds, only their closest known relatives. This is especially obvious with birds where the reptile group most like them in fact lived tens of millions of years after the earliest birds.

Cladistics presumes phylogeny

Which brings us to the most fundamental criticism of cladistics. Its cardinal premiss is that species (and higher taxonomic groups) are actually related in the sense of being descended from a common ancestor. That is, cladistics is based on the assumption that there is a real phylogeny, and the aim of cladistics is to discern that phylogeny as closely as possible. And cladistic methodology, because it allows multiple gain and loss of characteristics, can always come up with a phylogeny that will fit the available data, possibly with a few that fit the data equally well. Just as it is possible to devise a tree-like structure to link two or more sequences of amino acids (or nucleotides) no matter how similar or dissimilar they are (see Ch. 7), so it is possible to relate organisms or groups of organisms in a notional phylogeny irrespective of whether they are really related in a phylogenetic or evolutionary way. So the fact that one or more can be devised in no way verifies that any phylogeny actually occurred. In other words, cladistics cannot be used to test whether two groups are related – it simply assumes that they are. This is why cladistics permits multiple acquisition and loss of characteristics: it is the only way in which phylogenies can be derived from real data, and cladistic rationale and methodology depend on the assumption that a phylogeny exists.

Hence, for example, no matter how similar or different birds and reptiles are, it is possible to identify some group of reptiles which is closest to birds; but that is no evidence that they are in fact related. Put another way, if there is a range of possible ancestors to birds, such as the various reptile groups, then cladistics can identify which is the most likely candidate, but it does not test the initial assumption that one of them is.

This criticism is also relevant to another major application of cladistics, which is to determine basal groups. As mentioned in Chapter 10, a common feature of the fossil record is that many new types of organism – such as a vertebrate class or plant subphylum – appear in different lines, often at about the same time, without evident link between them. Some have concluded from the abrupt appearance of different lines that the groups concerned have evolved independently. Others, recognising the improbability of this, assume that the groups in question had a common source and seek to identify the basal group – the one from which the others diverged – by cladistic analysis of the characteristics of the groups concerned. This approach is used widely, and the results from such analyses may be announced with great conviction, such as claiming that the riddle of the origin of angiosperms has been solved.

However, we must recognise that no matter how similar or different the groups concerned are, cladistic analysis of their characteristics will be able to identify a possible phylogeny to link them. So the fact that a phylogeny can be devised, which identifies a basal group (or perhaps the analysis indicates that the basal group lies between two known groups) and links all of the others to it in an evolutionary manner, is no evidence at all that the proposed evolution actually occurred, or, indeed, that any such evolution occurred.

Note that I am not here criticising cladistic methodology: if the assumption is correct that the groups concerned evolved from a common stock then it is a reasonable way to proceed. But I am emphasizing that cladistics does not test the assumption. The methodology is sure to produce a possible theoretical phylogeny, so the fact that it does so does not show that the assumption is correct.

Coelum formation

The acceptance of cladistic thinking – allowing gain and loss of characteristics – can be taken to remarkable lengths in order to make the facts fit an underlying evolutionary hypothesis, and a prime example of this comes from the chordates.

All but the simplest animal phyla have an internal cavity called a coelum which in vertebrates forms the abdominal cavity or peritoneum. In the course of early embryological development, immediately following gastrulation (see above), the coelum arises in one of two quite different ways. In most invertebrate groups – collectively known as protostomes because the mouth forms from the first opening in the blastula – it arises as a cavity within the mesoderm. Because it arises by splitting of the mesoderm, this process of coelum formation is called schizocoely. In other invertebrate groups – known collectively as deuterostomes because the mouth forms from a second opening into the gastrula, notably the echinoderms – the coelum arises from outpocketings of the cavity (primitive gut) within the gastrula, and the process is called enterocoely.

Because characteristic features of protostomes and deuterostomes – including how the coelum forms – arise so early in development and, presumably, in the course of evolution, they are seen as important for defining relationships between animal phyla. Deuterostomes are generally considered more advanced than protostomes, and it is thought that, as the early deuterostomes evolved from some protostome group, their method of coelum formation changed from schizocoely to enterocoely. The chordates, including vertebrates, are deuterostomes and, consistent with this, in the cephalochordates and lampreys the coelum forms by enterocoely. Not surprisingly, these facts are interpreted as showing that chordates (including vertebrates) evolved from some deuterostome group, inheriting their distinguishing characteristics such as enterocoely.

However, in all other vertebrates the coelum forms by schizocoely! Despite this anomaly, in view of the similarities among vertebrates, it is thought that they must have evolved from a common source and, indeed, that vertebrates share a common ancestor with other chordates. Consequently it is concluded that chordates evolved from an early deuterostome, retaining deuterostome characteristics, as in amphioxus; subsequently vertebrates emerged from chordates, again retaining deuterostome characteristics, as in lamprey; but, at about the time of the early jawless fish, the main vertebrate line reverted to schizocoely which has been retained in all vertebrates since then. This view is maintained even though, as mentioned previously, it is difficult to envisage what might drive significant reorganization of early development, especially in view of the disruption it would entail.

In the first place, this example clearly illustrates how an evolutionary explanation is forced upon facts which *prima facie* do not support a common ancestry. It also reinforces that phylogeny is the premiss of cladistics, and any combination of characteristic can be explained in terms of gain and/or loss of characteristics. In particular, it is deemed preferable to accept that coelum formation changed from enterocoely to schizocoely in the main vertebrate line, no matter how impracticable that would be, in order to support the evolutionary hypothesis that links vertebrates to invertebrates through the collective phylum of chordates, rather than accept separate origins.

Secondly, the radically different methods of coelum formation provide a further example of non-homology, and at a deep level. It clearly separates other vertebrates from lampreys and the cephalochordates.

Incongruent data

It is widely thought that cladistics provides a way of unravelling the maze of characteristics occurring within groups of organisms; and of doing so objectively – avoiding the criticism of subjectivity levelled against the early taxonomists. Indeed its practitioners aim to use as many defining characteristics as possible, in this way hoping to maximise the objectivity and reliability of their conclusions. And – if the cladistic rationale is sound – we would have expected that the more data we use, the clearer would be the picture that emerges. However, contrary to expectations, exactly the opposite happens; and I can illustrate this no more clearly than with the following brief exchange which took place at a recent symposium on the subject of homology (Carroll).

Wilkins: If one has a lot of characters in the fossil evidence, each one of which is a bit dicey, how reliable are the cladograms at the end of the day, and how large are the data sets?

Carroll: In general, the more data you have the less well established they are.

Meyer: I agree. The consistency decreases with increasing numbers of characters and taxa.

That is, with limited data it is possible to produce well-defined trees to link characteristics from various groups of organisms, but the trees become less clear as more data are included in the analysis.

It is recognised that the link between genotype and phenotype can be indirect and complex, such that the degree of change in phenotype may not closely reflect the degree of change of genotype and, therefore, that morphological characteristics may be misleading as a basis for determining phylogenies. Consequently, biologists have turned increasingly to molecular data which is perceived as being more objective and less open to misinterpretation. However, we have already seen in Chapter 7 that very much the same sort of picture emerges for molecular data too: with limited data it is possible to produce well-defined trees to link the amino acid sequences of equivalent proteins from various groups of organisms, but the trees become less convincing as more data become available. In the words of Colin Patterson (1933–98), who was Director of the British Museum (Natural History), 'the difficulties of assigning homology to molecules parallel many of the difficulties of assigning homology to morphological structures' (1988).

Surely results such as these should make us question the underlying evolutionary rationale of cladistics. Unfortunately, few of today's scientists are prepared to be so radical – a theme I shall take up in Chapter 14. But first a little digression.

13

Chicken and Egg

In Chapter 10 I considered only the development of life from the Cambrian period onwards; it is now time to turn the clock further back and look at the enigma of the origin of life. For a while I was undecided about including this because I think the argument confronting evolution based on the improbability of biological macromolecules is compelling even with life's mechanisms available for generating them efficiently, so how much more hopeless is the prospect of obtaining something useful from a mere concoction of simple organic compounds. However, a book such as this would seem incomplete without at least some comment on this subject; and it provides further opportunity to point out how in this area, too, the realities of biochemical complexity all too often are glossed over.

We have seen how, from the classical Greeks if not before, the seemingly uncaused growth of moulds and the like led to belief in spontaneous generation. Such observations and the apparently simple nature of 'lower' life forms, especially before the development of microscopy, masked the gulf that actually exists between inanimate things and living organisms, and misled people into thinking that one could easily lead to the other. It was typified by Aristotle's *scala naturae* in which he believed that inanimate matter merged into the simplest or lowest forms of life. This view was still prevalent in the 18th century, with both Buffon and Lamarck, as part of their evolutionary schemes, maintaining that simple forms of life emerged in this sort of way; and it persisted well into the 19th, for example with Robert Chambers who cited what he considered to be experimental evidence for spontaneous generation in his *Vestiges*. It was not until the work of Louis Pasteur (1822–95) in the second half of that century that such ideas were finally shown to be false.

Along with the idea of spontaneous generation, though arguably inconsistent with it, was the philosophy of vitalism – that life cannot be reduced to mere physics and chemistry but requires some sort of animating force, and early vitalistic views even included that organic compounds could be formed only by living things. However, in 1828 the German chemist Friedrich Wöhler (1800–82) synthesized urea. Although urea is now regarded as a very simple organic compound, especially compared with biological proteins, at the time it marked a significant breakthrough, not just in terms of chemistry, but in breaking down the mystique around biology.

So, with this as backdrop, when the theory of evolution took shape in the middle of the 19th century it was not difficult to imagine that simple life could have emerged from some prebiotic medium. Darwin said little regarding the origin of life, even allowing in the *Origin* (Ch. 14) for a supernatural formation of the earliest forms of life from which evolution proceeded. But Thomas Huxley (1870) speculated as follows:

> ... if it were given to me to look beyond the abyss of geologically recorded time to the still more remote period when the earth was passing through physical and chemical conditions which it can no more see again than a man can recall his infancy, I should expect to be a witness of the evolution of living protoplasm from non-living matter. I should expect to see it appear under forms of great simplicity, endowed, like existing fungi, with the power of determining the formation of new protoplasm from such matters as ammonium carbonates, oxalates and tartrates, alkaline and earthy phosphates, and water, without the aid of light.

These ideas were developed by the Russian biochemist Aleksandr Oparin (1894–1980) in the early 1920s who argued that, because most organic compounds are readily destroyed by oxidation in the modern environment, the primeval world must have had a reducing environment (in which oxygen was substantially absent) in order for prebiotic molecules to form and survive. Haldane published similar ideas a few years later and was the first to refer to a primeval soup.

In 1953, the same year that Watson and Crick determined the structure of DNA, the American chemist Stanley Miller carried out his famous primeval soup experiment. Within a sealed vessel, he put together a proposed primitive atmosphere comprising water vapour, hydrogen, methane and ammonia and discharged electric sparks through it to simulate energy provided by lightning. After a week he analysed the contents of a distillate from this atmosphere and found an array of organic compounds, notably including amino acids. Similar experiments by Miller and many others after him have produced various other compounds important in biological systems such as sugars and the nitrogenous bases used in nucleic acids.

The timing of these experiments could hardly have been better. Just when basic genetic mechanisms were being unravelled, here was the first clear evidence of abiotic sources of organic compounds – an indication of how life might have started from an inorganic source. It was recognised that amino acids and nitrogenous bases are only rudimentary building blocks, but biologists were sure the rest would follow. It was all coming together: Darwinism had successfully assimilated Mendelian genetics, and now, just as the theory of evolution was finding its molecular basis, here was evidence for extending evolution back to the earliest times. Biologists were convinced it was just a matter of time before the whole story, from prebiology through simple organisms to the full array of present-day life, would be unravelled. It was even speculated that some simple forms of life may eventually be produced in the soup experiments. So by the 1960s, along with complete acceptance of Neo-Darwinism as explanation for how life had diversified, came confident belief that the origin of life itself could be accounted for solely in terms of physics and chemistry.

However, the fact is that nearly half a century after Miller's first experiment, and despite enormous effort by umpteen laboratories worldwide, the answers are not forthcoming. Stanley Miller has said that the biologists of the 1950s expected to find the explanation of the origin of life within 25 years, but it hadn't worked out that way; in fact 40 years after his first soup experiments, he admitted that solving the riddle of the origin of life had turned out to be much more difficult than he or anyone else had envisioned (Horgan, Ch. 5).

Their optimism was due in part to the rapid advances being made at the time, especially in discovering the basic biochemical processes of DNA replication and protein synthesis. And it was reinforced by the elegance of these processes – this elegance gave the mistaken impression that biological systems are simple, with biologists at the time not realising what intricate mechanisms are actually involved in implementing them. Now we know much more, such as the details of replication and transcription mentioned in Chapter 8. Also, the soup experiments have highlighted the amazing ability of enzymes to carry out reactions with a specificity and efficiency that organic chemists can only dream about.

Biologists have become increasingly aware that the real stumbling block to the origin of life is its complexity – complexity in terms of the interdependence of molecules and biochemical pathways within cell metabolism, and complexity at the molecular level of the individual components. The combination of complexities at these different levels presents insurmountable difficulties to getting anything that is remotely life-like. Lynn Margulis (a prominent figure in the field, who we will meet again later) has commented that, because of the complexity of even the simplest forms of life, a bacterium is much closer to a human being than it is to any cocktail of organic compounds in some putative primeval soup; this would certainly not have been the view in the 19th century (for example, Thomas Huxley's comment above) and probably not in the first half of the 20th.

A clear indication that the problem is proving insoluble is that there is no generally accepted coherent theory for the origin of life; different groups of investigators have their preferred options, and point out the failings of others'. For example, in a recent review of research on the origin of life, Leslie Orgel wrote:

> it would be hard to find two investigators who agree even the broad outline of the events that occurred so long ago and made possible the subsequent evolution of life in all its variety.

PRIMITIVE MACROMOLECULES

As I have just mentioned, the core of the problem is the considerable complexity of even the 'simplest' forms of life, or even of some notional system that is stripped down to the theoretical bare necessities of life. And, of course, central to the dilemma is the interdependence of proteins and nucleic acids – that each relies on the other for replication – a classic chicken-and-egg scenario. There is general agreement that it is too improbable for co-operative proteins and nucleic acids to arise together, and consequently that the first step must have been some sort of self-replicating molecule, though there are differing views as to whether this was more likely to have been protein or nucleic acid, or even if we should be looking for something completely different. Obviously, a prerequisite for any such molecule is that it be a polymer of reasonable size, so the first aspect to consider is how such polymers might have arisen from a soup of monomers (amino acids and/or nitrogenous bases). Some accounts of the origin of life assume that, given the monomers, the polymers will happen automatically – but that is just ignoring the facts.

Polypeptides

At first it seemed most likely that the earliest replicative units would have been based on proteins because of their evident catalytic capability. This view was supported by the fairly ready appearance of amino acids (rather than nitrogenous bases) in the soup experiments; although it should be noted that the only common ones are the two smallest, glycine and alanine, and these are only minor constituents, with the majority of the organic matter being an indeterminate tarry residue. The point has also been made that, if the experiments are prolonged, in the hope of increasing the yield and/or variety of amino acids, the actual result is to increase the quantity of tar.

When it comes to the polymerization of amino acids, it is important to note two things about peptide bond formation. First is that it is a condensation reaction, involving the net removal of a water molecule (a hydrogen atom (H) from the amino group of one acid and a hydroxyl (OH) from the carboxyl group of the other, together constituting H_2O, see Box 6.4); so it will not occur spontaneously in an aqueous medium, in fact the presence of water inhibits the reaction. Second, peptide bond formation is endothermic, i.e. it absorbs energy, so it does not occur spontaneously but needs a source of energy.

To satisfy these two requirements, initial polymerization experiments were based on the heating of dry amino acids. The result was substantially a tarry mass, but with sufficient identifiable 'proteinoids' of comparable size to small biological proteins to encourage further work along these lines, at least for a while. It should be noted however that many chemical groups on the various amino acid side chains reacted with each other (i.e. there were many unwanted side reactions as well as the desired formation of peptide bonds), so the protenoids contain multiple branchings of the amino acids and many of the side chains (which are of course crucial in terms of polypeptide folding and enzyme activity) are corrupted. (Attempts have been made to demonstrate that the protenoids have some measure of catalytic activity: but such as has been shown is entirely non-specific – such as promoting the degradation of hydrogen peroxide into water and oxygen, which all sorts of compounds can achieve, and even occurs spontaneously.)

Later attempts have employed compounds which are able to effect condensation reactions within an aqueous medium. Probably the leading contenders are cyanide derivatives which have been found in some soup experiments and which can condense various organic compounds, including amino acids. Although some success has been obtained in this way with combining two or three amino acids, the basic problem of unspecificity remains i.e. any reactive part of the side chains of the amino acids gets in on the action and the end result is a disorganized morass which bares no relation to the ordered structure of biological proteins.

But the difficulties involved in *synthesizing* peptides are only half of the problem: the other concerns breaking the peptide bonds, which is necessary in any scenario which recycles amino acids in order to generate a range of amino acid sequences – to try to find one with useful activity. As the formation of peptide bonds requires an input of energy and loss of a water molecule, the corollary of this is that the breaking of these bonds by hydrolysis is energetically favourable and involves the addition of a water molecule; so it might have been thought from this that peptide bond hydrolysis occurs readily in an aqueous environment, but that would be wrong. Although there is a net release of energy, the reaction involves a high activation energy: it is like going from a given height in one valley to a lower one in the next, but having to climb over a high ridge in between to get there. The activation energy for hydrolysis of peptide bonds is such that there is virtually no spontaneous hydrolysis at all under ambient conditions – something which makes proteins stable molecules and so suitable in biology for enzymes and building tissues. The procedure used in the laboratory to break down proteins into their component amino acids is boiling in 6M hydrochloric acid (about the same strength as the sulphuric acid in a car battery) for about a day. (In contrast, each molecule of the digestive enzymes trypsin or chymotrypsin, acting in our stomachs which contain about 0.1M hydrochloric acid at body temperature, can hydrolyse about 100 peptide bonds per second.)

So, taking together the above points regarding the making and breaking of peptide bonds presents a very bleak picture for exploring amino acid sequences abiotically. It is not just that the reactions themselves are likely to be slow and inefficient in any sort of soup environment, it is the fact that the conditions for making and breaking peptide bonds are so different, quite inconsistent with each other. What this means is that any scenario for trying out amino acid sequences requires a location in which the conditions can vary significantly, alternating between conditions suitable for peptide formation and breaking. Now it is not too difficult to envisage possible situations (e.g. using volcanoes and ocean vents) that might have given the required changing conditions, but there are substantial practical drawbacks. The first is that the volume of medium where both reactions can take place will be restricted to where the appropriate varying conditions exist – which must surely be of limited extent, so we can no longer think of large bodies of water (e.g. primeval 'oceans') being involved. And second, the rate of producing different polypeptides cannot be greater than the rate of change of the environmental conditions. Clearly, these factors severely limit the number of polypeptides that could be generated at any one time, and the rate at which they could be recycled. This scenario also supposes the fortuitous circumstance that the amino acids and polypeptides were somehow retained within this favourable location and not flushed out by whatever caused the changing conditions.

Faced with these daunting practical difficulties, it has to be concluded that the number of polypeptides that might have been produced in a prebiotic world must have been tiny. And it beggars belief to think that such hopeless odds could somehow strike it lucky and produce a polypeptide with significant 'biological' activity such as the ability to replicate itself.

One way around the hopeless situation with proteins is to propose that the first biologically significant molecules were not proteins but nucleic acids. So let's now take a look at those.

Polynucleotides

Interest in the possible role of RNA as the earliest macromolecules arose in the 1980s when it was realised that RNA not only has an information-mediating role within cells, but is also involved in carrying out catalytic functions. For example, RNA comprises about 60% of ribosomes and evidence emerged indicating that RNA has a primary role in the synthesis of peptide bonds. (Actually, this is an intriguing facet of the chicken-and-egg relationship of proteins and nucleic acids, with each actually synthesizing the other.) Secondly, you will recall that most structural genes in eukaryotes have introns which are included in the synthesis of mRNA and then spliced out before translation. RNA molecules are involved in many of the RNA-splicing processes, and it was especially interesting to find that some RNA introns are capable of self-splicing i.e. they can catalyse their own excision, albeit at a much slower rate than the proteins which usually do it. The particular significance of this is, first, that RNA is having a catalytic role without the aid of proteins and, second, that it is acting on itself – a possible indication of the sort of activity that would be required for self-replication.

Further support for the idea of RNA as having been the primitive biological molecules comes from the existence of RNA viruses. Most viruses use DNA which is transcribed into RNA before translation to produce proteins, as occurs in higher organisms; but RNA viruses use RNA as their genetic material, and this can be used directly for protein synthesis. They are seen as possible relics from an early RNA-based biological world.

All of this sounds very promising in theory, but whilst generating polypeptides in a soup environment is difficult, obtaining polynucleotides is virtually impossible. The reason is that each nucleotide comprises three chemical groups – a sugar (ribose) and phosphate as well as a nitrogenous base (see Box 6.1). These three components need to be present and react together in the right way in order to produce a single nucleotide, and then the nucleotides need to be polymerized. Each of these steps is an endo-thermic condensation reaction, so requires the presence of some high-energy condensing agent to carry it out. Advocates of the prebiotic RNA world point out that there is evidence for all the necessary ingredients be-ing available: it is argued that the nitrogenous bases and sugars have been found in rudimentary soup experiments based on inorganic compounds, that phosphate could be available from the weathering of rocks, and that cyanide-type condensation compounds have been found in other experi-ments. And they will point to their successes – that each step of polynu-cleotide synthesis has been demonstrated in a sort of soup environment. But sceptics respond that in order to achieve this, the right precursors and high-energy compounds need to be provided, and on a step by step basis. That is, to obtain nucleosides (base + ribose) requires starting with a mix-ture of nitrogenous bases and ribose and a suitable condensing agent; and to obtain nucleotides requires mixing nucleosides with phosphate and a different condensing agent. In other words, only by breaking down the synthesis into these discrete steps, and supplying appropriate intermedi-ates in large amounts, and under suitable reaction conditions, can the de-sired products be obtained. And there has been very little success in polymerizing nucleotides even when they are provided ready-made. So, although there are claims that nucleotides and polynucleotides have been found in soup-type experiments, these successes are unconvincing and widely criticised. Stanley Miller dismissed many of these claims as mere 'paper chemistry' – perhaps possible in theory, but they do not happen in practice. Any realistic primeval soup is such a cocktail of compounds that any desired products are swamped by undesired ones, and destroyed be-fore they can proceed to the next step.

Finally, of course, the production of a range of polynucleotides encoun-ters the same sort of physical constraint as for producing polypeptides, in terms of requiring very different reaction conditions for the various reac-tions. In fact it is even more problematic with polynucleotides because of the different bonds that need to be made and broken.

Replicating units

So there are immense obstacles in the way of obtaining any sort of poly-peptide or polynucleotide in a soup environment. But the difficulties do not stop there.

First, as we have seen already for polypeptides, to have any hope of biological activity it is necessary for a macromolecule to fold, and this im-poses a minimum size on the compound concerned: for polynucleotides it is estimated at 40 to 60 nucleotides. Also, just as with polypeptides, we cannot assume that any polynucleotide of appropriate length will fold, but the nucleotides must be able to be packed together in a way that is analogous to the amino acids of a protein. Even though there are only four bases in polynucleotides rather than the 20 amino acids of proteins, the odds are against it. As I said in Chapter 7, this criterion alone is probably too difficult to meet, regardless of any possible biological activity.

Second, even supposing that such prebiotic polymers could be produced, there are other fundamental problems. Notably, it is not just about finding a useful sequence (though that is unlikely enough), it is being able to reproduce it reasonably reliably.

It is at this point that proteins run into what seems to be an insurmountable hurdle because there appears to be no way in which a polypeptide can determine a peptide sequence, i.e. nothing analogous to the base pairing of nucleic acids. This makes it very difficult to see how a protein could act alone to replicate either itself or some other polypeptide. Some researchers have suggested that a group of proteins could cooperate in the replication of each other; but these ideas have not received much support because of the evident improbability of obtaining two or more cooperating macromolecules at the same time.[1]

Even for polynucleotides, which have the potential for replication by base-pairing, it is far from straightforward. Darwin realised that evolution (at the morphological level) must strike a balance between reliable reproduction of a species – to ensure continuity – on the one hand, and opportunistic variation on the other. The same sort of idea applies to prebiotic molecules, and poses a dilemma for the production of a primitive replicator, because a poor replicator is far more likely to degrade through miscopying than to improve its performance. So, although the common presumption is that a crude replicator can gradually improve its performance through a natural selection sort of process, in fact there is a threshold before that could take place. That is, a replicator must already have a reasonably good performance in order to be able to improve on that performance.

Further, we cannot expect short sequences to provide good performance, but the threshold level of performance might be achieved only with a sequence that is much longer than the minimum required for folding. Then, of course, the longer the sequence, the less likely that it will arise at random; and we should note that, before we have a replicator, there is not even a theoretical role for natural selection, so it can arise only at random.

---

[1] Some organisms produce short polypeptide antibiotics by enzymatic joining of amino acids, i.e. independently of any nucleic acids. But this requires several enzymes (which are coded for by DNA and synthesised in the normal way) to provide a polypeptide of just a few amino acids; so the information content of the amino acid sequences of the enzymes vastly exceeds that of the polypeptide product, so this is not seen as a possible model for prebiotic replicating units based on proteins.

Because of the link between size and performance and that there is no realistic possibility of obtaining a replicator randomly, another way of seeing
the dilemma is that we would need a replicator already in place to have
any hope of generating a replicator.

Finally, we should not forget that it is mere speculation that an RNA-
based replicase can exist at all.

Qβ virus

The possibility that the first biological molecules might have been RNA
has led to various studies on RNA viruses, looking for clues to the origin of
life. In particular, certain experiments on the RNA virus Qβ have aroused
considerable interest in some quarters and have been used to bolster various prebiotic evolutionary scenarios. As I explain below, I do not think
they actually shed any light on the origin of life, and when applied in this
way are being misapplied if not misrepresented; but because they are
quoted in various discussions relating to the subject, some readers may
have come across them, so it seems appropriate to mention them here. If
anything, the fact that these experiments have been press-ganged into the
origin of life debate shows how desperate some workers in this field are to
find something – anything.

Qβ is an RNA virus that infects bacteria. Its RNA is about 4500 nucleotides long and encodes four proteins. One protein encapsulates the RNA to
form an infective particle, one enables injection of the RNA into the host
cell, one disrupts the host cell, and one combines with three ribosomal
proteins of the host cell to form an RNA replicase which identifies and duplicates the viral RNA. There are two sets of experiments, using this Qβ
replicase, that have been used to try to support theories regarding the origin of life.[2]

It has been possible to isolate the Qβ replicase and viral RNA; and if
they are mixed in a suitable medium, and provided with a supply of ribonucleotides (nucleotides containing ribose instead of deoxyribose), the
replicase readily produces copies of the viral RNA. The initial strand of
RNA acts as a template, and the replicase builds a complementary strand
by base-pairing, though the two strands immediately separate – they do
not form a double helix.

In the first experiment, after a limited reaction time a sample of the reaction medium was transferred to a fresh solution containing ribonucleotides and replicase but no RNA, i.e. the only RNA entering this second
solution was carried over in the sample from the first. After a further limited time, a sample from the second was transferred to a third reaction
solution, and so on ... After several such transfers, it was found that the
RNA now differed from the original RNA, in particular in being more rap-

---

[2] These experiments and the evolutionary conclusions based on them are summarised
by Eigen (1983) which also gives references to the original work.

idly duplicated. These results are portrayed as demonstrating molecular evolution – that, via a selective process, the RNA has adapted to its new circumstances of successive transfers, and has reduced its replication time in order to improve its chance of survival.

But let's try to understand what's going on. Clearly, it takes a finite time to make a copy of RNA and, whatever the reaction time available, there will always be some RNA incompletely copied and transferred to the next reaction vessel as an incomplete copy. The shorter the reaction time is, the higher will be the proportion of incomplete to complete copies, and of course if the reaction time is less than that required for making a copy then no complete copies will be made. Also, because shorter strands will take less time to replicate, in a given time more copies of these will be produced than of longer ones, so this will increase their proportion of the overall RNA as well. Consequently, it is inevitable that serial transfers will ultimately result in the RNA being progressively shortened, and hence take less time to duplicate. In these experiments the RNA was reduced to about 500 nucleotides – and had lost its ability to code for the proteins. And this loss of useful genes is portrayed as evolution – as support for natural selection operating at the prebiotic, molecular, level to generate biologically relevant macromolecules! The significant point is, of course, that in these contrived experiments the RNA no longer has a biological function to maintain.

In a second set of experiments, it was found that when replicase is added to a medium containing ribonucleotides but no RNA template, (and using unusual reaction conditions and a very long reaction time), some strands of RNA up to about 220 nucleotides in length could be produced. These experiments are paraded as template-free synthesis of RNA, demonstrating that RNA strands could have arisen spontaneously in a prebiotic world. But the 220 nucleotide-long strands are remarkably similar to 'midivariants' which had been isolated previously and found to contain the Qβ replicase recognition site for initiating replication. It is quite evident that, far from being template-free, that part of the replicase which normally recognises an RNA sequence is, under the unusual reaction conditions used in these experiments, acting as a template on which the recognition sequence is constructed.

What should also be borne in mind when considering any possible application of these results to the origin of life is that in all the experiments the complete enzyme is provided, as is a supply of ribonucleotide triphosphates, and the reaction conditions are deliberately chosen to facilitate RNA synthesis. Note also that, although much is made of this as a replicase system, the replicase enzyme cannot arise directly from the RNA it produces, but requires the host cell's translation machinery (with its complement of tRNAs, tRNA activating enzymes, ribosomes etc.). The RNA does not even code for all of the replicase, but for just one of its four subunits.

One might have thought that, at least in theory, the RNA of Qβ could encode all of the replicase, but in fact that is not the case because it would make the RNA too long. The amount of information that can be securely encoded by a system depends on the quality of the coding system. (This is related to the fact that a reliable replicase would probably require a longer sequence than is required merely to fold.) Because it lacks base pairing, single strand RNA is not a very secure coding system, and the information required to code for a complete replicase probably exceeds what can reliably be encoded by it. This, of course, is why living systems, which require a huge amount of information to be encoded reliably, use the much more secure double stranded DNA.

If anything, far from giving any support to the origin of early simple biological systems, these experiments in fact underline the mutual dependence of proteins and nucleic acids and the complexity of the molecular machinery connecting them.

Great Expectations

Before leaving the subject of early replicating systems, there is one more point to emphasize. The current goal of origin of life studies is a basic replicating system – whether it be a self-replicating molecule or a small collection of cooperating molecules. The wide-spread belief is that this is *the* hurdle: that once this is achieved then the rest will follow relatively easily. This reasoning stems, of course, from the success of natural selection in segregating genes, and the extrapolation from this that natural selection can direct the progressive development of useful macromolecules. But we have already seen that this does not work because so much of their sequence (amino acids or nucleotides) is specific – so much needs to be right for a macromolecule to have any meaningful activity.

Which brings us back to what I said at the start of the chapter: even with an efficient replicating system in place there is no way biologically significant macromolecules can arise.

THE GENETIC CODE

Given that current biological systems are based on the mutual dependence of polypeptides and nucleic acids then, whichever might have come first, at some stage a link would need to have arisen between them. This link now requires many biochemical components – involved in DNA replication, transcription and translation – and at the heart of it is the genetic code. The origin of the genetic code, and how it is implemented, is acknowledged to be a major hurdle, and I shall mention just one or two of the main problems.

To try to explain the source of the code various researchers have sought some sort of chemical affinity between amino acids and their corresponding codons. But this approach seems misguided. It is not just that the code is mediated by tRNAs which carry the anti-codon rather than the codon, but the amino acid has no role in identifying the tRNA or the codon. This can be seen from an experiment in which the amino acid cysteine was bound to its appropriate tRNA in the normal way – using the relevant activating enzyme (see Ch. 6), and then it was chemically modified to alanine. When the altered aminoacyl-tRNA was used in an *in vitro* protein synthesizing system (including mRNA, ribosomes etc.), the resulting polypeptide contained alanine (instead of the usual cysteine) corresponding to wherever the codon UGU occurred in the mRNA. This clearly shows that it is the tRNA alone with its appropriate anticodon that matches the codon on the mRNA.

More importantly, this experiment emphasizes the crucial role of the activating enzymes. Interest in the genetic code tends to focus on the role of the tRNAs, but that is only one half of implementing the code. Just as important as the codon-anticodon pairing is the ability of each activating enzyme to bring together an amino acid with its appropriate tRNA, which clearly requires specific recognition sites for both molecules. It is evident that implementation of the code requires two sets of intermediary molecules: the tRNAs which interact with the ribosomes and recognise the appropriate codon on mRNA, and the acylating enzymes which attach the right amino acid to its tRNA. This is the sort of complexity that pervades biological systems, and which poses such a formidable challenge to an evolutionary explanation for its existence. It would be improbable enough if the code were implemented by only the tRNAs which have about 70 to 80 nucleotides, but the equally crucial and complementary role of the activating enzymes, which are hundreds of amino acids long, excludes any realistic possibility that this sort of arrangement could arise opportunistically.

As well as these intermediaries, there are of course the ribosomes. These are required to enable activated tRNAs to associate with an mRNA, to do so in phase with the codons and sequentially, and to catalyse formation of the peptide bonds. All of which are indispensable steps in the link between nucleic acids and proteins.

In view of the many components involved in implementing the genetic code scientists have tried to see how it might have arisen in a gradual, evolutionary, manner. But this encounters all sorts of difficulties with something as fundamental as the genetic code. For example, it is usually suggested that to begin with the code applied to only a few amino acids which gradually increased in number, but there are obvious drawbacks to such a scenario.

First, it would seem that the early codons need have used only two bases (which could code for up to 16 amino acids); but a subsequent change to three bases would seriously disrupt the code. Recognising this difficulty, most researchers assume that the code used 3-base codons from the outset; which was remarkably fortuitous or implies some measure of foresight on the part of evolution.

Much more serious are the implications for proteins based on a severely limited set of amino acids. In particular, it must be presumed that early activating enzymes comprised fewer types of amino acids and yet had the necessary level of specificity for reliable implementation of the code. There is no evidence of this; and subsequent reorganization of the enzymes as they made use of newly available amino acids would run into the same sort of improbability that I mentioned in Chapter 7 in the context of short primordial proteins. Similar limitations would apply to the protein components of the ribosomes which have an equally essential role. Further, recall from Chapter 6 that tRNAs tend to have atypical bases which are synthesized in the usual way but subsequently modified. These modifications are carried out by enzymes, so these enzymes too would need to have started life based on a limited number of amino acids; or it has to be assumed that these modifications are later refinements – even though they appear to be necessary for reliable implementation of the code.

And finally, what is going to motivate the addition of new amino acids to the genetic code? They would have little if any utility until incorporated into proteins, but that will not happen until they are included in the genetic code. So they must be synthesized (by enzymes which lack them), and all of the necessary machinery for including them in the code (dedicated tRNAs and activating enzymes) put in place – and all done opportunistically!

EARLY ENZYMES

It is usually proposed that the first biochemical pathway to evolve was that converting glucose to pyruvate, known as glycolysis. There are several reasons given for this: sugars such as glucose have been found in soup experiments, so glucose is thought to have been available; glycolysis occurs in almost all living organisms, so it is assumed to have arisen before divergence of the various groups; and it is an anaerobic reaction, i.e. does not require oxygen, so this would be consistent with an absence of oxygen on the early earth.

However appealing this may appear at first sight, as with so many proposed evolutionary scenarios, it does not stand up once we look more closely, especially at the realities of the biochemical implications.

First to consider are those relevant to an evolutionary origin of any biochemical pathway. That is, the conversion of glucose to pyruvate involves ten separate enzymes (see Fig. 12), most of which have between 300 and 500 amino acids. Even committed evolutionary biologists accept it is out of the question for ten enzymes with complementary functions to arise at more or less the same time and place; so it is necessary to consider ways in which the pathway might have arisen incrementally – forwards and/or backwards, as explained in Chapter 8.

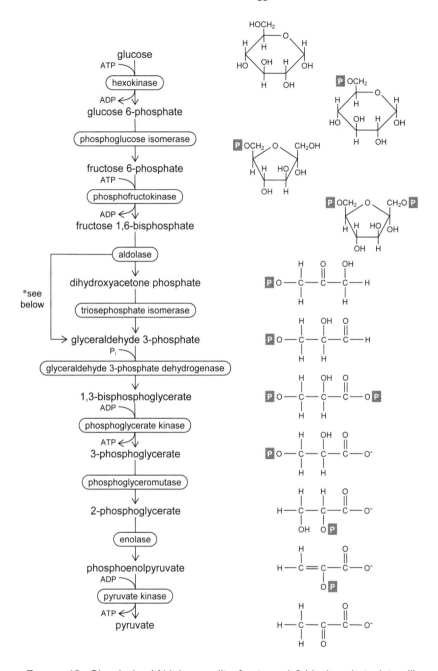

FIGURE 12. Glycolysis. *Aldolase splits fructose 1,6-bisphosphate into dihy-droxyacetone phosphate and glyceraldehyde 3-phosphate; the former is con-verted to the latter by triosphosphate isomerase, resulting in two molecules of glyceraldehyde 3-phosphate (and of each subsequent compound) for each molecule of glucose metabolised. So four molecules of ATP are generated in the second half of the pathway. P_i represents inorganic phosphate.

It is not generally thought that glycolysis arose backwards because sugar rather than the oxidized pyruvate[3] would have existed in the presumed reducing environment of the early earth, but it is appropriate to add some further comment. Recall that this sort of scenario would assume that pyruvate was available in the environment, then its precursor, and so on. However, we should note that all of the intermediates between glucose and pyruvate are phosphorylated, i.e. they have one or two of the sugar's hydroxyl groups replaced by phosphate. Replacing a hydroxyl group with phosphate is a condensation reaction (with elimination of a water molecule) comparable with those mentioned above in connection with the formation of peptide bonds, nucleotides etc. In view of the difficulties with this sort of reaction it is doubtful that any of the intermediates could have arisen abiotically in significant quantities, certainly not enough to enable evolution of the glycolytic pathway.

The more widely held view is that glycolysis was built up in the forwards direction. As we saw previously this relies on the intermediates having utility in their own right. But in the case of glycolysis, especially in the context of a pathway arising very early such that additional uses of the intermediates would not have developed, it must be concluded that the intermediates had little if any utility *per se*. Especially significant is that the value of glycolysis is not so much in its intermediates or end-products as such, but that it is a means of extracting energy from glucose – of generating the high-energy compound ATP (see Box 8.1). However, although there is a net production of ATP, before any is generated in glycolysis some has to be consumed. Specifically, two molecules of ATP are required to add phosphates to a molecule of sugar (with two others being added from inorganic phosphate after the six-carbon sugar has been split into two three-carbon compounds), before a total of four molecules of ATP are generated when the phosphate groups are removed. Indeed, after using up the first ATP there are at least five more steps (each requiring a separate enzyme) before any ATP is generated, and nine before there is a net production of ATP. Given that evolution does not have foresight, so it cannot anticipate future benefits, this casts grave doubt on the possibility that enzymes early on in the pathway could have had any utility in the absence of the later ones – utility that would have been essential for them to have been selectively retained.

As well as these general problems with any sort of evolutionary scenario for the origin of a biochemical pathway, there is a particular difficulty with glycolysis. Of the ten enzymes in the pathway six catalyse reactions involving the transfer of a phosphate group. Chemically, it is just as favourable for phosphate to react with the hydroxyl of water to form phosphoric acid as it is with the hydroxyl of a sugar or ADP. But such a

---

[3] In chemical terms, the conversion of glucose to pyruvate is considered to be oxidation even though it does not involve the addition of oxygen.

reaction would be totally useless biologically. Consequently, for these transfer reactions it is essential that water be excluded from the active sites of the relevant enzymes to prevent the abortive hydrolytic reactions. This is achieved by some part of the polypeptide being hinged: when the hinge is open the reactants can gain access to the active site, but once they are in place the hinge closes over them, completely excluding water while the reaction takes place, and opening again afterwards. In some cases, such as hexokinase, the whole enzyme is in two hinged halves, somewhat like a clam. Clearly this is quite sophisticated molecular engineering, and – as complete exclusion of water is essential for the appropriate reaction to occur – is a good example of where a part-formed enzyme (e.g. catalysing the phosphorylation reaction but not excluding water) would be of no use at all, and emphasizes the degree of specificity required of the amino acid sequence.

Finally, if glycolysis were one of the first metabolic pathways to arise, then it is reasonable to presume that there must have been only a limited repertoire of enzymes available at the time, so we might have expected the enzymes to be related and to be able to discern something of the relationships between them. For example, because the enzymes act on compounds of similar structure (phosphorylated sugars and glycerols), once one enzyme of the pathway had arisen, then perhaps it would not be so difficult for another to evolve from it, making use of common features of e.g. the active site. Alternatively, enzymes catalysing the same sort of reaction might have evolved from a primitive version – diverging from it to adapt to a particular reaction, but their ancestry being evident in their primary and/or tertiary structures, as is believed to be the case with the globins (Ch. 7). One biochemist, Linda Fothergill-Gilmore, specifically investigated an evolutionary origin of the glycolytic enzymes with these sorts of possibilities in mind. She concluded that the glycolytic enzymes did not show signs of such relationships, but 'By and large the glycolytic pathway appears to have resulted from the chance assemblage of independently evolving enzymes.' In particular, within glycolysis there are groups of enzymes that catalyse the same sort of reaction, e.g. the kinases; but, far from appearing to be related, she concluded that 'overwhelmingly the crystallographic and sequence information show that the enzymes within each group are not closely related structurally, and it appears most unlikely that each group has diverged from a primitive form.' In fact the amino acid sequences are so different that it was impossible to see how to begin to try to align them. All of which casts further doubt on the supposed evolutionary origin of glycolysis.

EUKARYOTES FROM PROKARYOTES

It was in the 1960s, as electron microscopy was beginning to reveal intra-cellular structure and biochemistry was emphasizing an underlying unity of all organisms, when Lynn Margulis proposed that eukaryotes had arisen from symbiotic aggregations of prokaryotes. The suggestion arose largely from the fact that eukaryotic organelles are comparable in size to prokaryotes. Some of these, notably mitochondria and chloroplasts (where photosynthesis takes place), even contain DNA which codes for some of their proteins (and RNAs) and have the means to synthesize them – clearly indicating a degree of autonomy and suggesting a prior independent exis-tence. Also, the DNA in these organelles is circular like that of prokaryotes rather than the linear DNA of eukaryotic (nuclear) chromosomes; and their ribosomes are similar in size to those of prokaryotes, which is smaller than those in the cytoplasm (on the endoplasmic reticulum, ER) where most eukaryotic protein synthesis occurs.

Genetic implications

The idea of prokaryotes cooperating (or, a related idea that eukaryotes arose by some prokaryotes cells being engulfed and assimilated by others) may seem an appealing scenario. However, when considering possible evolutionary progress it is all too easy to lapse into an 'inheritance of ac-quired characteristics' way of thinking. That is, just as I emphasized when discussing the idea of feathers having evolved from frayed scales, it is im-portant to recognise that, even if a cooperative assemblage of prokaryotes did arise, it is of no evolutionary significance unless there is a genetic ba-sis to ensure that the arrangement is propagated. And there are substantial genetic implications for this sort of scenario.

For example, although organelles have DNA to encode some of their proteins, most are encoded in the cell nucleus and are synthesized by ri-bosomes on the ER along with other cell proteins. So one of the genetic implications is that much of the DNA of the assimilated prokaryotes has to be transferred to the nucleus. But there are considerable difficulties with this.

First, there are a few but significant differences in the genetic code as operated by mitochondria compared with the 'universal' code. This sug-gests that the contributing prokaryotes used different genetic codes before they formed their cooperative unit; but that would require that the genes were recoded in the course of their transfer to the nucleus – and it is hard to envisage how this sort of recoding might have arisen. The situation is even more mysterious – and difficult to explain in evolutionary terms – because the differences in the code vary for different groups of organisms. In particular, most plant mitochondria use the universal code; but it is thought that plants and animals diverged after eukaryotic organisms were established, to account for their many similarities as eukaryotes. So an alternative, to accommodate this, is that in at least some organisms the mitochondrial and/or nuclear codes have changed since formation of the early eukaryotic cells; but it is widely accepted that changes in the genetic code are fraught with uncertainty – far more likely to produce functionless polypeptides than retain useful proteins.

Second, when such as a mitochondrial protein is synthesized in the eukaryotic cytoplasm there needs to be some way of identifying that it is a mitochondrial protein and ensuring it is despatched accordingly. This is usually achieved by the relevant protein being coded with an extra length of polypeptide which acts as a label and which is removed on safe arrival at the correct destination. Modification of the structural gene in this way would by itself require a remarkable coincidence, but what is even more remarkable is the molecular machinery required to implement the translocation.

In the case of most proteins destined for a mitochondrion, once synthesized in the cytosol (including their extended amino acid sequence) they are bound by other proteins, a particular group of chaperone molecules. Molecular chaperones are frequently used to protect proteins from unfolding, as with the lens crystallins (Ch. 11), but in this case their role is to prevent the mitochondrial proteins from folding prematurely, and require energy from ATP to do this. Chaperones bind at various points along the polypeptide to prevent its folding, but leave the label sequence free. The latter is identified by a receptor protein in the outer mitochondrial membrane, and thereby attaches the polypeptide to the mitochondrion. At a few places the inner and outer mitochondrial membranes are in very close proximity, and at these sites a collection of various proteins form a 'channel' which passes right through the membranes. It is to these channels that the polypeptide is guided and through which it passes. The polypeptide needs to be in its unfolded state so that it is thin enough to pass through the channel. Transfer into the mitochondrion also requires energy, but this time derived from the same gradient of hydrogen ions that is used to synthesize ATP (Box 8.1). Once inside the mitochondrial matrix, first a set of molecular chaperones similar to those used outside prevents premature folding, and then a different set catalyses the correct folding of the mitochondrial protein, again requiring energy from ATP. And finally the polypeptide 'label' is enzymatically removed, to leave the finished protein.

Variations on this theme are used for proteins that are located in the inner or outer mitochondrial membranes or in the space in between. An interesting case of the latter is cytochrome c whose polypeptide is synthesized in the cytosol, but without its haem group it does not fold properly and is able to pass through a channel in the outer mitochondrial membrane. Once in the intermembrane space, not only is haem available, but so is an enzyme that bonds it to cytochrome c. This enables cytochrome c to fold properly, and it is kept in place because in the folded state it cannot pass through the membranes on either side of it.

Hence, although transferring DNA from organelle to nucleus may seem straightforward, in fact there are considerable biochemical implications, involving at least some changes to the gene itself and the origination of completely new ones. Also, we must not overlook the need for control proteins and sequences to ensure the right proteins are synthesized when appropriate. In view of the numerous biochemical changes required, one wonders what advantages would have been sufficient to promote them (from a natural selection perspective) – or how they might actually have been implemented.

Biochemical novelties

This leads on to the fact that there are many other substantial biochemical novelties implicated in the supposed transition from prokaryotes to eukaryotes. When the scenario was first proposed, knowledge of intracellular structure was limited and the obvious feature of eukaryotic cells was their subdivision into organelles. However, since then we have discovered that eukaryotic cells are characterized not only by their organelles and the direct implications – such as directing proteins to the right destination – but also that they have a much more complex internal structure which requires an array of molecular machinery that is completely absent from prokaryotes.

Cytoskeleton

We have already met some of these, such as the cytoskeleton based on the highly specific actin which so pervades eukaryotic cells that it is their most abundant protein, and the array of other proteins which are essential for combining the basic actin filaments into a structural framework (Ch. 8). The actin cytoskeleton also provides the infrastructure, used in conjunction with myosin motor proteins, for cell motility, including moving organelles around the cell.

A second class of eukaryotic macromolecules are the intermediate filaments, which is a collective term for various groups of proteins, intermediate in size between actin filaments and microtubules (see below), with various structural roles associated with eukaryotic cells. They include the keratins and some of the lens crystallins, mentioned previously. An important group of intermediate filaments in the present context are the lamins which form an internal supporting structure for the nucleus of eukaryotic cells, and which require additional specific proteins, analogous to those used for the actin cytoskeleton, for interconnecting filaments and attaching them to the internal nuclear membrane. (The nucleus has inner and outer membranes, with nuclear pores passing right though both. The pores are not simply holes, but contain sophisticated molecular machinery which controls the transport of compounds between the nucleus and cytosol.)

Microtubules are the other group of major structural components of eukaryotic cells. Their core structure consists of small tubes built from alternating proteins of α- and β-tubulin, but there are several different types with particular functions. The individual microtubules are usually wound together – somewhat like strands of a rope – to provide larger structures, for example in the eukaryotic flagellum (cf. bacterial flagellum, Ch. 8). The length of microtubules is dynamic, under close molecular control, and is used to provide motive force. A particularly important application of this – especially significant in the present context – is that microtubules form the mitotic spindles. It is these that spread out from the centrosomes (stationed at the cell poles, either side of the chromosomes, see Box 5.1), locate and bind to the chromosomes, and then shorten, thereby separating the chromatids. The microtubules attach to a dedicated region of each chromatid, where there are specific DNA sequences, which are located by specific proteins, which are bound by the microtubules. Without the DNA sequences or the binding proteins, a chromosome cannot be captured by the spindle fibres, and is lost at cell division. Not only is this yet another example of several components being necessary for an essential biological function, but also clearly illustrates the additional complexity of eukaryotic organization.

Telomeres

This also leads to other implications of eukaryotic genetic material, such as the histones, with their highly specific amino acid sequences, which are required for tight packing of the long eukaryotic DNA in the cell nucleus. Of particular significance here are the problems posed by the linear eukaryotic chromosomes compared with circular prokaryotic DNA.

One problem, mentioned in Chapter 8, is that it is not possible to copy the last few bases on the lagging strand because they need to be approached from upstream, but there is no 'upstream'. You may recall that this is overcome by a telomerase which incorporates a length of RNA from which it copies to provide the last few bases of the DNA; in fact, the ends of eukaryotic chromosomes consist of repeated copies of the base sequence derived from the telomerase. Further, we now know that the telomerase does not act in isolation but, typical of most biological systems, involves the cooperation of several other proteins. It is evident that a telomerase is necessary to avoid the gradual loss of genetic material and therefore essential for long-term viability of eukaryotic organisms; conversely, a telomerase has no value for a prokaryotic organism with its circular DNA. So it is hard to see how the transition from circular to linear DNA could occur gradually – it seems to require yet another incredible coincidence.

There is another problem with the ends of linear DNA. Cells have the means to rejoin lengths of DNA that have been broken, and if a linear chromosome came to an abrupt end, the cell's machinery would treat it as a break and join it to the end of another chromosome. It appears that this possibility is avoided by the free end – where there are multiple repeats of the same sequence derived from the telomerase RNA – being looped back and spliced into itself. It is thought that the telomerase and various other proteins are involved in implementing this.

As I mentioned at the start of this chapter, scientists researching the origin of life are well aware that it is fraught with difficulties, of which I have outlined only a few. However, most popular books about evolution, and even textbooks on the subject, include a section on the origin of life – and usually indicate that life emerged fairly readily from some form of primeval environment. The many substantial difficulties are minimised or not even mentioned at all. So it is not surprising that so many people believe life did originate in this sort of way. This is one facet of the fact that evolution is the underlying thesis of biology – the prevailing paradigm – and it is to this that I now turn.

14

Sense and Sensibility

In the preceding chapters I have argued that the theory of evolution is fundamentally flawed at the biochemical level. Whatever other difficulties there may be, such as gaps in the fossil record, or whatever aspects of evolution may be substantiated, such as morphological change due to the selection and segregation of genes – whatever the other arguments for or against, an overriding factor is that biological macromolecules have proved to be much too complex, much too specific, and much too numerous to be accounted for by the current theory of evolution. It is not just that they are so improbable that they could not possibly have arisen by chance, but there is no evidence that they have acquired their complexity gradually, and no plausible means by which they could do so.

Natural selection was, of course, the principal contribution made by Darwin in the formulation of his theory, and it is still seen as the key to evolution – to progressive incremental development. So it is not surprising that a great deal of weight is given to the proven operation of natural selection at the morphological level, such as the spread of melanism in peppered moths. Not only is this seen by many as proof of the theory of evolution as a whole, but it is also presumed that the demonstrable action of natural selection at the morphological level can confidently be extrapolated to the molecular level, i.e. that a similar selection process can account for the formation of useful biological macromolecules by progressively improving their amino acid or nucleotide sequence. However, in Chapters 7 and 8 I have shown that this extrapolation fails because macromolecules cannot arise gradually: a large part of their sequence must be correct for them to have any utility at all – so much needs to be right that it is too improbable for this to happen, so any possible selection process is unable to get started. Then, in Chapter 9, we saw that the demonstrable operation of natural selection can be accounted for exclusively in terms of the segregation of genes. However, as we cannot account for the formation of those genes, this leaves a gaping hole in the theory of evolution, on which I elaborated in Chapter 11.

And when we look at the fossil record, although we can see some gradual evolution there, this too is only in a limited morphological sense such as may be accounted for by gene segregation. In contrast, probably the most striking feature of the fossil record is the abrupt appearance of distinct new groups. This corresponds with the occurrence of substantial new genetic material; but it is sudden, not in a gradual evolutionary manner. So the principal problem posed by the fossil record is consistent with the fundamental problem arising from biochemistry.

Taken together, these are substantive reasons to question the reality of evolution, at least in the overall sense of being a sufficient explanation for the existence of the various forms of life as we know it – which, after all, is what it's all about.

However, I would be the first to admit that much of what I have presented is not essentially new – for example, the abrupt appearance of new fossil groups has been known since Darwin's day, the improbability of biological macromolecules was recognised at least a generation ago, and the suggestion that some 'evolutionary' change may be due entirely to gene segregation rather than constructive mutation has been suggested before – although I am not aware of the arguments having been brought together previously in a comparable fashion. When I set out to write this book my sole aim was to present the scientific challenges to evolution; but, as my research uncovered various precedents, it became clear that a separate issue also needs to be addressed: that of the reluctance of the scientific community to take seriously the significant criticisms that are raised regarding the theory of evolution. This reluctance is the main subject of the present chapter, an important aspect of which is the nature of scientific enquiry, so it is here that I start.

THE SCIENTIFIC METHOD

The hypothetico-deductive method

We saw in Chapter 1 that the first milestone for science was to shake off the classical philosophical notions about the natural world and start to acquire an empirical approach. Aristotle led the way with induction – the derivation of general patterns from a set of diverse observations – as the basis for scientific laws. This was re-emphasised nearly two millennia later by Francis Bacon who maintained that scientists should consider a body of data without preconceptions and let the facts speak for themselves.

However, not long after him, Newton pioneered the hypothetico-deductive method where a scientist formulates a hypothesis (which may or may not be clearly derived from the known facts) and then sees how well the data fit. As more and wider-ranging facts are accumulated that are consistent with the theory then it gains credibility and becomes generally accepted. In particular, predictions are deduced from the theory, and then investigations carried out to see if the predictions are confirmed. Some suggest that after much corroboration the hypothesis becomes established as a scientific law. But that is just semantics: it is maintained that all scientific theories are held in a tentative way, always open to scrutiny.

What is perhaps less widely appreciated is that scientific theories are not provable. No matter how many times a scientific 'law' seems to hold, and even if we know of no exceptions, we cannot be sure it always will. It is recognised that there could be some unexamined, even unforeseen, circumstances in which it does not apply. Or maybe we are being misled into believing the theory by coincidental combinations of events.[1]

However, whilst we cannot prove a theory is true, we can show it to be false. Indeed, the 20th-century philosopher Karl Popper argued that the test of whether a theory is 'scientific' is whether or not it could (at least in theory!) be falsified. That is, if a theory cannot be used to make some sort of prediction that can be tested experimentally, then it is not to be regarded as a proper scientific theory. And where observations arise which are not consistent with the theory, known as counter-instances, the theory is to be rejected and another more suitable one found.

It is generally agreed that this is how science should proceed – proposing theories, testing them, and where necessary modifying or even rejecting them in the light of observation – and this falsifiable view of science was written into 20th-century philosophies of science. And it is widely thought that this is how science is actually carried out: the common present-day belief is that a scientist looks at the facts dispassionately and construes the theory that best makes sense of them all, rejecting a theory when any facts contradict it. Even many scientists believe that this is how they work. But probably very few do. This has been pointed out by some recent philosophers of science, and notably by Thomas Kuhn in his *Structure of Scientific Revolutions* (1962).

Science and paradigms

Kuhn was initially a physicist but is best remembered for his study on the history of science which led him to conclude that it does not advance in quite the objective fashion that is generally supposed. Rather, he showed that a more realistic picture is to see science as proceeding by a succession of what he called paradigms. Although there is debate over some aspects of what Kuhn said, by and large his concept of scientific paradigms has been widely accepted as a fair reflection of how much of science is carried out.

In any field of enquiry there is a prevailing paradigm which is the word Kuhn adopted to describe the underlying explanatory scheme for that particular branch of science. Most scientific research takes place within and consistent with that paradigm, and Kuhn called this 'normal science'. Much of this is directed to making increasingly accurate observations relevant to the paradigm or testing predictions arising from it.

---

[1] This is known as the problem of induction, which is fundamental to the validity of scientific theories. Whereas we can be confident of the conclusion from a deductive argument (provided the premises are true and our line of reasoning is sound), we cannot be sure that a conclusion from induction will necessarily be valid. Although many eminent scientists and philosophers of science, in an effort to make scientific theories more certain, have grappled with this perplexing quandary, no solution has yet been forthcoming, and it appears to be an insoluble problem.

Take, for example, the geocentric system and epicycles of Ptolemy's *Almagest* which was the prevailing astronomical paradigm in the West for such a long time (see Ch. 2). For centuries, first the Alexandrians, followed by the Arabs, and then astronomers in western Europe, this was the basis of their cosmological system, and they assessed all of their data in the context of this model. Much of their work consisted of taking observations, comparing these with their models, and adjusting the epicycles to give better agreement, or developing better mathematical methods for calculating the epicycles.

Paradigm puzzles

However, it is not uncommon for there to be some observations which do not seem to fit the paradigm. Strictly speaking, at least in the orthodox view of science, these anomalies are counter-instances in the sense that they are inconsistent with the prevailing theory and *prima facie* invalidate or falsify it. However, unlike the popular view of science, on coming across these counter-instances (which apparently disprove the accepted theory) scientists do not immediately abandon their theory or paradigm. This may come as something of a shock to some readers, but really it is entirely reasonable: It is quite possible that the offending observation itself is in error, so it needs to be checked out; the interpretation of the observation may be at fault, so that it does not actually contradict the paradigm, even though at first it seems to; or it may just be that some small modification to the paradigm could accommodate the anomalous result. So there is much to be said for not abandoning a useful theory too readily. In fact, Kuhn maintained that a significant part of 'normal science' is the addressing of these anomalies – 'puzzles' as he called them. Indeed, he went so far as to say that the prevailing paradigm is not only explanatory, it also sets problems for scientists to work on.

In other words, scientists are used to the idea that there are some observations which cannot be explained by the prevailing paradigm, may even appear to contradict it. But, for most, their belief in the essential truth of the paradigm is such that they expect an explanation will be found that can resolve the conflict, consistent with the paradigm. And it should not be assumed that puzzles are soon resolved – they may remain outstanding for decades or even centuries! How comfortably a scientific discipline can live with its outstanding puzzles depends on how they are viewed – whether they are deemed to be peripheral or central to the prevailing paradigm, or perhaps if it is just felt that current methodology (e.g. instrumentation) is not up to resolving the issues. It also depends on how firmly established the paradigm is.

For a long time there was satisfaction with Ptolemy's system – most astronomers accepted the reality of the epicycles, even though they did not account for observations particularly accurately. Discrepancies were seen as due to errors of measurement or a need to improve the definition of the epicycles, but not to question the epicycle system itself.

On the other hand, a persistent puzzle was uncertainty as to what physical reality lay behind the epicycles. No physical arrangement was ever proposed that was felt to be satisfactory. This bothered some more than others.

Paradigms in crisis

When a paradigm is well established scientists are fairly happy to assign *ad hoc* explanations for apparent anomalies, and/or keep some of them as puzzles to be worked on. But there is a deep-seated belief on the part of most scientists that the universe is orderly, which means they expect 'true' theories to be elegant; and the accumulation of special cases (*ad hoc* explanations) and unresolved puzzles detracts from that elegance. If the accumulation persists then at least in some quarters there is a growing sense of dissatisfaction with the paradigm, and eventually someone will come up with an alternative which they feel provides a better overall explanation of the available data. They will present this to the scientific community, usually through a relevant scientific publication, and a general debate will ensue as to its merits *vis-à-vis* the old paradigm. During this debate, whether one sees the anomalies as puzzles or counter-instances will depend very much on one's perspective – which side of the debate you are on. Eventually, one side will win. Either the old paradigm will be affirmed, emerging all the stronger because it has shaken off a contender, or the new one will take over – a paradigm 'revolution' will have taken place. But to win over the scientific community, the new paradigm must clearly have superior explanatory power compared with the old one. That does not mean it has to answer all the problems, it may even introduce some of its own; but overall it is considered to explain more of the facts, and appear to have the potential to resolve outstanding problems.

Which brings us to Copernicus. Over the preceding couple of centuries, since the *Almagest* was adopted in the West, there had been growing dissatisfaction that the epicycles lacked a physical basis, and that even their mathematical representation was continually having to be adjusted to accommodate new observations. All of this detracted from their elegance and confidence in them. The chief attraction of the heliocentric system was the elegant way it explained retrograde motion. However, the central position of the earth was so engrained in the minds of most medieval scientists that, in order to overthrow this concept, any alternative theory would have to offer substantial compensating advantages. But the Copernican system did not give a more accurate account of observations unless various epicycles were included, so it won few converts.

Even Kepler's correct determination of elliptical orbits (free of epicycles) did not win the day. But when Newton came up with a fundamental explanation for the elliptical orbits, and the variations in speed around the orbit – that was elegant, and convincing. Yet even Newton's laws did not explain everything. It was left for Laplace to resolve the interactions of the planets on each other's orbits, stellar parallax was observed only from the 19th century, and not until Einstein's theory of relativity in the 20th century was an anomaly of Mercury's orbit resolved.

The prevailing paradigm

Most of what we know, at least in an academic sense, we know because we have been taught it, not from our own investigation. For example, we know that $2 + 2 = 4$ from personal experience, but probably most of us know that $12 \times 12 = 144$ because we have learned it, though some will have checked it out for themselves. Many know that the chemical formula for water is H_2O because they have been taught it, but only a few will have carried out any experimentation to demonstrate the truth of it. And almost all who know that proteins are composed of amino acids, or DNA of nucleotides, will know it only because they have been taught it.

Perhaps the prime distinguishing feature of a paradigm – the feature which clearly demonstrates that it is the prevailing paradigm – is that it is the one that is taught. It is taught to those training to be its practitioners and, depending on the extent of interest in the subject, it is taught to the wider public. It could hardly be any other way: the paradigm is firmly believed to be true, so it is taught as such. Just as the geocentric system was taught before Copernicus, and the heliocentric one is taught now.

Although at the time of a paradigm revolution there is much debate about the pros and cons of the competing theories, once the dust has settled and a new paradigm is established, it simply becomes part of the accepted system of knowledge. It is interesting to note that, even as early as the 18th century, the historian cum philosopher David Hume (1711–76) commented:

> In reality [...] the modern system of astronomy is now so much received by all inquirers, and has become so essential a part even of our earliest education, that we are not commonly very scrupulous in examining the reasons upon which it is founded. [*Dialogues*, Part 2]

Social factors

Finally, one of the important points that Kuhn emphasized about science is that scientists are human and their views about scientific theories are influenced by non-scientific factors. For example, at a personal level, we can readily understand that a scientist who advances a new theory will be inclined to champion his ideas and be biased in their favour. It is to be expected that he will seek to defend it against attack, rather than have a completely neutral stance towards it. Similarly, how many are willing to question, even to themselves, the fundamental tenets of their science if it may undermine many years', perhaps a lifetime's, work? Kuhn even went so far as to say that:

> Normal science, the activity in which most scientists inevitably spend almost all their time, is predicated on the assumption that the scientific community knows what the world is like. Much of the success of the enterprise derives from the community's willingness to defend that assumption, if necessary at considerable cost. Normal science, for example, often suppresses fundamental novelties because they are necessarily subversive of its basic commitments. [*Scientific Revolutions*, Ch. 1]

Related to this is the fact that most scientists do not normally work in isolation but as part of a scientific community, and their reputation with peers is important to them. Similarly, although scientists today are mostly free of fear of sanction by church or state, the need for continued funding or employment can exercise considerable constraint on freedom of expression. There are few scientists of the stature of Galileo who are prepared to stand up against the body of scientific opinion of their time.

Paradigms of geology

It is all too easy for us to think that the science of our own day has got it 'right', but a recent major paradigm revolution in mainstream geology should caution us against overconfidence.

Up to the beginning of the 20th century geologists believed the world to be essentially static, especially that the continents are more or less where they have always been; they may have been submerged due to changes in sea level, but on a global scale the continents are fixed. However, in 1912, a meteorologist named Alfred Wegener (1880–1930) proposed that the continents are not fixed but drift slowly around the globe. There was a very strong reaction to this suggestion by most geologists, partly, no doubt, because it seemed to contradict their hard-won doctrine of uniformitarianism, and very few took the idea seriously. Only in the 1960s, when evidence emerged of the spreading of ocean floors, did the idea become generally accepted. Now, of course, almost everyone knows of continental drift, and plate tectonics is the established paradigm. It is used to explain the occurrence of earthquakes, volcanoes, mountain ranges and, from an evolutionary point of view, migration of species between land masses which were connected in the past though now separated.

THE PARADIGM OF EVOLUTION

So how does this understanding of how science proceeds apply to the theory of evolution?

Rise of the evolutionary paradigm

Before the 19th century, undoubtedly the prevailing paradigm relating to the biological world was creationism; in particular, it was believed that all living organisms were created substantially in their present form by a supernatural designer and maker. This paradigm increasingly came under attack from various observations which included the enormous variety of organisms (it was deemed improbable that any creator would specifically make so many individual types of creatures), adaptation (again, it seemed unlikely that all of the many intricate adaptations would be customised designs) and fossils (especially of extinct species, because, for many, part of the creationist view was that a benevolent creator would not allow extinctions). Further, because the paradigm of creation was associated with a recent time-scale but the fossils indicate an ancient earth, this also counted against creation. However, creationism was not superseded until there was a workable alternative: although general evolutionary ideas were suggested from the mid 18th century onwards by the likes of Buffon and Lamarck they were insufficient to displace the creation paradigm. Darwin's success was to present evolution as a coherent theory, by providing an explanatory mechanism for how evolution could occur: the action of natural selection on variations could explain the adaptation and divergence of species.

The paradigm established

Although the theory of evolution was accepted by most naturalists well before the end of the 19th century, it was by no means firmly established. An important reason for this was the lack of a coherent theory of inheritance. And when it came, because of the quantum (particulate) nature of inherited characteristics, at first it seemed to militate against Darwin's gradual evolution. The major importance of the early population geneticists was their successful synthesis of Darwinism with Mendelian genetics to provide a consistent overall theory, Neo-Darwinism, well supported with experimental evidence for natural selection and speciation. The fact that Darwinism faced a serious challenge and won through substantially increased its credibility and the theory of evolution emerged all the stronger.

The elegance and proven success of Neo-Darwinism meant that the evolutionary paradigm was firmly in the driving seat by mid 20th century. By then it had become (and still is) the unifying explanatory thesis of biology. The dominance of Neo-Darwinism means that any new knowledge to emerge is interpreted in the light of the theory of evolution. A point I have repeated is that the evolutionary paradigm was firmly established before there was any significant understanding of molecular biology; consequently, as the knowledge of biochemistry emerged, it had to be fitted into the evolutionary framework, and there was profound reluctance to allow the new knowledge to challenge the established paradigm.

I suspect that many readers, convinced of the basic objectivity of science, will doubt the extent of the truth of that last statement. But a clear example arose in a symposium that took place at The Wistar Institute of

Anatomy and Biology, Philadelphia, in 1966. This was perhaps the heyday of Neo-Darwinism, and the implications of molecular biology were just beginning to be appreciated. The symposium was specifically convened to enable leading evolutionists to discuss 'Mathematical Challenges to the Neo-Darwinian Interpretation of Evolution', so it was clear that the purpose of the meeting was to take these mathematical challenges seriously – they were not some sideline at a conference on another issue.

One of the mathematicians, Dr Stanislaw Ulam, presented a paper in which he outlined various difficulties, especially of there not being enough time available for evolution to occur – even to accumulate a series of advantageous changes (i.e. not trying to get there in one jump, which had been the scenario of the preceding paper). In the following discussion, one of the biologists, Prof. C. H. Waddington, said:

> You are asking, is there enough time for evolution to produce such complicated things as the eye? Let me put it the other way around: Evolution has produced such complicated things as the eye; can we deduce from this anything about the system by which it has been produced? [Ulam]

Followed by Sir Peter Medawar:

> May I make a point here in support of what Waddington says? I think the way you have treated this is a curious inversion of what would normally be a scientific process of reasoning. It is indeed a fact that the eye has evolved; and, as Waddington says, the fact that it has done so shows that this formulation [i.e. the mathematical argument presented by Ulam] is, I think, a mistaken one.

And, in presenting the next paper, Ernst Mayr said that they should approach the subject from the point of view that evolution has happened.

These comments clearly illustrate what it means to be operating within a paradigm. For the biologists it was not a question of how good or bad the mathematical argument was – they did not consider that. As far as they were concerned their paradigm was true and they rejected anything that would not fall in line with it. Rather than allow anything to challenge their own beliefs, their firm presumption was that there must be something wrong with the supposed challenge, even if they had no idea what it was. The symposium was convened specifically to consider mathematical challenges to Neo-Darwinism, but for the biologists it could be construed only as biological objections to the maths!

Evolution: the prevailing paradigm

As mentioned earlier, most of what we know, we know from being taught. Many people believe we have evolved, and for most it is solely because they have been taught it. And people believe it to be true as firmly as they believe water to be H_2O, because they have been taught it with equal conviction. Some readers may protest that it is not just a question of indoctrination – that they can point to proof of evolution. For example they believe it is shown to be true in the fossil record, or by demonstrations of natural selection. But the reason why they think these observations constitute proof of evolution is simply because they have been taught that this is evidence that proves evolution to be true. Similarly, many people believe that the struggle for survival portrayed in many nature documentaries is ecological proof of evolution because they are told repeatedly that this is relevant proof. However, as I have explained in the preceding chapters, on closer inspection these 'proofs' are nothing like as convincing as is generally believed: at best they demonstrate a limited aspect of evolution. Because evolution is the prevailing paradigm, all too often only the supporting evidence is taught, and the problems and contrary evidence are glossed over or ignored altogether.

We are taught most of what we 'know'. Realistically it has to be this way. One of the reasons for the success and continued growth of science and technology is the ability to transmit information from one generation to the next without it having to be reacquired from scratch. So do not misunderstand me – I am not criticising the fact that what we know is mainly from being taught. I am not even criticising that most people believe we evolved only because they have been taught it. But I am stressing that we need to recognise that this is the case. And when we are presented with arguments that challenge the theory of evolution, we should not just retort that we know evolution is true so there must be something wrong with the supposed contrary arguments.

Evolution's "puzzles"

So now let's turn to the puzzles, anomalies or counter-instances (depending on one's outlook) of the theory of evolution.

The origin of life

Probably the major puzzle, the one that is acknowledged even by a significant number of evolutionists, is the problem of the origin of life. Many of those who have not looked at the subject will simply accept the popular myth about primitive life emerging readily from a primeval soup, which became widespread after Miller's experiments. But any who have looked at the problem at all seriously know it is a major stumbling block, as I have indicated in the preceding chapter.

It is perhaps in the context of the origin of life that biologists are most keenly aware of the improbability of useful amino acid or nucleotide sequences: that even the simplest self-replicating systems, whether involving one molecule or a few, require macromolecules that are large enough and specific enough to be able to perform their required role sufficiently reliably. Once even a basic 'life' is under way, the improbability of macromolecules is largely overlooked, because another myth is that with a self-replicating system in place then some sort of natural selection can begin and further useful macromolecules may be found relatively readily – though I have pointed out what a misguided and forlorn hope this is.

The abrupt appearance of major groups

In Chapter 10 we saw that most if not all of the major groups of organisms appear in the fossil record abruptly, without transitional forms to link them with previously existing organisms. Clearly this is inconsistent with the gradual evolution proposed by Darwin or, for that matter, of Neo-Darwinism. The situation is exacerbated by the fact that in so many instances the major new groups – ranging from various invertebrate phyla to vertebrate classes and the higher plants – arise as multiple distinct lines. These lines are so distinct, without evident connecting family tree, that they appear to have separate origins; but even advocates of evolution recognise that it is unacceptably improbable that a major advance – whether it be an insect's wing, polydactyl limb, amniotic egg, mammary gland, leaf or flower – could evolve independently multiple times, especially at about the same time.

The gaps in the fossil record are seen as a separate problem from that of the origin of life. What I have shown here is that they are but a different facet of the same fundamental problem: the impossibility of obtaining biological macromolecules by chance, even progressively.

It is illuminating to consider the phylogenies based on comparative amino acid or nucleotide sequences in the context of the fossil record. The molecular phylogenies were at first hailed as a triumph – as independent corroboration of evolution based on morphology. But as more sequences have become available they are seen to show the same pattern as the fossil record: that, whilst closely related species may have similar sequences, different groups of organisms are quite distinct from each other. That is, both the fossil record and molecular cladograms point away from a gradual evolutionary origin of the major groups of organisms. Even though gradual morphological or molecular change may occur subsequently, any attempt to retropolate, using either morphological or molecular data, is found to be highly inconsistent with the abrupt first appearance.

Of course, various 'explanations' are offered to account for the abruptness of the fossil record and the discontinuities of the molecular phylogenies, for example that at such times evolution proceeds exceptionally rapidly. However, to say that there are different rates of evolution is merely to impose an evolutionary explanation on the facts, it certainly does not explain them.

Lack of constructive mutations

Although the Neo-Darwinists thought they had ample evidence of mutations that were relevant to evolution, modern knowledge of molecular biology has completely undermined this. As biochemists unravelled the nature of genes and how they are expressed, it showed that the mutations the Neo-Darwinists had found were due solely to the corruption of genes, not to the production of new ones. Hence, whilst they had thought that constructive mutations could arise readily, and that it was just a matter of time before advantageous ones would turn up, we now know otherwise. Despite extensive efforts to generate and identify mutations, none has been found that offers constructive morphological change such as is required to fuel evolution. Even the mutations which confer resistance to insecticides etc., cited as evidence of advantageous mutations, are losing their impact as we uncover what is happening at the molecular level. A century after de Vries with his variations in the evening primrose and recovery of Mendel's work on genetics, and nearly half a century after Watson and Crick determined the molecular structure of genes, the best evolutionists can offer is the circular argument that constructive mutations must arise sometimes otherwise evolution could not happen.

Molecular evolution

At the heart of this absence of advantageous mutations is, of course, the specificity and hence improbability of biological macromolecules. This was recognised to some extent quite early on, but it was presumed that it could be overcome by macromolecules evolving at the molecular level in a similar way to morphological evolution; and, as the comparative amino acid sequences became available, at first these were seen as showing this sort of evolutionary change. But in due course it became apparent that almost all of the inferred 'evolutionary' changes in sequence of biological macromolecules are in fact neutral – most of the variation we see in amino acid sequences are of little if any selective consequence. Even if one accepts the molecular phylogenies, the message they clearly convey is that proteins (and RNAs) were fully functional when they first appeared on the scene: all we see are variations on a theme which was already successful; we do not see evidence of a trial and error process by which a crude protein or RNA was gradually improved.

Homology

Most of the morphological features that are regarded as homologies had been identified as such in a non-evolutionary context and used as the basis for classification. Subsequently, evolution – inheritance from a common ancestor – appeared to provide a convincing (and elegant) explanation for them. If phylogeny is the reason for homologies then we would have expected at least to see homologues arise from equivalent embryological sources, and probably by comparable developmental processes. However, as embryological investigations have proceeded they have turned up an increasing number of substantial anomalies – marked incongruence between homology and supposed phylogeny; especially significant are those stemming from early development. From an evolutionary point of view this is very perplexing. There is a profound reluctance to deny the long-standing 'obvious' homologies, but there is an equal or even greater reluctance to abandon evolution.

Some biologists have presented homologies in a way that is rather interesting in the present context. They suggest that when morphological features are described as homologous it is, in effect, proposing a thesis – that they are derived in some way from a common source – and that the thesis is open to scrutiny in the light of other information:

> Homology after all is a conjecture of similarity to be explained by common descent. The more characters (conjectured similarities) congruently supporting a hierarchy of relationships, the stronger the evidence for a regularity of character distribution among taxa, suggestive of an underlying cause – which is evolution. What matters is not so much how we get to conjectures of similarities, but that these hypotheses of homology be tested against all other characters known in the search for regularity i.e. congruent character distribution. [Rieppel]

Where other features (e.g. developmental source and/or process) are congruent then it supports the original thesis; but where they are incongruent than it casts doubt on the thesis, even falsifies it. This is pertinent here because, although there are marked anomalies – anomalies which should challenge the evolutionary explanation – most biologists are unwilling to accept this conclusion. Instead they regard the incongruities as puzzles awaiting a satisfactory explanation consistent with evolution, or settle for *ad hoc* explanations such as advocating that homology may exist at different levels which are not necessarily congruent. Most biologists would rather modify what we mean by homology than abandon evolution; in the light of which Colin Patterson (1982) commented:

> The paradox is what we add to homology to justify phylogeny. Is it stratigraphy, biogeography, or simply the belief that organic phenomena must have a natural, historical explanation?

Homology is a further example of where an early or superficial understanding appeared to support evolution, but as we uncover more of the facts they even contradict it.

Sex

There are two aspects about sex that are contrary to the rationale of evolution and, therefore, why the presumed evolution of sex presents a problem.

First is the waste of resources in producing males. To appreciate this, compare a sexually reproducing female with one that reproduces asexually, i.e. parthenogenetically, which is not uncommon in various lower plants and animals. Assuming that the sexual female produces equal numbers of males and females, then only half of her offspring can go on to have further offspring, whereas all of the parthenogenetic offspring will be females which, in due course, may reproduce. Hence, if both types of female produce the same number of individuals (corresponding to a comparable economic cost of reproduction), the parthenogenetic female will, overall, proliferate at twice the rate of the sexual one. In other words, the species reproducing sexually has only 50% of the fitness of the parthenogenetic species (compare this with a typical difference in fitness of about 1% or less, mentioned in Chapter 5), and being at a substantial selective disadvantage should, by the operation of natural selection, be ousted quite rapidly by the parthenogenetic species. To compensate for this selective disadvantage of sex the male would need to make a substantial contribution to the overall fitness of the species, for example by collecting food for and/or protecting the female and her offspring. But this occurs in only some sexual species.

The second reason strikes at the heart of evolutionary rationale. The whole point of natural selection is that those organisms which survive better reproduce preferentially, and hence tend to perpetuate their favourable genetic composition. However, for thriving sexually reproducing individuals, rather than perpetuating all of their successful genotype they pass on only half of it. Whilst it can be argued that individuals having advantageous genotypes will arise in both sexes and, overall, fitter males will generally mate with fitter females, there is no getting away from the fact that sexually reproducing individuals are gambling with 50% of their successful genotype. In contrast, thriving parthenogenetic females pass on 100% of their successful genotype. So clearly sexual reproduction conflicts with conventional evolutionary wisdom.

It is obvious that the advantage of sexual reproduction is the genetic flexibility which arises from the shuffling of genes through recombination and fertilization. This shuffling provides the opportunity for genotypes to be generated that are more successful and, especially important, for the species to adapt fairly rapidly to changing circumstances. Conversely it is thought that asexual species must be less adaptable and more susceptible to extinction. This is supported by their sparse and sporadic occurrence within most taxonomic groups. In other words, it is suggested that if asexual species arise (which is usually assumed to be by reversion from a sexually-reproducing species), they do not last long because of their lack of genetic variability, and this explains why sexual reproduction is favoured in the long term.

However, this does not explain how sexual reproduction arose. Genetic flexibility is of benefit only to future generations, not to the present population; but natural selection has no foresight – long term survival does not carry any weight in terms of short term selective advantage. The two penalties mentioned above are significant short term disadvantages which should have acted against sex evolving in the first place. If the implementation of sexual reproduction could be attributed to a simple mutation then it might be argued that it could arise repeatedly – so often that at least some, by chance, survived the short term disadvantages and became established as the norm due to long term advantage. But all are agreed that implementing sexual reproduction is a sophisticated affair, requiring many biochemical novelties, notably meiosis, and is unlikely to arise readily or frequently. Needless to say, my view is that there is no possibility that the biochemical complexity of sexual reproduction could ever arise in an evolutionary manner.

John Maynard Smith, who was one of the first to recognise that the evolution of sex is a problem, wrote:

> We are therefore driven to the conclusion that the early stages in the evolution of the sexual process took place under the influence of selective forces quite different from those which are responsible for the maintenance and spread of sexual processes once they were erected. [*The Theory of Evolution*, Ch. 12]

It is very puzzling indeed if different circumstances are required for a feature to evolve than are required for its evolutionary preservation.

Living with the puzzles

In view of these substantial difficulties one might wonder why the theory of evolution is still so firmly and widely held. So far as the layperson is concerned, because evolution is taught, directly and indirectly from our earliest school days if not before, and taught as a coherent theory with little or no mention of the problems, most people have no reason to doubt the paradigm, and believe unquestioningly that it is true.

But what of the scientists? Surely they should be more familiar with the facts, and be objective enough to recognise that there are some serious unresolved problems? I think that for these, too, the primary reason is that evolution is the prevailing and pervasive paradigm. The only way significant scientific progress can be made is by building on earlier work, so, necessarily, much of what scientists know, even in their own area, is learned from others rather than first-hand experience. In their training scientists are taught the fundamental concepts of their discipline and, as Hume commented regarding astronomy, once a paradigm is established its tenets are accepted as true and generally not scrutinized. And Kuhn added that there can be a profound reluctance to question them. Then, so far as evolution is concerned, there are other factors as well.

Limited evolution

First – and this is probably relevant to the layperson too – is that some aspects of the theory of evolution have been shown to be true. In particular, natural selection has been amply demonstrated in various situations, famously in the rise and fall of melanism in the peppered moths. Then, some of the examples outlined in Chapter 9 of adaptation and diversification are reasonably well known, such as the closely related species of Galapagos finches which can confidently be inferred to have descended from a common ancestor. Similarly, as mentioned in Chapter 10, there are a few examples in the fossil record, notably the ancestry of the horse, where there is good evidence of gradual morphological change over a period of time. Examples such as these are seen by many as adequate justification for the whole evolutionary process, not appreciating that they are limited in extent, explicable solely in terms of gene segregation, and that full-scale evolution requires substantial, but unsupported, extrapolation. This is reinforced by the fact that many, like the 19th-century naturalists, still think that the only alternatives are evolution and a rigid fixity of species; so, as it is evident that species are not fixed, it follows that the theory of evolution must be true in its entirety.

The multidisciplinary nature of science

One reason for scientists not appreciating the significance of the problems arises from the subdivision and specialisation of modern science. Science today is so extensive that no individual can hope to have an in-depth knowledge of more than a small area, and knowledge outside of a scientist's sphere of activity may be little different from the popular understanding. However, the problems with evolution arise in diverse scientific disciplines. Hence, whilst biologists are aware of the problems in their own area of work, they are unaware of those in others.

For example, the population geneticists are probably the most content. They know evolution occurs because they can demonstrate the action of natural selection conclusively, and go a long way towards showing how speciation occurs. They are happy provided they are given the raw genetic material, and do not worry too much about where it has come from. Although they are aware of the lack of evidence for advantageous mutations, they know that mutations do occur and the wealth of genetic variability assures them that constructive genes must arise sometime, somehow or other; and they can at least point to the mutations which convey resistance to antibiotics etc. Most are largely unaware of the biochemical difficulties, and dismiss reports of gaps in the fossil record as nothing but creationist propaganda.

Then there are the biochemists who are unravelling the intricacies of molecular biology. They are impressed with the elegance of subcellular systems, and wonder at how evolution has wrought such complexity. It is not uncommon to come across comments such as this:

> Much of the basic fabric of living organisms now stands revealed, with its enigma of how the extraordinary interlocked system of nucleic acids and proteins (both of which, in different ways, are necessary for the synthesis of each) first evolved. [Whitehouse, Preface]

But most are immersed in the evolutionary paradigm, believing that natural selection and the fossil record prove evolution has occurred, so do not doubt there must be an evolutionary explanation for the biochemistry. Indeed many point to the similarities of protein structure as their contribution to the evidence for evolution. However, whilst the importance of a correct 3D structure for proteins to function is universally acknowledged, and emphasized in biochemistry textbooks, it seems few biochemists recognise that 'a good fold is a rare fold' or the minimum size required of a macromolecule to adopt a stable 3D structure; and even fewer appreciate the significance of these facts in relation to the supposed evolution of macromolecules.

The palaeontologists are the least happy. Although, in the popular perception, the fossil record is one of the key 'proofs' of evolution, the palaeontologists know otherwise. It is they who are most keenly aware of the substantial gaps between major groups, and are unhappy with many phylogenies that are based on cladistics (whether using morphological and/or molecular data) with scant regard to the fossils. Nevertheless, despite the difficulties, most palaeontologists are satisfied with the evidence for evolution they can see, such as of the horse, and modern evidence for the action of natural selection, are unaware of the biochemical issues, and hence believe that satisfactory evolutionary explanations will be found to bridge the gaps.

Probably most biologists think homology is good evidence of evolution. Only those studying the subject or comparative embryology are likely to be aware of the non-homologous development of apparently homologous morphological features, i.e. of the substantial inconsistencies between homology and supposed phylogeny.

Finally, for each of these groups, being relatively unaware of problems in other areas means they are inclined to see their own problems as puzzles to which they believe an answer will eventually be found, rather than as counter-instances for which there may not be an evolutionary explanation.

Fact versus mechanism

Some biologists, recognising there is evidence for some evolution but also that there are substantial difficulties, adopt the stance that the puzzles relate only to mechanism. That is, there may be debate about *how* evolution has occurred, but not *that* it has occurred. This view carries quite a lot of weight with many because it appears to face up to the facts and adopt a scientific attitude to the outstanding problems.

However, the important point is, of course, that any demonstrable evolution is limited and can be attributed to gene segregation, whereas new groups of organisms appear abruptly, not in a progressive or evolutionary manner. Proof of the former is no evidence for the latter.

On the one hand, many do not recognise this important distinction. But on the other one cannot help feeling that some advocates of this view are primarily trying to obfuscate the issues (rather like to trying justify the whole evolutionary scenario on the basis of industrial melanism) – it provides a way of making light of the problems.

Cladistics

One of the lines of work supporting an evolutionary outlook is cladistic analysis. The resulting cladograms, showing how species are believed to have derived from an ancestral stock, carry significant weight because of their visual impact – apparently demonstrating an actual phylogeny. In fact, some, e.g. relating to species within a genus, may well give a fair indication of how they have arisen through gene segregation.

However, a major application of cladistics is to bridge the substantial gaps in the fossil record. Because the analysis is carried out objectively it conveys a sense of reliability, even reality. But, as we have seen, this perception is misplaced because cladistics is based on the premiss of evolution – it does not test it. The cladograms are no more than a best consensus of the data: the premiss is that evolution is such a good explanation for the congruent characteristics that it must be right, and there must be an explanation consistent with evolution for the incongruent ones. If the resulting cladogams were self-consistent and clearly supported by the data, then this would provide circumstantial support for evolution. But the fact that the cladograms are not well defined by the data, compounded by the inconsistency of cladgrams based on morphological and molecular data, clearly shows they are not as reliable as they appear.

Ad hoc explanations and further research

Because evolution is the prevailing paradigm, the response of biologists when they come across difficulties is to propose explanations that are consistent with it, and this automatically minimises their potential to challenge the paradigm. A prime example of this is to permit secondary acquisition and/or loss of characteristics in cladistic analysis, with little or no consideration of the practicalities of doing so, even when they relate to very early stages of development, such as coelum formation. Related to this, biologists propose that there are different levels of homology, and the concept of partial homology, in order to accommodate the inconsistencies between adult morphology, embryological development, and genetic mechanisms. Or to account for the fossil record it is proposed that evolution must proceed at different rates. And, at the heart of the problem with evolution, to circumvent the improbability of biological macromolecules it is proposed that they must have evolved from simpler molecules having partial activity, even though there is no evidence for them and that the suggestion contradicts what we know of protein structure and function.

In addition, as indicated above, scientists do not expect theories to answer all of the questions. So the fact that there are problems, far from being seen as fatal to the theory, may even be one of its attractions, because it sets puzzles for scientists to work on. That is, it stimulates further research to find elegant solutions to the problems, such as, how could natural selection drive substantial changes in embryological development, especially whilst retaining similar adult structures? or how could a primitive protein evolve progressively into an efficient one?

A VIEW FROM OUTSIDE THE PARADIGM

One of the points made by Kuhn is that it is hard for those caught up in a paradigm to assess it objectively. Some beliefs are all but unshakable whatever the contrary evidence. Indeed, as we have seen, for those operating within the evolutionary paradigm, there can be a profound reluctance to expose its basic tenets to scrutiny. Commitment to a paradigm can be so deep-rooted that this reluctance is not even perceived as such and, contrary to normal thinking, it is considered unreasonable to question the essential truth of the paradigm rather than not to. And, of course, for most there is no motivation to do so – they have no reason to doubt the truth of the paradigm. Also, for those immersed in the paradigm, especially for those participating in the relevant science, they can be so taken up with their own area of work, so focused on addressing their particular puzzles, that they miss the wider picture – a case of not seeing the wood for the trees.

What I aim to do here is to bring together various points that have been raised in preceding chapters, to try to present a coherent view of the theory of evolution as seen from outside of the paradigm. Of course, not for a moment do I pretend to be commenting from a completely neutral or paradigm-free stance. On the contrary, I recognise that we all have a worldview, that I perceive the theory of evolution from my particular standpoint, and that some of my fundamental premises and beliefs would be unacceptable to many evolutionary biologists – but more on that in the final chapter.

I think one of the weaknesses of some previous critiques of the theory of evolution is their tendency to find fault with just about everything about it – they are so eager to condemn evolution that they are inclined to 'throw out the baby with the bathwater'. In contrast, as should be apparent from earlier chapters, I believe various aspects of the theory of evolution are correct, indeed are valuable insights into the workings of biology, and I will start my comments by recognising these.

Variation and natural selection

First is an awareness that, contrary to the traditional view, species are not fixed. They have genetic variability which leads to variations within a species and which, under appropriate conditions, can lead to distinct varieties, subspecies, and ultimately species. Darwin recognised it is the existence of variations that makes evolution of any sort possible, and I think he was also right to perceive varieties as potential incipient species. Further, Darwin correctly identified the occurrence of natural selection and its role in guiding substantial morphological change through the accumulation of small variations.

Natural selection acting on randomly occurring variations explains a wide range of biological observations, and is a particularly valuable insight

for ecology. First might be mentioned the conservative role of natural selection – that of maintaining the general health of a species by weeding out
its less fit individuals, including those carrying detrimental mutations.
Then, whilst recombination results only in a random mix of genes, natural
selection perpetuates those combinations which are beneficial and rejects
the unfavourable. One could say that natural selection is able to identify
coordinated combinations of the available characteristics. This allows a
species to adapt to changing conditions, typified by the development and
then recession of melanism in the peppered moths, or to a range of environments such as in the yarrow plant. Natural selection is also at least part
of the explanation for the co-adaptation of multiple species, such as in one
mimicking the other, or the co-adaptations of some insects and flowers for
pollination. And, finally, there is the refinement of characteristics, such as
we see in the cheetah, where natural selection has matched and possibly
exceeded man's planned breeding for extreme characteristics.

All of this I agree with, and see as a valuable contribution to our understanding of biology. It is the unlimited extrapolations from these observations that I contest.

Evolution, but within limits

It has been said that, whereas before Darwin evolution was seen as possible, by identifying the role of natural selection Darwin made evolution not
only probable but certain. The argument is based on two premises: first
that species produce more offspring than survive to reproductive maturity,
and second that individuals within a species exhibit heritable variations
which affect fitness or survivability. Given these, it follows that there is
competition between individuals, and those with advantageous variations
will generally survive and reproduce more than those without them; hence
the genes that confer the favourable variations will be preferentially perpetuated, and evolution will have taken place. This is a simple deductive
argument based on premises which we know from observation to be generally true; so we can conclude that evolution must occur. This conclusive
deductive argument has carried much weight in some quarters.

However, the argument fails as a proof of evolution in its widest sense
because it does not consider how variations arise, and hence ignores the
possibility that there is a limit to the variations that can occur. In other
words, and what is not generally recognised, the truth or applicability of
the argument is limited to the extent of available variations. The abundant
variations occurring in natural and cultivated species led Darwin to believe
that they would always be available. He saw no reason to doubt that the
process of generating variations and selecting advantageous ones could go
on for ever, and, by successive accumulation of small changes, the whole
array of living organisms could have evolved from simple organisms. Even
though some of his contemporaries pointed to the limitations of domestic
breeding as an indication that there would be limitations to the possible

variations in nature, Darwin believed that such limits, over the eons of geological time, would relax, or that in some way nature was less restricted than man. Because, at the time, there was virtually no knowledge as to the basis of those limitations, Darwin and his followers, even the Neo-Darwinians, can be excused for hoping that there was some way around them. However, it is now abundantly clear that the limitations arise from the genetic material available; and, even though the genetic variability may be immense, it is far from infinite. Importantly, although the available variability permits variations within a group of organisms, it does not enable advancement to higher organisms.

In the light of modern biochemical knowledge, in order to justify extrapolation of the deductive argument to the evolution of more complex forms from simpler ones, it is necessary to suppose that meaningful genes can arise from essentially random variations of nucleotide sequences. However, the central point I have been making is that this presumption fails: First, there is a *prima facie* case of improbability, even taking into account the levels of non-specificity that we find in biological macromolecules. Then there is no way in which such macromolecules can develop their activity progressively, in a trial and error manner comparable with natural selection at the morphological level, because of the minimum number of nucleotides or amino acids that need to be right for the resulting macromolecule to have any utility on which a selective process can operate. This is compounded by the fact that most macromolecules are cooperative – which means that multiple macromolecules must reach some level of utility at more or less the same time and place – which is so improbable a scenario that it must be rejected, and is not even advocated by evolutionists. Further, there is no evidence that this theoretical objection has, in practice, been overcome: the abrupt emergence of new groups in the fossil record corresponds with the sudden, not gradual, occurrence of new genetic material; and the comparative amino acid sequences demonstrate that the essential structures of proteins were already determined by the time they arrived on the scene.

What all of this means is that, although Darwin was right to identify the existence of variations, the role of natural selection, and the consequent evolution this enables, he was wrong in his unlimited extrapolation. Unfortunately, not only the Neo-Darwinians, but also today's biologists who should know better, have followed his error. It is clear from our modern knowledge of genetics that, although some evolution is possible, even inevitable, any such evolution must be within limits that are defined by the existing genetic material. Put another way, in terms of the steps identified in Chapter 9, evolution is substantially confined to the second and third steps of recombination and selection. I recognise that this will be totally unacceptable to the ardent evolutionist; but it is more consistent with the facts. Extending the theory to include macroevolution – evolution that requires the formation of new useful genetic material – is mere speculation, and speculation in the face of a weight of contrary evidence.

Unfortunately, a conclusion such as this – which accepts some evolution but rejects the whole evolutionary account – runs into a perceptual difficulty which I need to mention before proceeding further:

When the theory of evolution was first propounded, the alternative was a rigid fixity of species, and in the second half of the 19th century many, including highly respected scientists such as Louis Agassiz (1807–73), still maintained that extreme position. However, as increasing evidence emerged for morphological change, notably in the ancestry of the horse and melanism in the moths, even staunch opponents of evolution had to concede that some 'evolutionary' change occurred. Further evidence of evolutionary processes that came to light through the 20th century led them to accept even more change. This progressive withdrawal from the fixity position has no doubt weakened the case against evolution in its widest sense, because any remaining objections are perceived merely as part of a continuing retreat on the part of anti-evolutionists, with no reason to believe there is any actual limit. In other words, it has led many evolutionists to believe it is just a matter of time before any opposition fades away and all adopt their position. This view was expressed by the historian of science, David Hull:

> As soon as a natural explanation was acknowledged for the origin and evolution of some species, nothing but time stood in the way of extending it to all species. [*Darwin and his critics*, Ch. 3]

I think this is a popular view, but that it is incorrect; and I do not contest it on the grounds of trying to defend some philosophical or religious position about essences or created kinds, but on biological facts. We now have a sound objective basis for recognising why some evolution occurs – because genetic variability exits, but also why this must be within limits – because the variability is finite. The boundary is essentially between segregating genes and originating them. Perhaps one of the reasons why the argument about evolution has gone on for so long is that evolution is not all or nothing. Because of traditional ideas, it seemed the choice had to be a rigid fixity of species or evolution of the whole tree of life, whereas the truth of the matter is somewhere in between. Although some have perceived this since even before Darwin's day, it has taken a long time for us to realise why it is so.

Evolution as a scientific theory

Another line taken by some opponents of evolution is that we need not take it seriously anyway because it is 'only a theory'. I explicitly distance myself from such 'arguments', which I think are vacuous. Whilst it is legitimate to point out that no theory is provable, and that includes evolution, it is inappropriate to dismiss evolution on this ground. Much of our modern everyday lives, and that includes those of evolution's critics, depends on the substantial truth of many scientific 'theories'; so to reject evolution as only a theory is certainly inconsistent, if not hypocritical.

So far as my view is concerned, I trust that comments earlier this chapter, indeed throughout the book, will have conveyed that I have much respect for the scientific method and its theories, and I think we need to take them seriously, even though they are not as objective or certain as many believe or we may wish them to be. Further, it was enlightening to read Kuhn and use his perception to help me see how I operate as a scientist; and his insight certainly helps to explain the entrenchment of the theory of evolution. However, my reason for rejecting the theory of evolution is not that it is just a theory or even the prevailing paradigm, but because of its anomalies and the major problems which, after much investigation, remain unsolved and seem to be without prospect of solution.

With that possible misunderstanding set aside, let's see how evolution fares as a scientific theory. You will recall that, in formulating the theory of evolution, Darwin followed Newton's hypothetico-deductive approach. That is, on the basis of various observations relating to the diversity, variability and adaptation of species he proposed that organisms evolve gradually through the progressive selection and accumulation of small changes. It seemed to Darwin that his theory explained a range of diverse facts – on the one hand the diversity of organisms, and on the other that they can be grouped into a hierarchical classification; the overall progression of fossil organisms, and the fact that many species became extinct; and some morphological similarities, or homologies, that are evident in embryos and/or mature forms. So clearly it was reasonable to propose this theory, but how does it stand up to scrutiny?

One of the tests of a scientific theory, one which confers confidence in it, is whether it can be shown to be reproducible, that its results can be repeated. Clearly this is not possible with evolution – we cannot turn back the clock. Darwin, of course, recognised this, and that is why, instead, he provided a wealth of other, circumstantial, evidence to support his theory. However, with the subsequent understanding of heredity, by uncovering part of the evolutionary process, it did become possible to reproduce under controlled conditions at least some aspects of evolution, such as selection and adaptation to living in a particular environment. Understandably, work such as this reinforced a sense that evolution must be true, completely true.

Falsification of evolution

However, the important prediction that Darwin's theory of evolution makes is that organisms will evolve – will become more complex – progressively. This should be discernible in the fossil record, but Darwin had to admit that it did not show the intermediates his theory requires. Even some of his contemporaries pointed out that new groups appear abruptly, and, as has been discussed adequately in Chapter 10, this pattern has been confirmed be the numerous fossils unearthed since then.

Darwin also recognised that his theory implied not only the gradual development of organisms, but also of their organs. In fact he provided this as a test for his theory, as a ground on which it could be falsified:

> If it could be demonstrated that any complex organ existed, which could not possibly have been formed by numerous, successive, slight modifications, my theory would absolutely break down. [*Origin*, Ch. 6]

The eye had long been recognised as a prime example of an 'organ of extreme perfection' – cited by Ray and Paley as clear proof of design – and Darwin outlined how he thought it could have arisen by a series of functioning intermediates. As elaborated in Chapter 11, it remains the focus of much current debate as to whether it could have evolved progressively, but, as I emphasized there, this debate completely ignores the biochemical implications of changes in morphology. When these are taken into account, *every* organ fails Darwin's test, because every organ is dependent on complex biological structures in order to function. Even the supposed evolution of eukaryotic cells from prokaryotic (Ch. 13) fails because the transition would involve multiple inter-dependent novelties of molecular biology. In fact, the same could be said of the origin of prokaryotic cells. Indeed, I suggest that every biological macromolecule is, in itself, of a complexity which cannot arise incrementally.

Which brings us back to the central failing of evolution: it cannot account for the complexity of molecular biology. This shows itself not only at the molecular level, but in the origin of life, of eukaryotic cells, sex, and the abrupt appearance of new forms in the fossil record; all of which are clear counter-instances of the overall theory of evolution. However, they are not generally recognised as such because of the fact that science no longer follows the popular view of falsifiability (if it ever did). Instead, scientists operate in paradigms which are maintained because of what they do explain, despite what they do not.

It is interesting to consider Darwin's scientific approach – or lack of it: On the one hand he declared that 'I have steadfastly endeavoured to keep my mind free so as to give up any hypothesis, however much beloved, as soon as facts are shown to be opposed to it' (Darwin, 1950), prompting comments from such as the well-known palaeontologist George Gaylord Simpson (1902–84) that 'Here in the flesh is the ideal and rather rare scientist who is truly dispassionate in search for truth, judicious beyond all emotion. He had no thesis to prove.' But, on the other hand, at the time the *Origin* was published, Darwin wrote to Asa Gray:

> I cannot possibly believe that a false theory would explain so many classes of facts, as I think it certainly does explain. On these grounds I drop my anchor, and believe that the difficulties will slowly disappear. [Darwin, 1950]

And that, I think, is why the theory of evolution is so entrenched today. The theory *does* explain various facts; and for some facts, notably those relating to the operation of natural selection at the morphological level, probably it is the correct explanation. However, what scientists in a paradigm are inclined to do is to give too much weight to what a theory does explain, mistakenly seeing this as proof, and insufficient weight to those facts that are not explained by the theory. But the test of a scientific theory is not those observations that support it, but the significance that should be attached to those that contradict it.

The theory of evolution explains many more facts today than it did in Darwin's time, and this encourages modern evolutionists to follow Darwin's belief in its truth. But there are also many more facts that it does not explain. In particular, some of the facts that are explained by natural selection at the morphological level, are not explained at the more fundamental level of molecular biology. And isn't science supposed to be about explaining at progressively deeper levels?

15

Pride and Prejudice

In Chapter 14 we saw that although there are substantial problems with the theory of evolution a key reason why these are not given their due weight is that evolution is the prevailing paradigm. Evolution is the accepted underlying explanatory thesis of biology, which means it is assumed to be true and it is confidently presumed that answers to the problems will eventually be found. I also commented on various subsidiary factors why the theory of evolution is retained, such as the multidisciplinary nature of science, and its value as a stimulus to scientific research.

However, in addition to those mentioned previously, I suggest there is an overriding reason why evolution remains the prevailing paradigm: it is, quite simply, that there is no viable alternative. With respect to this, it is especially relevant that Kuhn (and others such as Popper) observed that, contrary to popular ideas of how science proceeds, a paradigm is never rejected solely because it is seen to be false – it is rejected only when it can be replaced by an alternative that is considered to account better for the observed facts:

> once it has achieved the status of a paradigm, a scientific theory is declared invalid only if an alternate candidate is available to take its place. No process yet disclosed by the historical study of scientific development at all resembles the methodological stereotype of falsification by direct comparison with nature. The decision to reject one paradigm is always simultaneously the decision to accept another, and the judgement leading to that decision involves the comparison of both paradigms with nature and with each other. [*Structure of Scientific Revolutions*, Ch. 8]

In other words, even when attested facts are clearly in conflict with the paradigm – facts which show it to be false and should lead to its rejection – that alone is insufficient to cause the paradigm to be abandoned. The paradigm will be set aside only if and when a better alternative becomes available. Hence, in keeping with this perception of how science actually proceeds, no matter how deficient the theory of evolution may be, no matter how many aspects of biology it fails to explain, we cannot expect it to be rejected unless there is at least *something* to take its place. Indeed, Kuhn sees a replacement paradigm as an essential requirement of the scientific enterprise:

> To reject one paradigm without simultaneously substituting another is to reject science itself. [*ibid.*]

One reason for this is that science cannot operate in a conceptual vacuum: scientists need some sort of working hypothesis with which to view the world and assess their data. However, I believe it goes deeper than that: the requirement for at least some sort of paradigm is also a reflection of the current world-view that everything must have a scientific or natural explanation. That is, modern science is unwilling to admit that there may be some aspect of nature for which it cannot offer an explanation – and it would have any sort of explanation, no matter how defective it may be, rather than no explanation at all. And that is the important point so far as evolution is concerned. It is not that there is no conceivable alternative to evolution, but there is no viable *natural* alternative. It is recognised that we could revert to a supernatural explanation for life, but that is seen almost universally as a retrograde step, as unscientific, and consequently rejected out of hand by the main body of scientific opinion and, following their lead, in the public perception.

Over the last thousand years science has striven to throw off the old medieval world-view – with its philosophical approach to the universe which it inherited from the Greeks, and which became associated with authoritarianism and religious dogma – and to replace it with an empirical basis. Since the scientific revolution it has had outstanding success in discovering the laws of nature and replacing mythical notions with rational explanations based on the regular operation of those laws. Science has come to explain so much of the natural order that we have come to believe that everything is explicable in scientific terms – it has rejected all supernatural accounts in favour of exclusively natural ones – and this has become the modern world-view. Indeed, some go so far as to say that, ultimately, scientific explanations are the only ones worth having.

So it is that undergirding the paradigm of evolution is the paradigm of naturalism. Only naturalistic explanations are acceptable; anything else is seen as unscientific and therefore unacceptable, untrue, and not worthy of consideration. Consequently, as evolution is the only naturalistic candidate, it is concluded that we must have evolved. Given the major problems at the biochemical level with the theory of evolution, and the discontinuities of the fossil record, I suspect that if a naturalistic alternative could be proposed that offered a credible solution to these problems, even if it meant demoting the role of natural selection, then such an alternative would be enthusiastically embraced. But in the absence of such a candidate it is better to cling to evolution despite its evident deficiencies; not to do so would be to abandon the even more cherished paradigm of naturalism. For example, Julian Huxley wrote:

In any case, if we repudiate creationism, divine or vitalistic guidance, and the extremer forms of orthogenesis, as originators of adaptation, we must (unless we confess total ignorance and abandon for the time any attempts at explanation) invoke natural selection. [*Evolution*, Ch. 8]

In this final chapter I shall consider this naturalistic world-view.

THE MODERN SCIENTIFIC WORLD-VIEW

The rise of Science

We have seen how science began when the classical natural philosophers conceived that the universe is not erratic but rational and follows the consistent operation of natural laws, and they set about discovering those laws. At first their approach was much too philosophical and scarcely scientific at all. They thought that by reason alone (*a priori*) they could ascertain the fundamental principles of the universe, and deduce how the whole of nature must be in the light of those principles. However, this approach is fundamentally flawed – it just is not possible to deduce the laws of nature in this way and, because it is based on one's notions or presuppositions about how the world is or should be, it can lead to conclusions which are far from the truth. This is typified by the Pythagorean belief that nature is based on numbers and geometrical forms from which they assumed that the planets must follow circular orbits, and by Plato's Theory of Forms which led to belief in the fixity of species.

A much more 'scientific' attitude was advocated by Aristotle who emphasized the importance of observation to check conclusions from reason and deduction, and as the basis for elucidating nature's laws by induction. Unfortunately, his works were unavailable to the West until the second millennium AD. Even then, for a long time, scientific progress was retarded by the blinkered scholastic attitude to established authorities, coupled with the dominant position of the church, the close association of secular knowledge with religious beliefs, and the allegorising of nature.

However, in time, an increasing number of scientists, such as Galileo, were committed to the principle that theories about the universe must fit the observed facts, even if that meant rejecting established ideas. Inevitably this led to conflict with the establishment, first in astronomy where, as progressively more observations became available, it was increasingly evident that the old systems based on earth-centred circular motions were struggling to accommodate the facts. Eventually it was realised that the earth orbits the sun rather than vice versa, Kepler showed that the orbits are elliptical rather than circular, and Newton came up with a scientific explanation for why it is this way.

In the first place, the revolution in astronomy was a triumph for empirical science: it showed that the classical philosophical approach of deduction from subjective premises was unsound, and science had to be based on objective facts. It also helped to throw off the medieval view that the venerated philosophers had known it all, and scientists began to realise that there was a great deal of nature still to be explored and understood.

However, as amplified at the end of Chapter 2, it was much more than that: science began to emerge as a coherent discipline, distinct from philosophy and religion, and with an independent voice. Importantly, because science was increasingly based on demonstrable, and often reproducible, facts, and a systematic interpretation of those facts, it came to be seen as a reliable source of knowledge – more reliable than the deductions of philosophers or the revelations of religion. Along with this, scientists were seen as pursuing their work objectively and dispassionately, seeking after truth for its own sake, rather than trying to bolster any particular philosophical or religious position. This perceived objectivity and freedom from bias, especially compared with the dogmatic pronouncements from other quarters, many of which had now been shown to be false, gave science increasing credibility. Consequently, not only did science free itself from the traditional authorities, but itself became an authority: from then on, at least so far as the natural world is concerned, the voice that mattered was that of science.

Its standing has been enhanced by the evident effectiveness of science – not only in accurately describing nature, but also in finding out how it works. From the scientific revolution onwards, science has been very successful in deciphering the laws of nature – in physics, chemistry and biology – and at progressively more fundamental levels. Further, along with this increased scientific understanding of the universe, and arising from it, has come the ability to use nature for our own ends. This has resulted in accelerating technological progress which, as I commented right at the start of this book, has considerable impact on our everyday lives. In modern times, although there are some misgivings about the role of science in our society over issues such as nuclear power, exploitation and pollution of the environment, and genetic engineering, there is no doubt that most recognise the success of science and appreciate much of the material benefits this has brought. I think Bertrand Russell (1961, Introduction) summed up the current world-view when he wrote that 'all *definite* knowledge ... belongs to science'.

The growth of Naturalism

Along with this rise of science, and resulting from its success, not only has there been an increased understanding of the world around us, but also a fundamental change in our perception of it. Many primitive cultures had seen nature as personified or directly controlled by the gods; rather than perceiving it to be governed by laws, it seemed irregular and unpredictable, even capricious. However, as science developed, it began to show that much of the universe conforms to discernible patterns.

This could be seen clearly in astronomy, in the regular movements of the stars and planets, even though it took until Kepler before the correct form of planetary orbits was worked out. More far reaching than Kepler's work, of course, was Newton's which showed that the regular patterns were but the natural consequence of the consistent operation of the force of gravity and the laws of motion. Although Newton had thought God would still need to intervene from time to time to overcome the perturbations due to interactions between the planets, Laplace showed that even this was not necessary. Indeed, it was analysis of such perturbations in the planets' orbits that led in due course to the discoveries of Uranus, Neptune and Pluto. Not only was this further evidence of the efficacy of the scientific method, it also reinforced the concept that nature runs like a machine, following natural laws, without the need for any intervention from outside.

Whilst Newton and most of his contemporaries believed that nature's laws had been set in place and were maintained by God, as more natural laws were discovered and found to apply reliably and widely, any divine contribution was progressively marginalized. This was typified by the Deists who acknowledged the need of a First Cause, but one who thereafter did not interfere at all: day-to-day working of the universe was entirely due to the operation of autonomous natural laws. And Laplace clearly saw his work as supporting a naturalistic view of the universe: when Napoleon asked him where God fitted into his ideas, Laplace replied that he had no need of that hypothesis.

However, for a long time, a challenge to this growing naturalistic perception of the world came from fossils. In the early days, many thought they must have arisen by the action of celestial forces or even been placed in the ground directly by God. Eventually, it was realised that they are petrified remains of living things; but, to account for their occurrence in various distinct strata, over a wide range of depths, and sometimes at high altitude, seemed to require such catastrophic events that they must surely be extraordinary rather than natural. The possible explanation for fossils in terms of the biblical flood also served to reinforce the idea that they were evidence of God's intervention, transcending natural processes.

Ideas such as these were finally overturned by geologists in the early 19th century, especially by Lyell who emphasized that even geological structures were just the result of natural terrestrial processes acting in their normal way, but over immense periods of time. Adopting the uniformitarian principle in geology, accompanied by acceptance of a long age of the earth and surrender of the traditional biblical account, was a major turning point in the rejection of supernatural explanations and the growth of naturalism.

The revolutions in astronomy and geology were key milestones in the rise of a scientific and naturalistic outlook, but were far from being the only areas where science developed. From the scientific revolution onwards, and especially in the 19th century, ground-breaking progress was made in the basic sciences of physics and chemistry. Scientists began to understand the nature of light, magnetism, electricity and the composition of matter; and with this knowledge many of the phenomena which had previously appeared mysterious and baffling (e.g. lightning) were recognised to be merely the outworkings of natural processes. All of this served to reinforce the growing naturalistic outlook; but mystery continued to surround biology.

The importance of biology

Part of the attraction of Darwin's theory of evolution was that it provided an underlying explanatory thesis for biology – one that could account for adaptation and specialisation, the natural groupings of plants and animals, the progression of fossil organisms, and extinctions. But for many scientists, and others, the major importance of evolution was that it provided a natural explanation for biology which for so long had been the ultimate evidence for God's hand in nature, the last bastion of natural theology. It removed the need for a teleological explanation for the 'design' that is apparent in biology. Although the 19th-century biologists knew next to nothing of how biology works, the successive assimilation of small beneficial variations, by virtue of natural selection, seemed to them to be a convincing means by which the complexity of biology could arise progressively – simply through the application of natural processes. And the importance of evolution as a naturalistic account for biology is still recognised today.

> It was of course the great achievement of Darwin to show that there is a possibility of explaining teleology in non-teleological or ordinary causal terms. Darwinism is the best explanation we have. There are not, at the moment, any seriously competing hypotheses. [Popper]

A naturalistic perception of biology has been strengthened in recent decades as we have unravelled the fundamental processes of life. Whereas vitalistic ideas about life persisted well into the 20th century, these have all but disappeared as the molecular basis of biology has emerged. Whether it be routine metabolism or the action of hormones, the operation of sensory receptors or the conduction of nervous impulses, or how all of the necessary genetic information is encoded and decoded – although there is yet much to learn, we can now see how the mystery of life can be accounted for in terms of the operation of molecules, albeit very complex molecules. Life is made up of exactly the same stuff as inanimate matter; and, increasingly, we can see how it works too. This insight has removed much of the mystique of life and reinforced the view that everything – even biology – is reducible to the operation of natural processes.

Naturalism: the supposition of modern science

The success of science, coupled with the fact that so much of nature – especially biology – can now be explained in terms of natural laws, has led most to believe that everything must be explicable in these terms. That is, even when we come across phenomena for which we do not have a natural explanation, the presumption is that there must be one, and it will be found eventually. For example, in Leslie Orgel's review of research into the origin of life, which he portrayed as a detective investigation and in which he highlighted the many substantial difficulties, he concluded that 'we are very far from knowing whodunit. The only certainty is that there will be a rational solution.'

Not only is any sort of supernatural intervention considered to be unscientific, such ideas are also discredited because of their obvious association with religion which has itself been discredited in the eyes of many because of the antiscientific attitudes of some of its adherents. On various occasions in the past (and we have seen some notable examples in the early chapters of this book), a religious body or individual has issued pronouncements regarding the natural world – ranging from the movement of the planets to the formation of geological strata, from the age of the earth to the fixity of species – based on their interpretation of the Bible; and, time and again, such pronouncements have been found to be woefully wrong. Especially since the time of Darwin, religion has acquired a reputation for being reactionary or obscurantist if not directly opposed to science. Even in recent years we find comments such as the following from 'creation scientists':

> "Creation or Evolution?" Now this question is basically an issue of authority – the authority of God versus the authority of the scientists; the authority of the Word of God, the Bible, versus the authority of the words of the scientists in their textbooks. [White, Preface]

So it is not at all surprising that many scientists do not want to be associated with anything that might have religious connotations, which, of course, includes any sort of supernatural explanation for nature.

However, although we now understand so much about biology – indeed, I would say it is because of it – biology remains a challenge to this naturalistic paradigm. In fact, modern molecular biology, far from providing a more certain scientific explanation for biology, has highlighted that something is missing.

DESIGN

Towards the end of Chapter 8 I introduced the idea that the complexity of biology at the molecular level reinforces the perception of design in biology which has been perceived for a long time at the morphological level. It is time to look at this issue more closely. The question of design in nature is usually discussed in the context of evidence for there being a God, and is referred to as the argument from design. However, in the context of this book, more important is the argument *to* design. That is, how do we determine whether or not something has been designed? What distinguishes an object as being an intentional end result, rather than merely the outcome of undirected natural processes?

The argument to design

The first criterion we could consider is whether the object in question displays a specific or distinctive pattern, one that is not likely to arise at random. I rather like the example used by Richard Dawkins in his book *Climbing Mount Improbable* to illustrate this distinction: There is a rocky outcrop in Hawaii which, when viewed from the right angle, is remarkably suggestive of the face of John F. Kennedy. Despite this appearance, no-one thinks it has been intentionally fashioned that way: it is recognised as merely an accidental result of natural weathering, just as we all have seen familiar shapes in rocks, clouds, frosted windows and the like. He contrasts this outcrop with the faces of other US presidents at Mount Rushmore. These have clearly been designed and purposefully sculptured: it is evident because, as Dawkins says, of the number of features that are right in detail – giving a precise representation of the presidents' faces; and, whilst we know that erosion can sculpt rocks into all sorts of interesting shapes, it just is not realistic to think that the faces at Mt. Rushmore, with their close resemblance to specific recognisable people, could have been formed in that way.

Pattern alone can be a fair indication of design, but the case is greatly strengthened if we can also detect that the specific pattern is for a purpose, i.e. that the object, by virtue of its distinctive pattern, can serve a particular function. Purpose can, of course, be artistic or aesthetic rather than utilitarian, such as the carved figures at Mt. Rushmore. However, because aesthetic appreciation is subjective, there is a more reliable case for design if we can identify more of a utilitarian function. For instance, a naturally-occurring flint might be usable as an implement, perhaps as a knife or axe head; but we recognise man-made flint tools because of the detailed shaping, e.g. to fit into the hand or onto a handle, and/or the sharpness of its edge, and because we can see that the specific shape (pattern) is for a definite reason or purpose. The specific shape enables or at least facilitates its function: without that structure it would not be able to fulfil its purpose as well, if at all.

The archetypal example of design is, of course, Paley's watch. Pattern is apparent in the shapes of the individual components (e.g. regularly cogged wheels) and in the arrangement of the components in such a way that their interaction enables regular movements. The shapes of the components are right in many details, and the assembly is right in detail (how many have dismantled a watch or similar only to find they cannot put all the parts back together again?). The function of the whole watch is readily apparent, and this confers function to the components because of their subsidiary role. Hence it is evident that the watch as a whole, and its components, must have been designed.

We should, however, note an important point made by Paley: he did not presume that design was evident automatically, without any knowledge or understanding on our part. On the contrary, whilst to most people of Paley's time and culture (and ours), even if uneducated, it would have seemed obvious that the watch must have been designed, Paley recognised that, in fact, the conclusion is based on some measure of knowledge about the natural world and the watch. To conclude that the watch must have been designed, we need to know something of what natural processes can and cannot do, and something of the workings of a watch. Indeed he recognised that investigation may be required to discern the construction and function of the watch – to determine what distinguishes it from a naturally-occurring object (*Natural Theology*, Ch. 1).

'Design' in biology

Pattern and purpose

When it comes to biology, its specific patterns – from the plumage of a peacock's tail to the form of a flower – have long been recognised, as has the functionality of biological structures. Aristotle concluded from the adaptation of structure to function that animals must have been designed, and many after him such as Galen and Ray independently came to the same conclusion. As early as the 17th century Robert Hooke extended this perception of design in biology to the microscopic level – such as the barbules of feathers and the feet of insects. And the revival of microscopy in the 19th century led to further discoveries on an even smaller scale with, for example, the mechanism of mitosis to effect accurate segregation of genetic material between daughter cells.

Then, in the second half of the 20th century, with the development of electron microscopy, we began to see something of the internal organization of cells – the form and function of subcellular structures – getting down to the arrangement of individual molecules. One of the structures revealed quite early on was the sarcomere where, as described in Chapter 8, filaments of actin and myosin are intermeshed to form the basic unit of contractile tissue, as used in muscles. A recent example is the bacterial flagellum with its exquisite tiny motor – complete with rotor, stator and bearings – which surprised us all. In many examples such as these the interrelation of structure and function are unmistakable in biology at the molecular level: it is evident that there are many instances where the specific arrangement of individual molecules is essential for function.

Finally, not only is design apparent in the way biological macromolecules are pieced together to produce functioning biological units, but also in the structure of the macromolecules themselves. Here, as at the morphological level, we see clearly that pattern and purpose are intimately interlinked. One of the points I stressed in Chapter 7 is that, although the sequence of amino acids in a protein may, superficially, appear random, in fact it is anything but. The pattern may not need to be 100% specific, but it is not far off; just as some latitude, but not much, would be permissible in the shape or size of the components in a watch. In much the same way, the base sequences of tRNAs must be right in order for them to function correctly in their role of implementing the genetic code.

Design or apparent design?

So how do we account for 'design' in biology? Crucially, is it real or only apparent? The answer hinges on whether or not the form and function can realistically be accounted for by natural processes alone.

For most natural philosophers (but noting exceptions such as the Epicurians) and biologists before the 19th century, plants and animals were seen as so different from inanimate objects, having marvellous powers of growth, reproduction etc., that it seemed they could not possibly have arisen naturally. Hence, the purposeful structure and evident function of biology were accepted as a strong, even compelling, case for a designer: it was thought they must be the product of an intentional process, carried out with their end purpose in view; that is, a teleological explanation, implying a creator.

This interpretation was, of course, advocated by such as Ray and Paley. However what they did not realise sufficiently (Paley acknowledged it to some extent) was that in their day not enough was known about the fundamental workings of biology for their conclusion – that organisms must have been designed – to be reliable. Their knowledge was based almost exclusively on large scale morphological features, and they presumed that when more detailed information became available it would support their case. On the one hand they knew nothing of biology at the submicroscopic level, and on the other of biological mechanisms, such as variability and natural selection, which together might mould organisms in an adaptive manner.

The teleological interpretation of design in biology lasted for a long time but was eventually brought into question by Darwin. No longer was it necessary to believe that the final forms of organisms must have been preconceived, and that formative processes were directed to produce them. Complexity and adaptation could arise opportunistically provided they could be reached by a series of steps, each of which offered at least some advantage over its predecessor, because each step could then be selected naturally, and become a stepping stone to the next. The abundance of variations in nature, and the ease with which domestic breeders could select so many different features in their breeding programmes, persuaded Darwin that appropriate variations would always be available in nature to fuel this evolutionary process. After Darwin, the appearance of design in nature was seen to be just that – 'apparent design', not actually the product of intelligent conception.

This evolutionary account was reinforced with the discovery of genes and genetic mechanisms which could explain the source of variations, and the operation of natural selection in terms of changes in gene frequencies. And with the establishment of Neo-Darwinism, natural selection became the accepted explanation for all complexity and adaptation in biology. So much so that when the complexity of molecular biology began to emerge it was assumed that natural selection could account for 'apparent design' at this level too.

However – and this was the key point of Chapters 7 and 8 – although natural selection is widely considered to be the explanation for how the primary sequences of proteins and nucleic acids have arisen, and is taught as such in evolutionary text books, in fact it cannot work because so many amino acids or nucleotides need to be right before it can have any utility. There is no credible answer to this objection (in fact it is compounded by the interdependence of biological macromolecules), and there is no evidence that constructive evolution of macromolecules has actually happened.

Paley recognised that to assess whether an object is designed or not we need to know something of its construction and working. It is only with molecular biology that we have gained the necessary level of information for us to do this for biology. Along with this we also need to investigate what natural processes might occur and what they might be able to achieve; and in this context it was valuable to recognise the process of natural selection and appropriate to consider whether (in conjunction with some means such as mutation to vary the sequences) it could account for the formation of functioning biological macromolecules. However, in the light of information now available to us, an objective assessment should conclude that a natural selection sort of process cannot account for the formation of biological macromolecules. Consequently, from their specific pattern (the number of details which need to be right) and evident function there is a *prima facie* case of 'design'; and, despite extensive investigation into possible mechanisms, we do not know of any natural process which can account for them. Further, as I elaborated at the end of Chapter 8, whilst an argument for design based on morphology might be subjective – we just can't believe that such exquisite form and function could not be designed – the argument for design based on macromolecules is substantially objective.

Perhaps it is instructive to consider macromolecules in the light of Aristotle's four explanatory causes mentioned in Chapter 1. It is easy to see that their material cause is the constituent atoms (or we could take it to a lower level of subatomic particles if we wished), and that their formal cause is proteins or genes. Similarly, it is evident that they have a final cause or purpose in terms of their roles in biological systems – to code for a protein or tRNA, catalyse a reaction, or build a tissue. But what of an efficient cause? Of course we can say that they were made by other macromolecules, and so on...; but such a 'chicken and egg' explanation is clearly no longer adequate (just as we are no longer satisfied that the efficient cause for an organism is its parents). The only natural (or scientific) explanations that have been offered to account for macromolecules are a freak highly improbable event (which is hardly scientific, and very few advocate this) or a progressive selective process, which is generally accepted but I have explained at length why it is not viable. In other words, we cannot identify a competent efficient cause for biological macromolecules, and that is why they are at the core of the argument about design. Biological macromolecules, in themselves, present a case of design for which we do not have a natural or scientific explanation; they point clearly to there having been a purposeful designer. We can no longer evade or fudge the issue by referring to them as 'apparent design', 'designoid' or whatever. This is important because it is generally accepted that evolution provides an efficient casual explanation for biology; but we can see that at the fundamental molecular level it does not.

Further, once we recognise that we do not have an evolutionary or other natural explanation for biological macromolecules, there are direct implications for the concept of 'design' at the morphological level. As mentioned above, the old-fashioned, morphology-based, argument from design was undermined when it became evident that variations do arise in biology, that advantageous ones can be and are selected naturally, and hence that progressive change, including adaptation, is possible. In other words, evolution offered a powerful alternative explanation. However, we now know that those variations arise by the recombination of genes; which means the variations are entirely dependent on the existence of genes. But we have no satisfactory natural explanation for the origination of the genes. Put another way, the information to produce (potentially) useful variations resides in genes, but we have no adequate explanation for how that information – meaningful nucleotide sequences (and the decoding mechanisms) – arose. Consequently, the evolutionary response to the morphology-based design argument fails at a deeper level of explanation. That is, although apparent design at the morphological level can be explained in terms of selecting variations, we can now see that this is just putting the argument back a step. Because those variations depend on entities for which we have no explanation, it is no explanation at all.

The implications of design, or the argument from design

My prime objective here is to challenge the widespread belief that evolution provides a satisfactory (sufficient) explanation for design in biology. What we conclude from that is more in the realm of metaphysics – philosophy and theology – than science, so my comments here will be brief.

For something like a watch, no one seriously questions that its pieces must have been deliberately manufactured and assembled, the product of preconception and intelligent design, intention and purposeful action. Any other sort of explanation is not credible: it could not have 'just happened', the mere consequence of undirected natural processes. And it necessarily follows that there must have been a watchmaker.

Where biology is concerned, although most will adopt a similar conclusion – that there must have been a designer and maker – the situation is not quite so clear cut.

For example, some object that to posit a designer to account for the complexity of biology is no explanation at all because presumably the designer would need to be even more complex; so how do we explain the origin of that? But it seems to me that such a response is merely an aspect of the naturalistic way of thinking – insisting that for everything there must be a natural explanation – simply denying in principle or *a priori* (and regardless of any contrary evidence) that there could be an uncaused First Cause. Those with an exclusively naturalistic outlook will believe that there must be a natural explanation, that the argument to design must be wrong, even if they cannot see where it fails. Whereas for those who admit the possibility of a Creator it is acceptable that the Creator's existence might be evident in what has been created. On the other hand, I would accept that design need not have been by direct action of a supreme being but could have been mediated: perhaps through intermediate beings, as advocated by e.g. Fred Hoyle.

Which brings me to the important point that implementing design may not necessarily mean intervening in or suspending the laws of nature. Many have argued for design being implemented more subtly than that, e.g. through the indeterminacy of nature at the level of quantum mechanics. Readers interested in pursing issues such as these may find *Evolutionary and molecular biology: scientific perspectives on divine action* (Russell *et al.*) a useful starting point, even though most if not all of its authors accept evolution as a causal explanation for biology.

OBJECTIONS TO DESIGN

Hume's counter-argument

In debate relating to the argument from design it is often said that the argument was refuted by David Hume, one of the prominent figures of the Enlightenment, notably in his *Dialogues Concerning Natural Religion*, published posthumously in 1779. However, in the first place, an examination of his counter-argument shows that it is undermined in the light of modern biological knowledge; and, second, the conclusion he actually came to is all too frequently overstated or misrepresented. So it worth taking a little space to clarify these points.

The argument from design that Hume addressed runs as follows: Objects which we know are the product of human design and manufacture, such as watches and houses, clearly display pattern and purpose; and (based on our knowledge of these objects and of natural processes) it is inconceivable that articles such as these could arise other than by design and deliberate action.[1] We also see pattern and purpose (including adaptation of structure to function) on the small and large scale in the world around us. So, by applying the well-established analogy that we expect similar effects to result from similar causes, then, just as watches and houses must have had a designer and maker, we should infer that this is

---

[1] We recognise, of course, that some fabricated, even 'designed', objects do not have a purpose, and may not even display a pattern that is in any way out of the ordinary; but that is aside from the main issue.

also true of the universe.

Explanatory 'principles'

Whilst Hume accepts an analogy-based argument in principle, he questions how far it is valid, indeed if it is valid at all, to apply it in this case. In particular, being limited to an 18th-century understanding of nature, he suggests that pattern might result from other causes apart from intelligent design or 'reason', namely instinct, animal generation, vegetation or simply the self-organizing properties of matter itself (*Dialogues*, Part 7). And, he argues, because the world about us exhibits properties such as change and regeneration, similar to the properties of plants and animals, then perhaps it would be equally valid, or even more so, to compare the world with a plant or animal rather than a man-made machine. If that is the case, then the analogy to man-designed artefacts is weakened or fails completely.

However, even Paley, writing within 30 years of the *Dialogues* and without any more insight into the workings of biology than Hume, criticised the use of 'generation' or similar as an explanatory factor:

> The minds of most men are fond of what they call a *principle*, and of the appearance of simplicity, in accounting for phenomena. Yet this principle, this simplicity, resides merely in the *name*; which name, after all, comprises, perhaps, under it, a diversified, multifarious, or progressive operation, distinguishable into parts. The power in organised bodies [plants and animals], of producing bodies like themselves, is one of these principles. Give a philosopher this, and he can get on. But he does not reflect, what this mode of production, this principle (if such he choose to call it) requires; how much it presupposes; what an apparatus or instruments, some of which are strictly mechanical, is necessary to its success; what a train it includes of operations and changes, one succeeding another, one related to another, one ministering to another; all advancing, by intermediate, and, frequently, by sensible steps, to their ultimate result! Yet, because the whole of this complicated action is wrapped up in a single term, *generation*, we are to set it down as an elementary principle, and to suppose, that when we have resolved the things which we see into this principle, we have sufficiently accounted for their origin, without the necessity of a designing, intelligent Creator. The truth is, generation is not a principle, but a *process*. [*Natural Theology*, Ch. 23, emphases in original]

Modern biology has substantiated Paley's view that plant or animal reproduction or generation is indeed a process, comprising many discrete and complex steps, and that it is quite inadequate to explain such phenomena in terms of some nebulous principle of 'generation'. Still less are we now happy to accept as an explanatory principle for biology the self-organizing capacity of matter.

Further, what modern biology, especially in the last few decades, has clearly shown is that all of biology at the molecular level bears a striking resemblance to man-made devices. Whether it be plant or animal growth or reproduction, ultimately we now see all of this as being reducible to the operation of molecular machines. Not that we yet know it all – far from it – but from what we do now know it is certainly our expectation that it is all reducible in this way, and that in time we will be able to explain biology fully in these terms. This would include instinct and memory which we expect to be reducible to neuronal patterns based on biochemical features. Even if mind is a higher-level phenomenon – such that it is inappropriate to think of it being entirely reducible in this way – we expect its workings to be based on low level biochemical operations involving molecular machines.

So, in the light of modern biological knowledge, Hume's suggestion that it might be more appropriate to compare the world with a plant or animal completely loses its force. Instead, the analogy between nature and man-made machines is greatly strengthened. Although Hume, at best, accepted that there may be a remote analogy between the natural world and human artefacts, we should now conclude that the analogy is a great deal less remote than Hume supposed; in fact, especially if we think of recent advances in nanotechnology – the production of exceedingly tiny machines – it is a very close analogy.

Design in the universe

A second point relating to Hume's treatment of the argument from design is that he was considering design in the universe in general, or as a whole, rather than in specific objects such as biological organisms. Hence, in that context, examples of perceived *disorder* must be set against the examples of order (design) so as to gain an overall view of how ordered or 'designed' the universe appears to be. In Hume's view it was doubtful whether overall the universe displayed design rather than not (but, as we have seen, he did not credit much 'design' to living things).

Paley on the other hand (although he is sometimes misrepresented as following Hume's argument of overall design) saw each and every example of design in biology as a sufficient case. This is because his argument was based on the fact that everything must have an efficient cause; so, just one object for which there can unequivocally be no natural explanation is a sound case for a supernatural cause.

> Were there no other example in the world of contrivance [one of Paley's words for design], except that of the eye, it would be alone sufficient to support the conclusion which we draw from it, as to the necessity of an intelligent Creator. It could never be got rid of; because it could not be accounted for by any other supposition, which did not contradict all the principles we possess of knowledge; [...] even if other parts of nature presented nothing to our examination but disorder and confusion, the validity of this example would remain the same. If there were but one watch in the world, it would not be less certain that it had a maker. [*ibid.*, Ch. 6]

As indicated previously, given the state of biological knowledge in Paley's time, Paley could have been right or wrong as to whether, at bottom, organisms could be accounted for naturally. It might have turned out that the natural outworkings of physics and chemistry could produce the complexity of biology. Equally, Darwin could have been right or wrong about 'apparent design' being the result of accumulated variations – it depended on the fundamental cause(s) of variations. Molecular biology has shown that Paley's assessment, although somewhat subjective and, at the time, an unsubstantiated extrapolation from what he saw at the morphological level, was in fact right. We now know that all living organisms and, indeed, many of their parts considered independently, are unequivocal examples of design without a viable natural explanation.

The attributes of a designer

Finally, so far as Hume is concerned, the whole context of his *Dialogues* is not to question the existence of God but to debate what we can infer about God's attributes from what we know of the natural world. For example, one of his characters, Philo, says 'But surely, where reasonable men treat these subjects, the question can never be concerning the Being, but only the Nature, of the Deity' (Part 1). In line with this, a point Hume often makes is that, because of the constraints of an argument based on analogy, we cannot reliably infer infinite attributes for God from our limited experience of a finite world. In particular, he says, even if the argument from design is valid, we can infer from it only those qualities necessary for a Being to be able to fashion the universe, concluding 'That the cause or causes of order in the universe probably bear some remote analogy to human intelligence' (Part 12), – we cannot extrapolate even that He is omniscient or omnipotent, still less any moral attributes. Paley also recognised that 'Natural Theology' – what we can learn about God from nature – is limited. However, what we need to recognise is that, just because we cannot conclude from nature alone that the designer of nature necessarily has all of the attributes traditionally ascribed to God, this in no way weakens the argument from the design evident in nature that there must have been a Designer.

'Design' and modern science

The principal criticism levelled in modern times against teleological explanations or any sort of suggestion of a designer for biology is that it is ascientific, unscientific or even anti-scientific. That is, at best it is considered to have nothing at all to do with science, and at worst to have a negative impact on science.

Design is ascientific

The most common view is simply that any sort of supposed supernatural occurrence is clearly not susceptible to scientific enquiry, so science should have nothing to do with it. A typical expression of this is the following by Niles Eldredge:

> Adaptation is the very heart and soul of evolution. It is *the* scientific ac-
> count of why the living world comes in so many shapes and sizes: how the
> giraffe got its long neck, why porpoises look so much like sharks and the
> extinct ichthyosaurs, how birds fly, and literally millions of similar ques-
> tions. The only other account of this spectacular display of diversity is the
> creationist tale: that a supernatural Creator fashioned the world, including
> its organic contents, the way we find it. But that form of explanation, by its
> very nature, lies outside the bounds of the scientific enterprise.
> [*Reinventing Darwin*, Ch. 2, emphasis in original]

In keeping with this, most scientists just ignore any sort of possible supernatural account as simply not within their sphere of activity. And I think it
is quite satisfactory for a scientist, as a scientist, to do so, even perhaps
should do so. That is, in the course of their work it is appropriate that scientists neglect the possibility of any sort of supernatural explanation, and
search exclusively for natural causes – what is sometimes referred to as
methodological naturalism.

However, for most modern day scientists it does not stop at that. The
pervasive belief is that anything not susceptible to scientific enquiry is not
only outside of science, and not to be considered from a scientific point of
view, but implicitly has no part in reality and should not be taken seriously
from any point of view. That is, such is the confidence in the power of science and in an exclusively naturalistic outlook that anything that would lie
outside of this world-view must be spurious and should, therefore, not
merely be ignored, but categorically rejected. Julian Huxley, for example,
was particularly keen to outlaw any sort of explanation that was not
strictly 'scientific'.

> How has adaptation been brought about? Modern science must rule out
> creation or divine guidance. It cannot well avoid frowning upon entelechies
> and purposive vital urges. [*Evolution*, Ch. 8]

Indeed, regrettably, some writers maintain that the naturalistic, scientific
point of view is the only one that is intellectually respectable, and even
ridicule those who do not conform to their way of thinking.

But, clearly, what needs to be recognised is that this decision to reject
anything not susceptible to scientific enquiry is based on nothing more
than the modern supposition that everything must be explicable in naturalistic terms. Unfortunately, as with the paradigm of evolution, because
naturalism is the prevailing paradigm, it is taught as wholly true; so few
realise that it is but supposition, and most think and act as if it were certainly true, never questioning its validity.

'Design' inhibits scientific inquiry

From a scientific point of view, there is concern that admitting the slightest possibility of a non-natural explanation will have a negative impact on
the progress of science. The objective of modern scientists is to explain –

to find out how things work and how they have arisen, and motivation to pursue this objective comes from the conviction that a natural explanation, one that is completely comprehensible to science, exists and can be found. This firm belief that there must be a natural explanation means that when such an explanation resists elucidation, rather than discouraging scientists in their efforts, it is more likely to spur them on to greater effort to find that explanation. And undoubtedly this has led to substantial scientific discoveries.

Francis Bacon, in contrast to many of his predecessors who had advanced teleological explanations for the natural world, argued that such an outlook merely leads to verbal disputes and inhibits real scientific progress. Similarly, Lyell addressed this issue in his *Principles of Geology*: in the context of arguing for a natural uniformitarian explanation for geological structures, rather than any sort of extraordinary event, he commented that allowing the possibility of supernatural intervention would 'foster indolence and blunt the keen edge of curiosity'.

The concern is that allowing non-natural explanations is closing rather than opening up an area of nature for scientific study. So it is not surprising that some feel that admitting any possibility of design or creation is not only unscientific, but somewhat anti-science. However, whilst I accept these are legitimate concerns, some comment should be made in response to them.

First, although it is generally supposed that any non-natural cause must by definition be outside of science, I suggest that admitting the possibility of such causes is in fact consistent with the scientific approach in its widest sense. When scientists come across observations which cannot be accounted for by current scientific theories, they will explore various alternative hypotheses and may, as a result, discover a completely new type of phenomenon compared with preceding knowledge. Such was the case when Einstein developed his theory of relativity to account for various observations that were not consistent with Newton's mechanics. This illustrates that scientists should be open to the possibility that an explanation lies outside their current understanding; and probably most are. However that 'openness' is usually limited to exclusively natural phenomena. The reason for this is that natural phenomena have been found to explain so much that almost all scientists expect everything to be explicable in such terms. But that is not a reliable argument. A conclusion from induction should be open to scrutiny: it should not take precedence over observation but be subject to revision in the light of observation. A more scientific approach would be to have an outlook that is open to the unexpected.

Scientists should take Lyell's warning to heart and not make a teleological explanation an easy option, but, in the interests of good science, should search as hard as possible for a natural explanation (methodological naturalism). However, a good scientist also recognises that there are, or at least may be, limits to science, and we cannot safely assume that there are no other kinds of explanation. William Whewell grappled with this issue and concluded that, although scientists should exclude final causes in the course of their investigations, they were perfectly legitimate as the results of such investigation.

My second response is that even if the concept of teleology or design does not sit comfortably with scientists, even if admitting the possibility of supernatural explanations does have some negative impact on scientific investigations, these are not valid reasons for assuming such concepts are false. Most scientists would, I think, claim they are searching after a true account of the universe. The fact that some aspect of what we uncover might not be as we would wish it, is not a valid reason for rejecting it.

Today's scientists are in danger of following a similar error to that of the classical natural philosophers. Based on some knowledge of the world and their predilections, they decided what the basic principles of the world ought to be and deduced what nature should be like in the light of those principles. Unfortunately, all too often, they based their principles on inadequate information, so they were incorrect, and this led them to erroneous conclusions about the world. Their error was compounded because they were unwilling to submit their deductions to scrutiny – to compare conclusions with the facts of nature. Whilst the paradigm of naturalism is based on some facts – that much of nature can be explained in exclusively naturalistic terms – scientists have extrapolated from these facts the wider principle that everything must be explicable in these terms, and there is a profound reluctance to allow any observations to challenge that presumption. Some, for totally non-scientific reasons, want the world to be completely explicable in naturalistic terms; and are determined to maintain this view whatever the contrary evidence.

God of the gaps

Perhaps the most serious criticism that might be made of an argument from design based on the absence of a natural efficient cause (rather than based on overall order in the universe) is that it is merely postulating a supernatural cause to bridge a gap in our current knowledge of the world. That is, it displays a 'God of the gaps' mentality, which surely has been discredited by the advances of science.

As mentioned above when outlining the growth of naturalism, primitive cultures knew so little about nature that, it seemed, the gods must have had a hand in just about everything to keep it all going. But, as science uncovered the laws of nature, increasingly the world could be seen as a machine which simply follows natural laws, and in time the role for a God in nature was marginalized to the point of irrelevance. Not only was Newton instrumental in effecting this transformation, he also illustrates the change in attitude that was taking place. Although he discovered that the planetary orbits are elliptical as a natural consequence of the laws of gravity and motion, he felt that God would be required to prevent the system becoming unstable. Not very long after, this God-filled gap in his understanding was removed when Laplace showed that the perturbations in the orbits would sort themselves out naturally. Subsequently, as we have determined the laws of chemistry and physics – extending from quantum mechanics to relativity – most natural phenomena can now be seen as merely the outworkings of such laws.

Even though we now see most phenomena as just the operation of natural laws, many present day scientists, like Newton, maintain that there is clear evidence for a God in the overall design of the universe. This is evident in its rationality, in the institution of natural laws and, for modern-day physicists, in the assignation of appropriate values to various fundamental physical constants.

However, whilst the case for a Designer behind the universe as a whole is generally considered to be legitimate and is quite widely held, it is felt by most that we should no longer invoke a God merely to account for gaps in our current knowledge of how phenomena might be explained in terms of the operation of natural laws. This includes many of those who hold traditional religious views, who feel that a God of the gaps is, inevitably, a shrinking God, and they want no part of it. The widespread view is that, although there are still phenomena for which we do not have an adequate natural explanation, in view of the substantial and continuing scientific progress, we should expect that all these gaps will eventually be filled.

It is generally believed that this also applies to biology. Up to the 19th century, living things seemed so different from inanimate material that few doubted they must have had a supernatural origin. That view was challenged first by Darwin with the concept of natural selection and progressive development, then by the discovery of genetic mechanisms, and further undermined by modern molecular biology which has shown how the wonder of life is brought about through normal physical and chemical processes. To some, this knowledge of molecular biology, which has consigned vitalism to the history books, is convincing evidence that even in biology there is no longer a gap to be bridged by God. No longer do we need any sort of supernatural animating force, but – like everything else – biology is based on the ordinary operation of physics and chemistry.

However, although we can now describe much of how biology works – and we may confidently expect that most, perhaps all, of biology will be describable in molecular terms in due course – this knowledge of fundamental biological mechanisms does not provide any explanation whatever of how those mechanisms have arisen. In fact, it accentuates the problem, because the more of molecular biology that we uncover – whether it be how a fertilized egg develops into a mature individual, or how memory or mind works – the more do we realise how complex biology is. We do not have a credible natural explanation for even the simplest of biological macromolecules; and, as I have amply described previously, the problem is greatly compounded by the interactions of macromolecules – by their interdependence on which biological systems depend.

For example, whereas for Paley it was a mystery how muscles contract, we now know a great deal of how this is achieved. We are beginning to learn something of what determines that a cell will develop into muscle tissue, but currently know little of how synthesis of the relevant proteins is regulated, and nothing of how the proteins are assembled into a sarcomere. These latter features are still gaps in our knowledge, but we can be confident that, though it may take many years yet, these gaps will be filled; and I am not suggesting for a moment that we invoke a vitalistic watchmaker. But we can be equally confident that, when we do know how these things are done, we will find that they involve many genes and we will have no credible natural explanation for how the information content of those genes has arisen.

The popular perception is that increased scientific knowledge inevitably closes the gaps in our understanding, and progressively removes any need for non-natural explanations. However, biology is a clear example – perhaps it is the only example – where our increased knowledge has served to widen the gap rather than close it. Success in uncovering the molecular mechanisms of biology has misled scientists into thinking that they are explaining biology fully; but it does not explain how biology has arisen. In fact, the more of molecular biology we uncover, the stronger does the case for design become.

It would be rash, not learning anything from the history of science, to rule out the possibility of ever having a natural explanation for design in biology. Perhaps tomorrow someone will come up with a viable naturalistic account for the complexity and evident design of molecular biology. But, equally, it is nothing but blinkered naturalism to insist that one day we will find such an explanation. And, at present, the evidence is against it – the more we know of biology, the more formidable do the gaps appear.

I trust it will be apparent from the foregoing paragraphs that I am certainly not suggesting scientific research cease in a particular area in favour of a supernatural or teleological explanation. All I am challenging is the presumption that there *must* be a natural explanation. And, if persistent searching fails to come up with a natural explanation, it is entirely reasonable – and consistent with science – to leave open the possibility of a supernatural one. As Thomas Huxley (1860) wrote to Charles Kingsley:

> Sit down before fact as a little child, be prepared to give up every preconceived notion, follow humbly wherever and to whatever abysses nature leads, or you shall learn nothing.

Appendix 1. Stratigraphical column

Era	Period	Epoch	MY ago	First appearance
Cenozoic	Quaternary	Recent		
			0.01	
		Pleistocene		
			1.7	
		Pliocene		man
			5	
		Miocene		
			24	
	Tertiary	Oligocene		
			37	
		Eocene		
			58	
		Palaeocene		modern mammals
			67	
Mesozoic	Cretaceous			marsupials and placentals flowering plants
			140	
	Jurassic			birds modern amphibians
			245	
	Triassic			mammals dinosaurs
			285	
Palaeozoic	Permian			
			320	
	Carboniferous	Pennsylvanian		winged insects reptiles
			325	
		Mississippian		
			360	
	Devonian			bryophytes, gymnosperms insects amphibians bony fish
			410	
	Silurian			vascular plants land plants and animals jawed fish
			440	
	Ordovician			
			500	
	Cambrian			jawless fish invertebrates with exoskeletons
			540	
	Precambrian			invertebrates

Appendix 2. Classification of animals

This classification is very selective, substantially limited to taxonomic groups mentioned in the main text.

Phylum	Subphylum	Class	Subclass	Order	Suborder	Family	Genus
Arthropoda							
		Arachnida (e.g. spiders)					
		Crustacea					
		Hexapoda (insects)					
			Apterygota (wingless)				
				Collembola (springtails)			
				Thysanura (bristletails)			
			Pterygota (winged)				
				Coleoptera (beetles)			
				Dictyoptera (e.g. cockroaches)			
				Diptera (true flies)			
							Drosophila
				Ephemeroptera (mayflies)			
				Lepidoptera (butterflies and moths)			
				Odonata (dragonflies)			
				Orthoptera (grasshoppers)			
		Trilobita					
Brachiopoda (lampshells)							
Chordata							
	Cephalochordata (e.g. amphioxus)						
	Vertebrata						
		Acanthodii (spiny fish)					
		Agnatha (jawless fish)					
				Myxinida (hagfish)			
				Petromyzontida (lamprey)			
			Ostracodermi[a] (early agnathans)				
		Amphibia					
			Labyrinthodontia				
				Anthracosauria			
					Seymouriamorpha		
				Ichthyostegalia			
							Ichthyostega
			Lepospondyli				
				Aïstopoda			
				Microsauria			
			Lissamphibia				
				Anura (frogs and toads))			
				Apoda (caecilians)			
				Urodela (newts and salamanders)			
			Temnospondyli				
		Aves (birds)					
			Archaeornithes				
							Archaeopteryx
			Ornithurae (includes orders of modern birds)				
		Chondrichthyes (cartilaginous fish, e.g. sharks)					

---

[a] Now divided into several separate orders.

Phylum	Subphylum	Class	Subclass	Order	Suborder	Family	Genus
Chordata	Vertebrata						
		Mammalia					
			Eutheria (placentals)				
				Carnivora			
					Fissipedia (land-living)		
						Canidae (dogs)	
						Ursidae (bears)	
						Felidae (cats)	
					Pinnipedia (marine e.g. seals)		
				Cetacea (whales and dolphins)			
				Chiroptera (bats)			
				Perissodactyla (odd-toed hoofed animals)			
					Equoidea (horses)		
							Equus
							Hyracotherium
							(and others)
				Primates			
				Rodentia			
			Metatheria (marsupials)				
			Prototheria				
				Monotremata			
				Morganucodontia			
				Multituberculata			
		Osteichthyes (bony fish)					
			Actinopterygii (ray-finned fish)				
				Teleostei (modern ray-finned fish)			
			Sarcopterygii				
				Crossopterygii (lobe-finned fish)			
				Dipnoi (lung fish)			
		Placodermi (early jawed fish)					
				Arthrodira			
		Reptilia					
			Anapsida				
				Captorhinida			
					Captorhinomorpha		
				Chelonia (turtles and tortoises)			
				Mesosauria			
			Diapsida				
				Ichthyosauria			
				Ornithischia (bird-hipped dinosaurs)			
				Pterosauria			
				Saurischia (lizard-hipped dinosaurs)			
					Theropoda		
			Synapsida				
				Pelycosauria			
				Therapsida			
					Cynodontia		
					Theriodontia		
Echinodermata							
			Asteroidea (starfish)				
			Echinoidea (sea urchins)				
Mollusca							
			Bivalvia				
			Cephalopoda (inc ammonites)				
			Gastropoda				
Porifera (sponges)							
			Hexactinellida				

[about 25 more animal phyla]

References

Allard M W, Honeycutt R L and Novacek M J, 1999, Advances in higher level mammalian relationships, *Cladistics* 15:213–219.

Baba M, Darga L, Goodman M and Czelusniak J, 1981, Evolution of cytochrome *c* investigated by the maximum parsimony method, *Journal of Molecular Evolution* 17:197–213.

Behe M J, 1996, *Darwin's Black Box*, The Free Press.

Bendall D S (ed), *Evolution from molecules to men*, Cambridge University Press.

Bjerring H C, 1998, The fates of spiracular allostoses in mammals, *Acta Zoologica (Stockholm)* 79:51–67.

Bock W J, 2000, Explanatory history of the origin of feathers, *American Zoologist* 40(4):478–485.

Bowler P, 1992, *The Fontana history of the environmental sciences*, Fontana Press.

Brush A H, 1996, On the origin of feathers, *Journal of Evolutionary Biology* 9:131–42.

Brush A H, 2000, Evolving a protofeather and feather diversity, *American Zoologist* 40(4):631–9.

Buffon, 1780, *Natural History*; English translation of Volumes 1–9 by William Smellie.

Burnet T, 1684, *A Sacred Theory of the Earth*.

Carroll R, 1999, Homology among divergent Paleozoic tetrapod clades; in Hall *et al.*

Colbert E H, Morales M and Minkoff E, 2001, *Colbert's Evolution of the vertebrates*, 5th edn, Wiley-Liss.

Crane P, Friis E M and Pedersen K R, 1995, The origin and early diversification of angiosperms, *Nature* 374:27–33.

Darwin C, 1859, *On the Origin of Species*, Murray.

Darwin C, 1868, *Variation in Animals and Plants under Domestication*, Murray.

Darwin C, 1950, *Charles Darwin's Autobiography, with his notes and letters depicting the growth of the Origin of Species*, ed. Francis Darwin, Henry Schuman.

Darwin C, 1959, *The Autobiography of Charles Darwin 1809–1882 with Original Omissions Restored*, ed. Nora Barlow, Harcourt Brace.

Dawkins R, 1986, *The Blind Watchmaker*, Penguin Books.

Dawkins R, 1996, *Climbing Mount Improbable*, Viking.

de Beer G, 1958, *Embryos and Ancestors*, 3rd edn, Oxford University Press.

de Beer G, 1971, *Homology: an unsolved problem*, Oxford University Press.

de Beer G, 1975, *The evolution of flying and flightless birds*, Oxford University Press.

Dickerson R and Geis I, 1983, *Hemoglobin: structure, function, evolution and pathology*, Benjamin/Cummings Publishing Co.

Dobzhansky T, 1977, Populations, races, subspecies; in Dobzhansky T, Ayala F, Stebbins G L and Valentine J, *Evolution*, Freeman and Company.

Eigen M, 1983, Self-replication and molecular evolution; in Bendall.

Eldredge N, 1995, *Reinventing Darwin*, Weidenfeld and Nicholson.

Falconer D S, 1989, *Introduction to quantitative genetics*, 3rd edn, Longman.

Feduccia J A, 1996, *The origin and evolution of birds*, Yale University Press.

Fisher R A, 1930, *The genetical theory of natural selection*, Clarendon/Oxford.

Foote M, Hunter J P, Janis C M and Sepkoski J J Jr, 1999, Evolutionary and pre-servational constraints on origins of biologic groups: Divergence times of eutherian mammals, *Science* 283:1310–14,

Forsten A, 1989, Horse diversity through the ages, *Biological Reviews of the Cambridge Philosophical Society* 64:279–304

Fothergill-Gilmore L, 1986, Domains of glycolytic enzymes; in Hardie D G and Coggins J R (eds), *Multidomain proteins – structure and evolution*, Elsevier Science Publishers.

ffrench-Constant R H, 1998, Why are there so few resistance-associated mutations in insecticide target genes?, *Philosophical Transactions of the Royal Society of London, Series B* 353:1685–93.

Geist N R and Feduccia A, 2000, Gravity-defying behaviors: Identifying models for protoaves, *American Zoologist* 40(4): 664-675.

Goodwin B C, Holder N, Wylie C C (eds), 1983, *Development and evolution*, Sixth Symposium of the British Society for Developmental Biology, Cambridge University Press.

Grimaldi D, 1999, The co-radiations of pollinating insects and angiosperms in the Cretaceous, *Annals of the Missouri Botanical Garden* 86(2):373–406.

Haldane J B S, 1924, A mathematical theory of natural and artificial selection, Part 1, *Transactions of the Cambridge Philosophical Society* 23:19–41.

Haldane J B S, 1957, The cost of natural selection, *Journal of Genetics* 55:511–24.

Hall B K, 1992, *Evolutionary developmental biology*, Chapman & Hall.

Hall B K (ed), 1994, *Homology: The hierarchical basis of comparative biology*, Academic Press.

Hall B K, Bock G and Cardew G (eds), 1999, *Homology*, Proceedings of Novartis Foundation Symposium 222, 21–23 July 1998, John Wiley & Sons.

Hardison R C, 1991, Evolution of the globin gene families; in Selander R K, Clark A G and Whittam T S (eds), *Evolution at the molecular level*, Sinauer.

Hinchcliffe J R and Griffiths P J, 1983, The prechondrogenic patterns in tetrapod limb development and their phylogenetic significance; in Goodwin *et al.*

Homberger D G and de Silva K N, 2000, Functional microanatomy of the feather-bearing integument: Implications for the evolution of birds and avian flight, *American Zoologist* 40(4):553–74.

Hooke R, 1665, *Micrographia*; facsimile by Dover Publications Inc.,1961.

Horgan J, 1996, *The End of Science*, Little, Brown and Company.

Hoyle F and Wickramasinghe C, 1981, *Evolution from space*, J M Dent & Sons.

Hull D L, 1973, *Darwin and his critics*, Harvard University Press.

Hume D, 1789, *Dialogues concerning natural religion*.

Huxley J, 1942, *Evolution: The modern synthesis*, George Allen & Unwin.

Huxley T, 1860, Letter to Charles Kingsley of 23 September.

Huxley T, 1870, *Biogenesis and abiogenesis*, Presidential address to the British Association for Advancement of Science.

Jaçob F, 1983, Molecular tinkering in evolution; in Bendall.

Jarvik E, 1980, *Basic structure and evolution of vertebrates*, Academic Press.

Joysey K A, 1981, Molecular evolution and vertebrate phylogeny in perspective; in Ashton E H and Holmes R L (eds), *Perspectives in primate biology*, The Zoological Society of London.

Kellog E and Julian N, 1997, The structure and function of rubisco, *American Journal of Botany* 84(3):413–28.

Kettlewell, 1973, *The evolution of melanism*, Oxford University Press.

Kimura M, 1983, *The neutral theory of molecular evolution*, Cambridge University Press.

Kuhn T S, 1962, *The Structure of Scientific Revolutions*, University of Chicago Press.

Kyte J, 1995, *Structure in protein chemistry*, Garland Publishing.

Laurin M and Reisz R R, 1997, A new perspective on tetrapod phylogeny; in Sumida S S and Martin K L M (eds), *Amniote origins*, Academic Press.

Lyell C, 1830–33, *Principles of geology : Being an attempt to explain the former changes of the earth's surface, by reference to causes now in operation*, Murray.

Maynard Smith J, 1995, *The theory of evolution*, 3rd (Canto) edn, Cambridge University Press.

Mayr E, 1967, Evolutionary challenges to the mathematical interpretation of evolution; in Moorhead and Kaplan.

Mayr E, 1978, Evolution, *Scientific American*, Sep, W H Freeman & Co.

Meglitsch P and Schram F, 1991, *Invertebrate Zoology*, 3rd edn, Oxford University Press.

Mettler L and Gregg T, 1969, *Population Genetics and Evolution*, Longman and Todd.

Mills G C, 1991, Cytochrome *c*: gene structure, homology and ancestral relationships, *Journal of Theoretical Biology* 152:177–90.

Mixter R L (ed), 1959, *Evolution and Christian thought today*, Wm B Eerdmans Publishing Company.

Moorhead P S and Kaplan M M (eds), 1967, *Mathematical Challenges to the Neo-Darwinian Interpretation of Evolution*, Proceedings of a symposium at The Wistar Institute of Anatomy and Biology, 25–26 April 1966, The Wistar Institute Press.

Nelson G, 1978, Ontogeny, phylogeny, paleontology and the biogenetic law, *Systematic Zoology* 27:324–45

Nieukoop P D and Sutasurya L A, 1983, Some problems in the development and evolution of the chordates; in Goodwin *et al.*

Nilsson Dan-E and Pelger S, 1994, A pessimistic estimate of the time required for an eye to evolve, *Proceedings of the Royal Society of London Series B – Biological Sciences* 256:53–8.

Orgel L, 1998, The origin of life – a review of facts and speculations, *Trends in Biochemical Sciences* 23:491–5.

Paley W, 1802, *Natural Theology*, R. Faulder.

Panchen A L, 1992, *Classification, evolution and the nature of biology*, Cambridge University Press.

Panchen A L, 1999, Homology – history of a concept; in Hall *et al.*

Paterson C, 1982, Morphological characters and homology; in Joysey K A and Friday A E (eds), *Problems of phylogenetic reconstruction*, Academic Press.

Paul C R C, 1985, The adequacy of the fossil record reconsidered; In *Evolutionary case histories from the fossil record*, Special Papers in Palaeontology 33:7.

Pictet F, 1860, On the Origin of Species by Charles Darwin; in Hull.

Popper K, 1974, Scientific reduction and the essential incompleteness of all science; in Ayala F and Dobzhansky T (eds), *Studies in the philosophy of biology*, The Macmillan Press Ltd.

Raff R, Larval homologies and radical evolutionary changes in early development; in Hall *et al.*, 1999.

Ray J, 1678, *The Ornithology of Francis Willughby*, John Martyn.

Ray J, 1691, *The wisdom of God manifested in the works of the creation*; facsimile by Garland Publishing Inc., 1979.

Richardson M K, Hanken J, Gooneratine M L, Pieau C, Raynaud A, Selwood L and Wright G M, 1997, There is no highly conserved embryonic stage in the vertebrates: Implications for current theories of evolution and development, *Anatomy and Embryology* 196(2):91–106.

Ridley M, 1996, *Evolution*, 2nd edn, Blackwell Science.

Rieppel O, 1994, Homology, topology, and typology: the history of modern debates; in Hall.

Roth V L, 1984, On homology, *Biological Journal of the Linnean Society* 22:13–29.

Roth V L, 1988, The biological basis of homology; in *Ontogeny and Systematics*, Humphries C J (ed), Columbia University Press.

Ruse 1993 *The Darwinian Paradigm*, Routledge.

Ruben J A and Jones T D, 2000, Selective factors associated with the origin of fur and feathers, *American Zoologist* 40(4):585–96.

Ruben J A, Jones T D, Geist N R and Hillenius WJ, 1997, Lung structure and ventilation in theropod dinosaurs and early birds, *Science* 278:1267–70.

Russell B, 1961, *History of Western Philosophy*, 2nd edn, George Allen and Unwin.

Salvini-Plawen L v and Mayr E, 1976, On the evolution of photoreceptors and eyes, *Evolutionary Biology* 10:207–76.

Sarton G, 1975, *Introduction to the History of Science*, Vol 1, R E Kreiger Co.

Shankaranarayanan P, Banerjee M, Kacker R K, Aggarwal R K and Singh L, 1997, Genetic variation in Asiatic lions and Indian tigers, *Electrophoresis* 18(9):1693–1700.

Simpson G G, 1950, The meaning of Darwin; Introductory Essay in Darwin.

Strickberger M, 1990, *Evolution*, 2nd edn, Jones and Bartlett Publishers.

Sturtevant A H, 1965, *A History of Genetics*, Harper and Row.

Thomas A and Garner J, 1998, Are birds dinosaurs?, *Trends in Ecology and Evolution* 13:129–30, Elsevier Science.

Ulam S, 1967, How to formulate mathematically problems of rate of evolution; in Moorhead and Kaplan.

Valentine J W, Jablonski D and Erwin D H, 1999, Fossils, molecules and embryos: new perspectives on the Cambrian explosion, *Development* 126(5):851–9.

Vesalius A, 1555, *De fabrica corporis humani*, 2nd edition.

Wagner G P, 1989, The biological homology concept, *Annual Reviews of Ecology and Systematics.* 20:51–69.

Wagner G P, 1999, A research programme for testing the biological homology concept; in Hall *et al.*

Webster D and Webster M, 1974, *Comparative vertebrate morphology*, Academic Press.

Whewell W, 1833, Astronomy and general physics considered with reference to natural theology; Bridgewater treatises on the power wisdom and goodness of God as manifested in the creation, No. 3, William Pickering.

White A J M, 1978, *What about origins?*, Dunestone Printers Ltd.

Whitehouse H L K, 1973, *Towards an understanding of the Mechanism of Heredity*, 3rd edn, Edward Arnold (Publishers) Ltd.

Wolken J J, 1995, *Light detectors, photoreceptors, and imaging systems in nature*, Oxford University Press.

Index

Glossary: references to definitions of technical terms are identified in the index by page numbers in bold type. Bold type is also used to identify key texts on the relevant subject.